HANDBOOK OF INDUSTRIAL ENERGY CONSERVATION

A universal rule—after being specified—is no longer universal.

—Lao-tzu, the Book of Tao

HANDBOOK OF INDUSTRIAL ENERGY CONSERVATION

S. David Hu, Ph.D.

VAN NOSTRAND REINHOLD COMPANY
NEW YORK CINCINNATI TORONTO LONDON MELBOURNE

Copyright © 1983 by Van Nostrand Reinhold Company Inc.

Library of Congress Catalog Card Number: 81–15916
ISBN: 0–442–24426–6

Manufactured in the United States of America

Published by Van Nostrand Reinhold Company Inc.
135 West 50th Street, New York, N.Y. 10020

Van Nostrand Reinhold Publishing
1410 Birchmount Road
Scarborough, Ontario M1P 2E7, Canada

Van Nostrand Reinhold Australia Pty. Ltd.
17 Queen Street
Mitcham, Victoria 3132, Australia

Van Nostrand Reinhold Company Limited
Molly Millars Lane
Wokingham, Berkshire, England

15 14 13 12 11 10 9 8 7 6 5 4 3 2 1

Library of Congress Cataloging in Publication Data

Hu, S. David.
  Handbook of industrial energy conservation.

  Includes index.
  1. Energy conservation. 2. Energy policy.
I. Title
TJ163.3.H8   621.042   81–15916
ISBN 0–442–24426–6   AACR2

*To my family for their understanding and patience
in the course of completing this book*

Preface

This book presents a systems analysis approach to energy conservation problems. It is based on my past experience in modeling energy supply and demand at TRW, Inc., assessing conservation policies for the Department of Energy, and teaching systems analysis courses at the George Washington University. The experience of managing industrial R&D projects at the Electric Power Research Institute (EPRI) refined some of the concepts in this book.

This book can be used by energy planners for major corporations, large institutes, and government agencies whose concerns for energy conservation warrant comprehensive energy planning. Various parts of the book can be used in teaching courses in energy economics, management and policy, and systems analysis.

I am indebted to EPRI and TRW for supporting my research. Thanks also go to Milton Searl, Dr. John Chamberlin, Robert Mauro, James Eyssell, Rene Males, and Dr. Iraj Zandi for their constructive comments and support. Finally, I would like to express gratitude to my past graduate students at the National Aeronautics and Space Administration, Goddard Space Flight Center, Maryland, whose interaction formed the original concept for this book.

S. David Hu
San Jose, California

Energy Conversion Factors

Purchased Fuels

Coal	1 short ton	$= 26.2 \times 10^6$ Btu
Crude oil	1 barrel	$= 5.6 \times 10^6$ Btu
Residual oil (#5, #6)	1 barrel	$= 6.3 \times 10^6$ Btu
Dist. oil (#2)	1 barrel	$= 6.0 \times 10^6$ Btu
Natural gas, SNG, LNG	10^3 ft^3	$= 1.03 \times 10^6$ Btu
Electricity	10^3 kWh	$= 3.41 \times 10^6$ Btu

Other:

Steam (no condensate return)	10^3 lb	$= 1.3 \times 10^6$ Btu
Steam (condensate return)	10^3 lb	$= 1.1 \times 10^6$ Btu
Liquid propane gas	1 gallon	$= 0.092 \times 10^6$ Btu

Self-Generated Fuels

Bark (50% moisture)	1 short ton	$= 9.0 \times 10^6$ Btu
Hogged wood (50% moisture)	1 short ton	$= 7.7 \times 10^6$ Btu
Spent liquor (oven dry basis)	1 short ton (solids)	$= 13.4 \times 10^6$ Btu
Electricity	10^3 kWh	$= 3.41 \times 10^6$ Btu

Other Common Symbols

GJ	$= 0.95 \times 10^6$ Btu
Q or Quad (quadrillion Btu)	$= 10^9 \times 10^6$ Btu
MBD (million barrels/day)	$= 2.04$ Q

Contents

HANDBOOK OF INDUSTRIAL ENERGY CONSERVATION

1

The U.S. Energy Problem and Conservation of Industry

THE U.S. ENERGY PROBLEM

A stunning warning concerning the U.S. energy problem comes from Robert Mundell, a Canadian-born economics professor at Columbia University: The U.S. is "moving industrially into the situation Britain was in during the latter part of the 19th Century."[9] Much like Britain when it had only primitive factories geared to cheap raw materials from a colonial empire, the United States is encumbered a century later with an economy depending on inexpensive energy from obliging oil states. Whether Mundell's warning will come true largely depends on how we try to solve the energy problem.

It is not difficult to see the reasons for the comparison. For example, most of the U.S. facilities in the basic metal industries are more than 30 years old and were built at a time when energy was not a primary consideration in engineering design. Energy, particularly natural gas and oil, was inexpensive. Many of these facilities were modified in the 1960s and early 1970s to meet environmental requirements. For example, many coal-fired boilers were converted to natural gas and oil. These modifications significantly increased the energy required to run the same boilers. In the 1960s and early 1970s, environment, not energy, was the theme of government policy.

President Nixon's advisors debated about the amount of oil that should be imported from world oil-producing countries at the beginning of the 1970s, in order to reduce domestic oil prices. Then in 1973, the Arab oil embargo shocked Americans into an awareness of potential energy shortages, rising energy costs, and a lack of available solutions. The oil embargo was only the tip of the iceberg of energy-triggered economic problems that combine the result of energy-inefficient industrial plants and high energy prices. Since 1973, the U.S. economy has tried to adjust. Although a number of energy policies and legislative acts have been adopted, it will take a long time to see their effect. Consequently, the United States is experiencing double-digit inflation and an unemployment rate of more than 7%. The longer it takes the remedy to take effect, the more severe the inflation and unemployment will be.

One remedy would be conservation, which would have an immediate effect and yet maintain the American standard of living. Conservation in industry is particularly effective because industry is concerned about costs and profits, and it also makes equipment and appliances that use energy to produce goods and services. The remainder of this chapter will discuss the mystery of conservation, implications of industrial energy conservation, and the organization of this book.

THE MYSTERY OF CONSERVATION

The simplest definition of energy conservation is "using energy more efficiently." By extension, this means to substitute time, convenience, labor, and capital for energy. Following are some examples of energy conservation:

- It takes a little longer to plan your trip so that you can combine many shopping trips into one. You conserve energy by substituting time.
- It takes patience and inconvenience to carpool with other people. You conserve energy by substituting convenience.
- Instead of using an electric opener, you use a hand opener to open cans. You conserve energy by substituting labor.
- Because of the age of your refrigerator, you decide to buy a new, more energy-efficient one. You therefore substitute capital for energy.

It can be difficult to convey this definition of conservation. Conservation has usually been interpreted as a sacrifice in standard of living, but that is an incorrect interpretation. Conservation can both increase productivity and maintain our standard of living because more time and convenience are used for production. Conservation is thus a way of using energy more efficiently.

In particular, conservation in industry may mean making new investments in more efficient equipment to replace old, inefficient facilities. Thus conservation investments have the same results as other investments that bring more jobs, lower costs, better products, and more competition, if conservation investments are well planned.

Because of past low prices for oil and natural gas, conservation has historically been unable to compete with other industrial investments. This is particularly true when the cost of capital (interest rate) is high. Then, even rising energy prices may not induce industry to invest in conservation technologies.

OBSERVATIONS ON ENERGY CONSERVATION BY INDUSTRY

Because of industry's attention to costs and profits, continuing rising energy prices and uncertainty in energy supply may finally induce investment in conservation technologies when the cost of capital declines. Industrial conservation is the key to the nation's conservation efforts because industry produces equipment and appliances for itself and consumers. For example, if the automobile industry can produce cars that get 30% better mileage, the nation's energy savings will be substantial. In addition, many of these cars will be used to produce other goods and service. The following are some observations about industrial energy conservation.

The first observation on industrial energy conservation investment is that competition will increase because of the more energy-efficient production equipment installed in conservation programs. Investments in conservation are likely to be substantial and will be in the areas of large retrofitting of existing facilities or early retirement of existing equipment and implementation of new, energy-efficient processes. This is particularly true for the basic metal industries. Such improvements will reduce energy costs and thus increase industry productivity and competition.

The second observation is that, in order to justify expenses for energy conservation measures, industry must be able to expect an appropriate level of energy prices. Industry has to expect sufficient increases in energy prices in the future for future energy savings to compensate for today's capital investments. With today's high interest rates and tight money, it is difficult to foresee industry investing in energy-saving equipment if it expects energy prices to be only slightly higher than inflation. In other words, if energy prices were kept artificially low by either legislation or government policies, industry would find it impractical to voluntarily invest in energy-saving processes. It would then most likely use the capital to invest in more profitable facilities such as realty.

The third observation is that industry needs a constructive and cohesive government energy policy. For example, in the past two decades, the federal government's emphasis has shifted from energy production to environmental protection, to conservation. The shifting forced industry to change from coal burning to oil or gas burning in the 1960s and then to return to coal burning in the 1980s. Such shifting of government policies is costly.

The fourth observation is that coordination among various federal, state, and local agencies, both energy- and non-energy-related, is essential for the success of the nation's energy conservation efforts. In the past few years, the federal government were engaged in generating a series of energy laws and policies. However, most state and local governments have not acquired the same momentum. For example, most states are unprepared for emergencies and conservation, and DOE has provided little help to implement the state energy conservation mechanisms provided in the Emergency Energy Conservation Act of 1979.[10] It is questionable how much energy conservation can be accomplished without the cooperation of state and local governments.

Furthermore, if U.S. industry cannot produce energy-efficient equipment, its markets are in danger of being lost to foreign products that are more energy-efficient. The survival of U.S. industry depends on its ability to produce energy-efficient products. This point cannot be more clear than in the case of the automobile industry. The U.S. automobile industry has been forced by the consumer, not the government, to produce better, more energy-efficient cars, because one-quarter of its market has been taken over by small foreign cars, particularly Japanese cars.

ORGANIZATION OF THIS BOOK

To solve the U.S. energy problem, it is important for policy-makers to recognize the importance of all energy sources, particularly conservation.* Even though it is the general belief that slow growth appears inevitable for industrial nations over the next few years,[2] with wide-ranging, heavy dependence on oil imports in 1985, the United States still has to be prepared for the worst in regard to foreign oil supplies because of political uncertainty in the Persian Gulf region. Conservation, emergency oil storage, or even rationing appears to be taking a more important role in meeting future energy needs.** Because industrial energy conservation is essential to the success of U.S. conservation efforts, this *Handbook of Industrial Energy Conservation* intends to answer many questions in this area.

Morgan Guaranty Trust Co. reported that, during 1973–1979, U.S. energy use increased 0.3% for every 1% increase in the gross national product (GNP), compared with the historical pattern of 1% increase in energy consumption for every 1% increase in GNP. The U.S. energy/GNP ratio in 1979 was thus reduced by 10% from 1973 compared with a reduction of 17% in Japan; 10% in West Germany, the United Kingdom, and France; and almost zero in Canada.[3] Two questions are therefore raised: What is

---

* According to Bernard O'Keefe, Chairman of the Board of EG&G, Inc. of Wellesley, Mass., and one of President Reagan's energy advisors.[1] Others have indicated the same thing.[4,5]

** Morgan Guaranty has the same point of view.[3] A Stanford study also cited increased efficiency as a key to meeting future energy demands.[5]

the relationship between energy use in industry and economic growth? And how important is conservation in industry? The first question will be discussed in Chapter 2 and the second in Chapter 3.

According to many studies, executives in industry acknowledge that there are many energy-saving innovations in which they could have invested, but they have not done so because of other more profitable claims on capital.[6] The primary barriers are subsidized energy prices, which make conservation investments economically unattractive. Owing to price controls on oil and natural gas, the users of oil and gas have never paid full market prices for them. The consumer of electricity pays only average costs rather than the marginal costs of new generating capacity. Users of coal and nuclear energy do not pay the full prices because air pollution from coal and the fear of postulated nuclear accidents have never been fully included in the market price. Therefore, the major part of this book will focus on energy management and planning for industry, and the energy price level required for conservation investments to become economically attractive. Chapters related to this discussion are as follows:

- Chapter 4: Comprehensive energy conservation planning
- Chapter 5: Principles of energy conservation
- Chapter 6: Thermodynamics and energy efficiency
- Chapter 7: Economics of alternative energy conservation measures
- Chapter 8: Industrial energy conservation modeling
- Chapter 10: Economics of energy conservation policies—from industry's point of view

Chapter 4 outlines the basics of conservation planning, whereas Chapters 5, 6, 7, and 8 detail the major elements, and Chapter 10 presents an application.

Management Centre Europe surveyed 570 companies and found that 70% of European business executives faulted their governments for ineffectiveness in dealing with energy issues and conservation efforts.[7] Also a Department of Energy report, prepared by the Energy Information Administration,[8] concluded that government policies will have significantly little effect on the nation's energy prospects for the next 10 years. It is therefore important for us to examine the impact of government policies. Chapter 9 will present legislative, institutional, and environmental impacts on industrial energy conservation, and Chapter 11 will discuss energy management and policy from the nation's point of view.

Although most business executives are aware of conservation technologies and measures, we will review conservation technologies as follows:

- Chapter 12: Waste heat recovery and utilization
- Chapter 13: In-plant (industrial) cogeneration
- Chapter 14: Cogeneration: power plant reject heat utilization
- Appendix C: Overview of energy conservation technologies

Chapter 15 concludes the text with a discussion of the future of energy conservation.

REFERENCES

1. Maize, K. "Who are the Energy Stars in Ronald Reagan's Firmament?" *The Energy Daily,* Vol. 8, No. 176, September 15, 1980.
2. "Slow Growth Appears Inevitable," *The Wall Street Journal,* September 30, 1980.
3. "Caution Urged for U.S. Energy Policy," *Oil and Gas Journal,* October 20, 1980.
4. "Congress Expected to Direct DOE to Expand R&D in Industrial Conservation," *Inside D.O.E.,* July 4, 1980.
5. Institute for Energy Studies (IES). "Alternative Energy Futures: An Assessment of U.S. Options to 2025," Stanford, Calif.: Stanford University, June 1980.
6. Stobaugh, R. and Yergin, D. "Industrial Conservation Incentives," *Wall Street Journal,* July 31, 1980.
7. O'Neil, J. "Energy Problem," *Vision* (Business Monthly), Dublin, July/August 1980.
8. Energy Information Administration. "Energy Programs/Energy Markets," Department of Energy, DOE/EIA-0201/16, July 1980.
9. Janssen, R. "Some Overseas Fear U.S. is Following Britain Down Economic-Decline Path," *Wall Street Journal,* October 28, 1980.
10. "States Unprepared for Emergency, DOE Giving Little Help, Moffett says," *Energy Users Report,* September 11, 1980.

PART I
BACKGROUND AND IMPORTANCE OF INDUSTRIAL ENERGY CONSERVATION

This section presents an industrial energy consumption profile and discusses the importance of industrial energy conservation.

2
Industrial Energy Use and Economy

INTRODUCTION

Because of past abundance of low-cost energy, historically the rate of social progress among industrial societies has not been limited by energy availability. Although energy is essential for industrial production, energy cost has not been significant when compared with nonenergy costs. For example, energy cost represented only about 2% of 1975 value of shipments, compared with a nearly 21% cost for wages and salary.

Energy emerged as an important national issue at the beginning of the 1970s, when domestic supplies of inexpensive oil and natural gas dwindled, and imports of crude oil became necessary. The 1973 oil embargo produced an "energy crisis," which shocked the U.S. economy. The essence of the crisis has been an oil shortage, a result of the past consumption pattern depending on inexpensive oil and natural gas. The U.S. economy was further weakened when the oil cartel, the Organization of Petroleum Exporting Countries (OPEC), raised oil prices to levels quadruple those of 1973.

Conservation has become one of the most effective ways to provide immediate relief for the energy problem, to maintain economic growth and social progress. Because industrialists are concerned about profits, costs, and secure energy supply, energy conservation is especially important in the industrial sector. It is thus necessary to examine industrial energy use and economy in this chapter before the importance of conservation is presented in the following chapter. This chapter is divided into four sections: (1) industrial energy use, (2) importance of energy in production and employment, (3)

9

energy requirements for industry, and (4) comparison of energy uses among industrial societies.

INDUSTRIAL ENERGY USE

Industrial energy use includes energy used for fuel, energy used for nonfuel, and energy lost in power generation distributed to the industrial sector. Some may define industrial energy use as energy used only for fuel in industry, and report separately nonfuel use and energy loss. Unless specifically mentioned, the term industrial energy use in this chapter will include all three types of energy.

About 40% of the energy consumed in the United States is for industrial applications. Figure 2-1 shows the share of the 1980 industrial energy consumption: total industrial energy use was 30.6×10^{15} Btu (30.6 quad), which accounted for 39% of the 1980 total U.S. energy consumption. It was estimated that 39% of this industrial energy was consumed in process steam, 26% in direct heat, 19% in electric drive, and 16% in feedstock. Clearly, energy is essential for production processes, feedstock, and space-heating in order to maintain business operations.

Both fuel and nonfuel industrial energy uses have increased substantially since the 1950s, as shown in Fig. 2-2. The figure indicates that fuel and nonfuel industrial energy uses have almost doubled since 1947.

Industry has become increasingly dependent upon the two fuels in shortest supply—oil and natural gas. As shown in Fig. 2-2, coal contributed almost half of the industrial energy use in the 1940s and 1950s. However, since 1960, oil [petroleum and natural gas liquids (NGL)] and natural gas have overtaken coal. They account for more than 60% of industrial energy use.

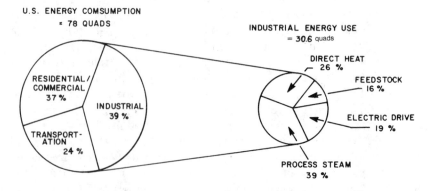

Fig. 2-1.   1980 industrial energy use. Source: Department of Energy. *Monthly Energy Review.*

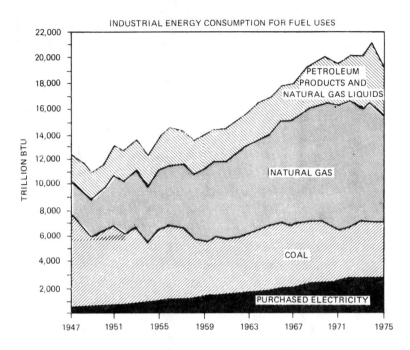

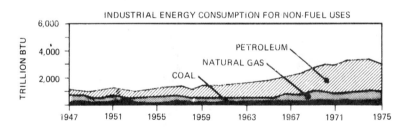

Fig. 2–2. Trends of industrial energy uses. Source: U.S. Department of Commerce.[1]

The importance of oil and natural gas in industrial energy use is reinforced in Fig. 2–3. The figure compares the proportion of each primary energy source in industrial consumption. It can be seen that oil and natural gas are more important as energy sources for industry than coal.

Industry consumes a substantial portion of U.S. natural gas and oil production. Figure 2–3 shows the detail of 1980 U.S. energy "flow" from the primary sources to the end-users. Without taking into account the natural gas and oil used to generate electricity for industrial use in 1980, industry used almost one-half of the U.S. gas supply and one-fifth of the U.S. petroleum and NGL supply.

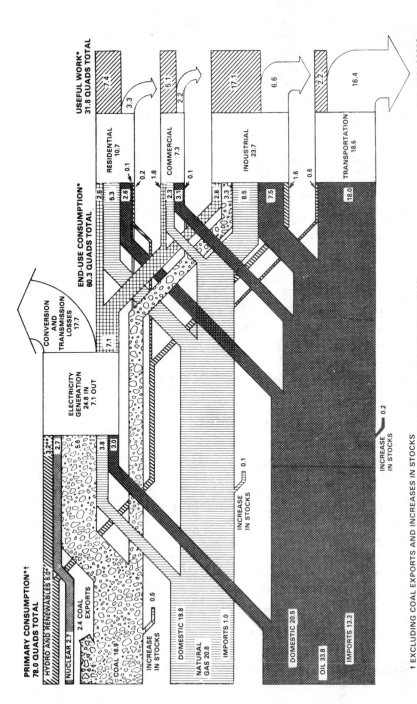

Fig. 2-3. U.S. energy flow, 1980 (quadrilion Btu). Source: DOE.[2]

PRIMARY CONSUMPTION*†
78.0 QUADS TOTAL

HYDRO AND RENEWABLES 5.0***

NUCLEAR 2.7

COAL 18.6**

2.4 COAL EXPORTS

INCREASE IN STOCKS 0.5

NATURAL GAS 20.8

DOMESTIC 19.8

IMPORTS 1.0

INCREASE IN STOCKS 0.1

OIL 33.8

DOMESTIC 20.5

IMPORTS 13.3

INCREASE IN STOCKS 0.2

CONVERSION AND TRANSMISSION LOSSES 17.7

ELECTRICITY GENERATION 24.8 IN 7.1 OUT

END-USE CONSUMPTION*
60.3 QUADS TOTAL

USEFUL WORK*
31.8 QUADS TOTAL

RESIDENTIAL 10.7

COMMERCIAL 7.3

INDUSTRIAL 23.7

TRANSPORTATION 18.6

7.4

5.1

17.1

2.2

3.3

2.2

6.6

16.4

END-USE LOSSES
28.5 QUADS TOTAL
(NET OF ELECTRICITY EFFICIENCY GAINS)

† EXCLUDING COAL EXPORTS AND INCREASES IN STOCKS
* INCLUDES 1.8 QUADS OF BIOMASS USE NOT CURRENTLY ACCOUNTED FOR IN DOE STATISTICS
** INCLUDES 0.2 QUADS OF IMPORTED HYDROELECTRIC POWER
*** BASED ON END-USE EFFICIENCIES FROM 1979 BROOKHAVEN DATA

12

Careful examination of the results of the energy flow in Fig. 2–3 pinpoints substantial energy waste in industry. In 1980, in addition to attributed energy loss of 6.9 quad for electric generation, transmission, and distribution, industry rejected 6.6 quad Btu into the atmosphere. The total industrial energy loss was 13.5 quad, out of a total industrial energy consumption of 30.6 quad in 1980. Clearly, energy loss accounts for more than one-third of industrial energy consumption. In other words, industrial production can be maintained even if this energy loss is reduced. Conservation can be practiced to reduce this industrial energy waste.

IMPORTANCE OF ENERGY IN PRODUCTION AND EMPLOYMENT

Energy in Industrial Production

The importance of energy in industrial production will be discussed in terms of energy input per dollar of value added to the gross national product (GNP), percentage of energy cost in production, and vulnerability of industrial production during energy shortages.

Energy Input per Dollar of Value Added. The trend of energy input per dollar of value added to GNP is shown in Fig. 2–4. Since 1947, while value added in manufacturing has increased at an annual average rate of 3.5%, the U.S. industrial fuel use of energy has increased at an average rate of 1.9%. Because of this fuel-efficiency improvement, fuel use of energy

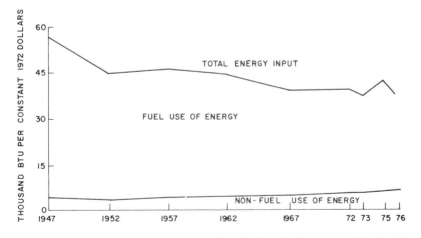

Fig. 2–4. Industrial energy consumption per dollar of value added to gross national product. Source: Bureau of the Census.[1]

per dollar of value added has decreased at an annual average rate of 1.0%, and total industrial energy input (including both fuel and nonfuel use) per dollar of value added declined at an annual average rate of 0.8%. Industry has shown a trend of slowly improving energy efficiency. For example, the steel industry used 47 million Btu per ton of finished steel production in 1950 and steadily improved energy efficiency to 35 million Btu per ton by 1973,* an average improvement of 1.1% annually.

Energy input per dollar of value added to GNP for most energy-intensive industries will be discussed in later sections.

Importance of Energy Costs in Production. Energy costs are relatively insignificant in most industries, compared with other production factors, such as wages and overhead. Table 2–1 shows that energy costs represented only 2.3% of manufacturers' value of shipments (or value added to GNP) for all industry in 1975. This percentage compares with nearly 21% for wages and salaries, and approximately 22% for overhead costs and profit. Analysis of 1976 industrial energy cost data shows similar results.

The importance of energy costs is further analyzed for selected energy-intensive industries in Table 2–2. The table indicates that, even after drastic

Table 2–1. Energy Cost Compared to Other Manufacturing Costs. (All Manufacturers, 1975)

Cost Category	Costs ($ Billions)	Fraction of Value of Shipments (Percent)
Purchased Energy:		
Electricity	$ 10.85	1.01
Natural gas	5.65	0.56
Fuel oils	3.52	0.35
Coal	1.31	0.13
Coke	0.82	0.08
Other Fuels	1.60	0.15
Energy Subtotal	$ 23.19	2.28
Wages and salaries	209.96	20.63
Materials	558.52	54.87
Other costs and profits	226.18	22.22
Value of shipments	$1,017.85	100.00

Source: U.S. Department of Commerce.[1]

---

* Owing to stricter pollution controls, some industries such as copper may require more energy to produce same units of products. See Chapter 10 for details.

Table 2-2. Importance of Energy Costs in Energy-Intensive Industries (1971 and 1976).

SIC Code	Industry	Energy Cost as Percentage of Value of Industry Shipments	
		1971	1976
20	Food and kindred products	0.85	1.19
22	Textile mill products	1.67	2.72
26	Paper and applied products	3.43	5.23
28	Chemical and allied products	3.22	5.17
286	Organic chemicals	6.75	7.09
287	Agricultural chemicals	1.89	5.92
281	Inorganic chemicals	—	14.49
282	Plastics	3.06	4.70
29	Petroleum and coal	2.36	2.55
2911	Petroleum refining	2.38	2.51
32	Stone, clay, and glass products	4.50	7.06
3211	Flat glass	4.19	7.42
322	Pressed and blown glass	4.78	8.25
	Glass	4.66	8.08
3241	Cement	15.58	23.70
33	Primary metals	3.98	5.82
3334	Aluminum	—	14.04
331	Steel	4.91	7.24
3331	Copper	—	3.13
34	Fabricated metals	1.05	1.47
35	Machinery (less electrical)	0.84	1.02
37	Transportation equipment	0.57	0.84
371	Motor vehicles	0.52	0.80
2851	Paint	0.60	0.78
34	Fabricated metals	1.05	1.47
362	Electrical industrial apparatus	1.28	1.57
	End-use products	1.04	1.44

Source: U.S. Department of Commerce.[1]

energy price increases after the 1973 Arab oil embargo, the energy costs in most industries have not changed significantly with respect to the percentages of values of industry shipments for energy costs. Except for cement, aluminum, and inorganic chemical industries, energy costs were less than 9% of the total value of industry shipments in 1976. Clearly, for many firms, profit margins are higher than energy cost percentages.

Vulnerability of Industrial Production. Although energy cost is not a substantial portion of production costs, industrial production is vulnerable to oil embargoes because of its increasing dependence on oil and gas. Figure 2-5 depicts energy sources for key industries. Most basic industries such as

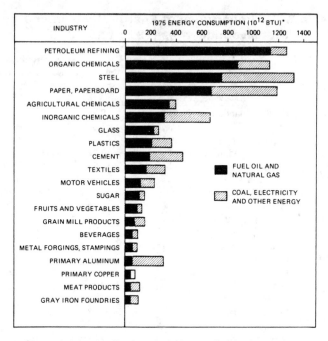

Fig. 2–5. Key industries vulnerable to oil and gas shortages. Source: U.S. Department of Commerce.[1]

petroleum refining, chemicals, steel, paper, glass, plastics, cements, textiles, motor vehicles, and grain mill products depend heavily on oil and natural gas. In particular, the following four major groups consume nearly 60% of all industrially used oil and natural gas:

- Four key chemical industries: 23%
- Petroleum refining: 15%
- Steel industry: 10%
- Paper industry: 9%

If there were an oil or natural gas shortage, the key industries producing goods for basic national needs would be affected significantly. The historical pattern of industrial fuel use is discussed below to emphasize industry's dependence on oil and natural gas.

Historical Industrial Fuel Use Pattern. Because of low prices of natural gas and oil and environmental regulations in the 1960s and early 1970s, the manufacturing and utility sectors converted coal-fired combustors into natural gas and oil-fired combustors. Part of this conversion was due to lower capital and operating costs.

According to the 1975 Federal Energy Administration survey of major industrial fuel-burning installations (MFBIs),* the percentage of total output capacity of major combustors by primary fuel is as follows:

- Gas: 53%
- Residual fuel oil: 19%
- Distillate oil: 2%
- Coal: 26%

Most small industrial and commercial boilers are designed to fire either oil or gas or a combination of the two.

More than 70% of the U.S. nonutility combustors use oil and gas as their primary fuel. They have a life cycle of about 30 years or more, although the Internal Revenue Service permits depreciation over 25 years. They will have to continue using increasingly expensive oil and gas, or be retired prematurely, or use oil and gas substitutes.

Regional Dependence on Oil and Gas. Primary fuel used for combustors varies over ten Environmental Protection Agency Regions, as shown in Table

Table 2-3. Regional Dependence on Fuel Types for Major Combustors.

EPA Region	Dominant Fuel Type	Percentage of Region's Capacity
1. (CT, MA, ME, RI, and VT)	Oil	93
2. (NJ and NY)	Oil	66
3. (DC, DE, MD, PA, VA, and WV)	Coal	51
4. (AL, FL, KY, GA, MS, NC, SC, and TN)	Gas	38
5. (IL, IN, MI, MN, OH, and WI)	Coal	47
6. (AK, LA, NM, OK, and TX)	Gas	94
7. (IA, KS, MI, and NE)	Gas	62
8. (CO, MT, ND, SD, UT, and WY)	Gas	51
9. (AZ, CA, and NV)	Gas	91
10. (ID, OR, and WA)	Gas	67

Source: EPA.[8]

---

* The MFBI is defined as an installation or unit other than a fossil power plant that has or is a fossil-fired boiler, burner, or other combustor of fuel or any combination thereof at a single site, and that has, individually or in combination, a design firing rate of 100 million Btu/hour or greater.[8]

2–3. For example, oil is the primary fuel for combustors in a New England Region (EPA Region 1), accounting for 93% of the region's capacity. Coal is the primary fuel in Pennsylvania–West Virginia (Region 3), accounting for 51% of the total region's capacity; natural gas is the primary fuel in the Gulf States* (Region 6). Details of primary fuel type and percentage are shown in Table 2–3. Six out of ten regions depend on gas as a primary energy source for combustors.

Energy and Employment

As shown in Fig. 2–6, the occupational mix in the United States has slowly shifted from that of an agrarian economy to a service-oriented economy in which government, service, and other nonagricultural activities require the most laborers. Employment in manufacturing has been steady, about 20% of the total civilian labor force.

The historical relationship between primary energy use and employment is shown in Fig. 2–7, and observed as the following equation:[9]

$$E = 13e^{0.0235 L} - 16 \qquad (2\text{–}1)$$

where E is the total energy use in 10^{15} Btu/year, and L is employed civilians in millions of persons.

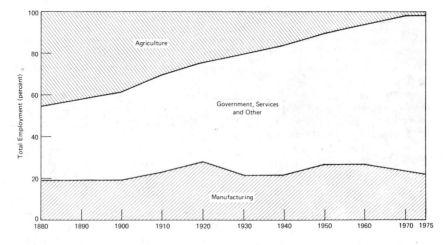

Fig. 2–6. Shift of the U.S. employment mix. Source: U.S. Department of Commerce.[3,4]

---

* Arkansas, Louisiana, New Mexico, Oklahoma, and Texas.

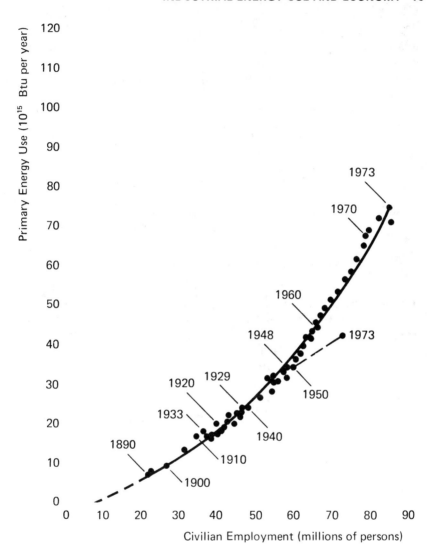

Fig. 2–7. Historical relationship between energy use and employment. Source: Starr.[9]

The exponential relationship between energy use and employment indicates that energy use increases more rapidly than does employment. On the other hand, the curvature shown in Fig. 2–7 is the result of a continuous increase in productivity and a continuous decline in annual work hours (i.e., labor being partially replaced by energy).

In the future, if full employment of the labor force, including a large number of female workers, is required, the need for ever-increasing energy use will be crucial. This pressure for more energy may be alleviated by conservation, which can improve energy efficiency and thereby reduce total energy requirements.

Because employment in manufacturing has been about 20% of the total force, Equation 2–1 can be approximated by the following equation:

$$E = 13e^{0.1175(IL)} - 16 \tag{2-2}$$

where E is total energy use in 10^{15} Btu/year and IL is employment in manufacturing, expressed in terms of millions of persons.

ENERGY REQUIREMENTS FOR INDUSTRY

Table 2–4 shows energy intensity (thousand Btu per 1976 dollars) for major energy-intensive industries in 1976. The ten most intensive industries are also ranked in the table. In order of energy intensity, they are: aluminum; cement; inorganic chemical; organic chemical; agricultural chemical; petroleum and coal (including petroleum refining); steel; copper; primary metal; and stone, clay, and glass products.

The pattern of energy requirements for each industry is analyzed in Tables 2–5 and 2–6. Table 2–5 shows the direct energy use coefficients by industry for 1947, 1958, 1963, and 1967. These figures indicate the amount of energy (in millions of Btu's) required to produce one dollar's output for each industry (in 1958 dollars). These coefficients summarize all the factors that determine how much energy is used to produce a given amount of output in an industry. The factors include the type of technology and production processes used and their efficiency, as well as the scale of individual operations, the degree of integration in the industry, the mix of products, and other factors that affect energy efficiencies.

Table 2–6 shows the total energy use per dollar of final demand. Total energy consists of direct (as shown in Table 2–5) and indirect fuel use required to produce products for industry. In other words, the total energy use is a product of direct energy use and the Leontief inverse. For example, the production of a car will involve the direct use of energy in operating machines, assembly lines, space heating, and so on. It will also require indirect energy use to produce materials, machinery, buildings, and so forth.

Examination of Tables 2–5 and 2–6 shows that energy coefficient improvements from 1947 to 1967 are not homogeneous among industries. For most

Table 2-4.   Energy Intensiveness for Selected Industries in 1976

SIC Code	Industry	Energy Intensity (10^3 Btu/1972 Dollar Value Added)	Rank
20	Food and kindred products	22.81	
22	Textile mill products	35.82	
26	Paper and applied products	77.2	
28	Chemical and allied products	77.99	
286	Organic chemicals	128.99	4
287	Agricultural chemicals	122.41	5
281	Inorganic chemicals	206.99	3
282	Plastics	76.95	
29	Petroleum and coal	112.4	6
2911	Petroleum refining	122.39	6
32	Stone, clay, and glass products	84.58	10
3211	Flat glass	73.46	
322	Pressed and blown glass	79.10	
	Glass	77.95	
3241	Cement	339.75	2
33	Primary metals	99.1	9
3334	Aluminum	476.51	1
331	Steel	109.76	7
3331	Copper	108.3	8
34	Fabricated metals	14.17	
35	Machinery (less electrical)	9.08	
37	Transportation equipment	10.45	
371	Motor vehicles	12.22	
2851	Paint	8.48	
34	Fabricated metals	14.17	
362	Electrical industrial apparatus	14.36	
	End use products	13.88	

Source: U.S. Department of Commerce.[1]

industries, total energy coefficients (shown in Table 2–6) have been improved, but improvements have not been substantial.

COMPARISON OF ENERGY USES AMONG INDUSTRIAL SOCIETIES

Energy is an essential input for production. If production in the United States could be maintained at a much lower level of energy use as is done in the other industrial countries, results of comparisons with other industrial societies could have major significance for the direction of U.S. conservation strategies.

Table 2-5. Direct Energy Use Coefficients.
(10^6 Btu's/1958 Dollar Output)

Sector	1947	1958	1963	1967
1. Coal mining*	.0086	.0222	.0212	.0179
2. Electric utilities*	.1732	.1263	.1063	.1038
3. Petroleum refining*	.0488	.1034	.1514	.1200
4. Gas utilities*	.2291	.2033	.2029	.1824
5. Crude petroleum	.0441	.0433	.0222	.0264
6. Agriculture	.0165	.0146	.0142	.0195
7. Metal ore mining	.0498	.0692	.0382	.0400
8. Non-metal mining	.1229	.1022	.0674	.0837
9. Construction	.0121	.0196	.0134	.0158
10. Food and kindred products	.0129	.0123	.0139	.0123
11. Tobacco manufacturers	.0025	.0020	.0024	.0031
12. Textiles and apparel	.0159	.0099	.0098	.0110
13. Lumber and wood products	.0139	.0127	.0142	.0127
14. Furniture and fixtures	.0097	.0073	.0080	.0081
15. Paper and allied products	.0672	.0611	.0665	.0602
16. Printing and publishing	.0043	.0047	.0079	.0049
17. Chemical, plastics, drugs, and paint	.0730	.0802	.0560	.0778
18. Rubber and misc. plastics	.1233	.0521	.0265	.0172
19. Leather and footwear	.0089	.0083	.0084	.0079
20. Stone, clay, and glass products	.1361	.0828	.0936	.0928
21. Primary iron and steel	.2337	.2140	.1656	.1567
22. Primary non-ferrous metals	.0529	.0558	.0808	.0549
23. Fabricated metal products	.0113	.0093	.0126	.0106
24. Machinery	.0117	.0078	.0102	.0079
25. Electrical machinery	.0091	.0073	.0110	.0065
26. Transportation equipment	.0094	.0063	.0065	.0075
27. Scientific instruments	.0072	.0071	.0087	.0064
28. Misc. manufacturing	.0127	.0111	.0108	.0062
29. Transportation	.1647	.1129	.1336	.1010
30. Communications	.0060	.0037	.0132	.0092
31. Water and other	0.0000	.0078	.0369	.0425
32. Commercial, financial, and services	.0202	.0179	.0185	.0166
33. Federal and local gov't enterprises	.0499	.0946	.1257	.0408
34. Imports	0.0000	0.0000	0.0000	0.0000
35. Dummy industries, misc. adjustments	0.0000	0.0000	0.0000	0.0000

Note: The number of decimal places shown should not be interpreted as a measure of accuracy.

* The units of the first four sectors are Btu's/Btu of total sector output. The remainder have units as indicated above.

Source: EPRI.[5]

Table 2–6. Total Direct and Indirect Energy Use Coefficients. (10^6 Btu's/1958 Dollar of Final Demand)

Sector	1947	1958	1963	1967
1. Coal mining*	.01737	.03309	.03115	.02497
2. Electric utilities*	.26101	.19332	.17771	.14516
3. Petroleum refining*	.12592	.17712	.22594	.18294
4. Gas utilities*	.30751	.27196	.27326	.23607
5. Crude petroleum	.06996	.06971	.04396	.04777
6. Agriculture	.06133	.04903	.04877	.05911
7. Metal ore mining	.08855	.12925	.07679	.07358
8. Non-metal mining	.19430	.15685	.11417	.12956
9. Construction	.08991	.08575	.07732	.07206
10. Food and kindred products	.06735	.05951	.06347	.06114
11. Tobacco manufacturers	.03866	.02934	.02478	.02710
12. Textiles and apparel	.07577	.05598	.05381	.05502
13. Lumber and wood products	.07500	.05775	.05438	.04877
14. Furniture and fixtures	.08551	.06791	.06659	.05542
15. Paper and allied products	.15526	.13057	.13942	.12095
16. Printing and publishing	.05251	.04772	.05638	.04370
17. Chemical, plastics, drugs, and paint	.17709	.16349	.11992	.14153
18. Rubber and misc. plastics	.20602	.11803	.08064	.06677
19. Leather and footwear	.06617	.04721	.05045	.04436
20. Stone, clay, and glass products	.22386	.15035	.16440	.15675
21. Primary iron and steel	.37877	.34781	.26353	.24178
22. Primary non-ferrous metals	.15529	.13924	.17991	.12502
23. Fabricated metal products	.14663	.12677	.11307	.09526
24. Machinery	.10535	.07836	.07501	.05979
25. Electrical machinery	.09949	.06460	.06192	.04160
26. Transportation equipment	.12060	.07680	.07119	.05810
27. Scientific instruments	.07036	.05182	.05473	.04345
28. Misc. manufacturing	.07931	.06836	.06466	.04899
29. Transportation	.20960	.15839	.19519	.14417
30. Communications	.02474	.01385	.02471	.01797
31. Water and other	.07971	.10733	.16589	.10242
32. Commercial, financial, and services	.04506	.04046	.04107	.03499
33. Federal and local gov't enterprises	.09727	.14057	.18574	.07787
34. Imports	0.00000	0.00000	0.00000	0.00000
35. Dummy industries, misc. adjustments	.02167	.01572	.01709	.01252

* The units of the first four sectors are Btu's/Btu of final demand. The remainder of the sectors have units as indicated above.

Source: EPRI.[5]

Relationship Between Energy and Output

As shown in Fig. 2–8, there is a strong multicountry correlation between national output (GDP)* per capita and energy consumption per capita. This relationship approximately indicates that, for countries of over $1,000 of GDP per capita, a 10% rise in output per capita would require approximately an 8% rise in energy consumption per capita.

Undoubtedly, energy used in productive activity is one of the components of economic growth, just as proceeds of that economic growth and rising income increase the consumption of energy-powered comforts and service. The application of fuel and power is a critical element in the state of a nation's economy. However, the question for our concern is whether the United States has used more than the necessary units of energy to produce the equivalent amount of goods, when compared with other nations.

Comparison of Energy Consumption

A convenient way of comparing energy uses among societies is to compare their energy/GDP or energy/output ratio. The ratio derived by dividing a country's energy consumption by its gross domestic product gives an indication of how efficiently energy is consumed to produce a given value of output. Conventionally, the higher the ratio, the more energy-intensive the economy is considered to be.**

Figure 2–9 shows trends of energy/GDP ratios in 1953–1976 for seven selected industrial societies. It indicates that the United States is the most energy-intensive of the seven countries. For example, in 1972 the United States used almost twice as much energy for a given value of output than the lowest user, which was France. This highest position in overall energy intensiveness of the U.S. economy has been maintained stably since 1953, with a good margin over other economies. This is in contrast to other nations' energy consumption. For example, the ratios of the United Kingdom, France, and Germany started the period relatively high and gradually fell off. After 1973, the energy/output ratios of all seven countries declined, owing to rising oil prices and resulting worldwide recession. This comparison seems to indicate that energy consumption per unit of economic output is not immutable, and it is possible in other societies to produce the same amount of output with less energy consumed.

---

* GDP (gross domestic product) instead of gross national product (GNP) is used because net factor income originating in overseas enterprises and investments that do not use domestic energy are excluded.

** However, note that the ratio does not take into account the difference of economic and geographic structures among nations.

Energy consumption per capita
(tons oil equivalent)

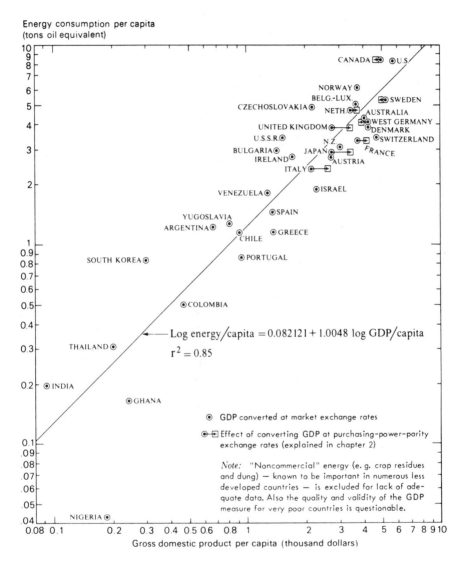

Fig. 2–8. Gross domestic product per capita versus energy consumption per capita for selected countries, 1972. Source: EPRI.[6]

In order to compare energy consumption among nations on a fairer basis, the industrial sector's energy consumption is chosen for further comparison. The reason for this choice is that industrial sectors are relatively homogeneous because of the nearly similar technologies and processes employed in the societies compared. Table 2–7 shows industrial energy consumption in nine

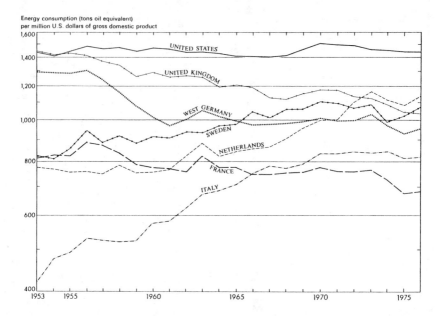

Fig. 2–9. Comparison of energy intensity among industrial societies. Source: EPRI.[7]

Table 2–7. Energy Consumption in the Industrial Sector Relative to GDP for Selected Societies, 1972.

Countries	Including Energy Sector* (Tons Oil Equiv. Per $ Million GDP)	Including Energy Sector (%)
United States	444	100.0
Canada	517	116.4
France	276	62.1
W. Germany	373	84.0
Italy	330	74.3
Netherlands	354	79.7
United Kingdom	399	89.9
Sweden	308	69.4
Japan	378	85.1

* Not including nonfuel energy use.
Source: Data obtained from RFF.[10]

selected societies. Except for Canada, the U.S. industrial sector consumed at least 10% more energy than the other industrial sectors to produce every dollar of GDP.

This energy intensity in the U.S. industrial sector might be attributed to relatively cheaper fuel prices. As shown in Table 2–8, 1972 industrial energy

Table 2–8. 1972 Relative Industrial Energy Prices* in Nine Selected Countries (U.S. = 1).

Prices	U.S.	Canada	France	W. Germany	Italy	Netherlands	U.K.	Sweden	Japan
Fuel									
Electricity	1.	1.61	1.99	2.66	2.47	2.37	2.26	1.36	2.31
Gas	1.	1.	1.82	1.94	1.76	1.24	1.35	NA	NA
Coal	1.	1.05	2.17	2.17	2.61	1.89	2.17	2.0	2.0
Petroleum products	1.	.79	1.32	1.04	1.25	.89	1.53	.96	1.14

NA: Negligible because of negligible consumption.

* Prices deflated by GDP rates of exchange.

Source: Data computed from RFF.[10]

prices other than petroleum prices in the United States were at least 80% lower than those for most other industrial nations (except Canada). It is difficult to suggest any correlation between high energy intensiveness and low energy prices. However, study in this area is needed so that conservation can be made more economically attractive for industry.

CONCLUSIONS

We have seen that energy use is essential for economic growth and employment. Improvement in energy efficiency over past decades has been slow because of an abundance of inexpensive energy. In order to maintain proper economic growth and employment, a sufficient energy supply is necessary, but energy has become scarcer and more expensive in recent years. More energy conservation is needed. In particular, conservation in industry seems to offer more opportunities technically and more incentives economically than in other areas.

REFERENCES

1. U.S. Department of Commerce, Bureau of the Census. *Annual Survey of Manufacturers, 1975,* September 1977.
2. U.S. Department of Energy. *The National Energy Plan,* A Report to Congress Required By Title VIII of the Department of Energy Organization Act (P.L. 95–91), DOE–0008, July 1981.
3. U.S. Department of Commerce, Bureau of the Census. *Historical Statistics of the United States, Colonial Times to 1970,* Bicentennial Edition. Washington, D.C.: U.S. Government Printing Office, 1975.
4. U.S. Department of Commerce, Bureau of the Census. *The U.S. Fact Book. The Statistical Abstract of the U.S. 1976.* New York: Grosset and Dunlap, 1976.
5. Electric Power Research Institute (EPRI). "An Input/Output Analysis of Energy Use Change from 1947 to 1958, 1958 to 1963, and 1963 to 1967," prepared by Battelle Pacific Northwest Laboratories, EPRI EA-281, November 1976.
6. EPRI. "How Industrial Societies Use Energy," prepared by Resources for the Future, EPRI EA-707-SY, February 1978.
7. EPRI. "Trends in Energy Use in Industrial Societies," prepared by Resources for the Future, EPRI EA-1471, August 1980.
8. U.S. Environmental Protection Agency. "Impact of Natural Gas Shortage on Major Industrial Fuel Burning Installations," Vol. III, EPA-450/3-77-017c, March 1977.
9. Starr, C. "Economic Growth, Employment and Energy," EPRI, March 1977.
10. Darmstadter, J. et al. *How Industrial Societies Use Energy,* published for Resources for the Future. Baltimore: Johns Hopkins University Press, 1977.

3

Energy Available for Industrial Use and the Role of Conservation

INTRODUCTION

In Chapter 2, Energy was shown to be an indispensable component of industrial production, employment, and economic growth. In particular, natural gas and oil are dominant in industrial energy use, and their shortages have caused local economic chaos in recent years. In this chapter, energy available for future industrial use and the role of conservation and other new energy technologies are discussed.

Recently, the industrial energy user has been encouraged to shift from oil and natural gas to coal and other fuels. This is the U.S. energy policy. However, coal may not be burned directly owing to environmental regulations, but needs to be converted to various types of synthetic fuels. The transition will take more than a decade; and conservation will have to play an important role in smoothing the transition, and in reducing future energy consumption even after the commercialization of synthetic fuels.

This chapter will consist of four sections: (1) methodology for forecasting industrial energy supply and demand, (2) the world and U.S. energy supply in 1990, (3) industrial energy availability and price in 1990, and (4) new energy technologies and conservation.

METHODOLOGY FOR FORECASTING INDUSTRIAL ENERGY SUPPLY AND DEMAND

In forecasting industrial energy availability and prices, two important factors must be considered: (1) energy demand is a derived demand, and (2) the

United States depends substantially on foreign energy sources, particularly oil. Energy itself is not final goods or service; its demand depends on the demand for goods and service it is used to produce. For example, the need for gasoline depends on the need for transportation service. When forecasting industrial energy, we also need to project the need for future services and goods that energy is used to produce. On the other hand, the United States has imported about one-half of its oil need from many countries, imported significant amounts of natural gas from Canada and Mexico, and exported coal to Western Europe and Japan. It is therefore necessary to consider the U.S. and world energy supply and demand to forecast energy use.

Forecasting energy outlook is a difficult and risky business. In the 1970s because of the energy crisis, a number of energy models were developed and updated to predict impacts of given energy policies. In particular, several large-scale computer energy models were developed. These models are complex and good only for the purposes they were designed for, at best. However, they are useful for repetitive energy projections. This chapter assesses and synthesizes some of the most recent projections.

Review of Alternative Approaches and Major Models and Studies

Alternative methods have been used in past years to forecast energy availability and prices. These methods range from a simple judgmental approach to a sophisticated engineering-economic method.

Some major approaches are as follows:

- Judgmental forecast
- Trend analysis
- Time series analysis
- Engineering analysis
- Economic modeling
- Mathematical program modeling
- Combination of any of the above approaches

Judgmental Forecast. A judgmental forecast is done by experts in the field. Although no formal sophisticated mathematical formulas are employed, this forecast can be very useful. Project Independence is an example of this approach.[4]

Trend Analysis. Trend analysis is a method that forecasts future events by extrapolating from historical trends.

Time Series Analysis. Time series analysis is similar to trend analysis. It analyzes the structure and patterns in the historical data and extrapolates from these structures and patterns for the future.

Engineering Analysis. Engineering analysis examines the characteristics of energy supply sources and estimates supply and cost of these sources based on engineering data. This approach is particularly useful when limited or no historical data exist. Examples are supplies and costs of new energy technologies such as coal gasification.

Economic Modeling. Economic modeling attempts to simulate future energy supply and price by using the relationships between the energy supplies/prices and other factors. These relationships are abstracted from available historical data. Often referred to as "econometric modeling," economic modeling can be expensive to develop and operate. Several energy models use this approach, including the Data Resources Inc. (DRI) and the Wharton energy models.

Mathematical Program Modeling. Mathematical program modeling attempts to simulate future energy supply by optimizing a social objective, subject to production and technical constraints. This approach includes linear, nonlinear, dynamic, and mixed-integer programming. This modeling is comprehensive and flexible but can be expensive to develop and operate. The Brookhaven Energy Systems Optimization Model (BESOM) is an example.[5]

Choosing an Approach. The choice of a particular forecasting approach depends on a number of factors: forecast purpose, data availability and quality, desired forecast reliability, and resources available to develop forecasts. For example, for some small-size producers, a combination of trend analysis and judgmental forecasts may be cost-effective in predicting their fuel price and supply.

Major Energy Models and Studies. During the past decade, a vast array of models and studies has been developed to forecast the U.S. energy future. The number of models and studies is so enormous that it is impossible to compile an exhaustive list. Since industrial energy use is a major component of energy, the supply and price of the aggregate industrial sector have been discussed in most models and studies. However, some of these models have not forecast the supply and price in detail for lower than two-digit SIC industries. Some of the energy studies and models are as follows:

- "Outlook for Energy in the U.S. to 1985," the Chase Manhattan Bank, 1972.[1]
- *A Time to Choose,* Ford Foundation, 1974.[2]
- "National Energy Outlook" [using the results of the Project Independence Evaluation System (PIES)], Federal Energy Administration, 1976.[3]
- "Energy Outlook," Exxon Company, U.S.A., 1977 and 1979.[6]
- "Workshop on Alternative Energy Strategies (WAES), Energy Global Projects, 1985–2000," G. Wilson, 1977.[9]
- "Energy Information Administration (EIA) Annual Report to Congress" [using results of the Midterm Energy Forecasting System (MEFS)], EIA, Department of Energy, 1978, 1979, 1980.[7]
- "Energy Review" (using results of the DRI Energy Model), DRI, 1978.[8]
- "Alternative Scenarios to Develop R&D Strategies" [using results of the Brookhaven Energy Systems Optimization Model (BESOM)], Brookhaven National Laboratory, 1978.[10]
- The Stanford University PILOT Model, developed by G. Dantzig.[11]
- The Stanford University ETA-MACRO model, developed by A. Manne.[12]
- The FOSSIL 2 model, developed by Dartmouth University and operated by the Office of Policy and Evaluation in the Department of Energy.[13]
- The DOE Long-Range Energy Analysis Package (LEAP), developed by Decision Focus, Inc. on the basis of the SRI-GULF Model[14] and operated by the Energy Information Administration (EIA) in DOE.[15]
- The Wharton Energy Model, developed by the Wharton Econometric Forecasting Associates.[16]

The DRI and Wharton energy models are the most comprehensive in detailing energy availability and prices for lower SIC level industries. However, they are proprietary and expensive to subscribe. The most accessible and updated model is the EIA's Midterm Energy Forecasting System (MEFS). The MEFS is the refined version of the early Project Independence Evaluation System (PIES). It is a national energy forecasting system used to make forecasts of energy prices and supplies at a regional level. Because the U.S. Congress requires EIA to present annual energy projections, the results of the MEFS are thus published annually in the EIA "Annual Report to Congress" and are widely disseminated. Their supporting data base models are well documented, and continually updated and refined.

Method for Forecasting Industrial Energy Price and Availability

Results of the EIA's Midterm Energy Forecasting System and other available projections will be compared to forecast industrial energy price and availability. EIA's results are available in the 1979 EIA "Annual Report to Congress"; other important sources for energy forecasts are the recently published "Exxon World Energy Outlook"[17] and "Exxon Energy Outlook."[16]

In using these projections, we recognize some difficulties observed from past forecasting experience. Three major difficulties are overly optimistic energy projection, shortsightedness of U.S. energy policy and planning, and uncertainty in world oil prices and supply.

Overly Optimistic Energy Projection. Comparisons of earlier energy projections indicate that the earlier the forecast, the more optimistic the U.S. energy supply prediction. Table 3–1 shows three sets of forecasts of the U.S. energy outlook for 1985. The forecasts were made in 1972, 1974, and 1976 by the Chase Manhattan Bank, Ford Foundation, and Federal Energy Administration (FEA), respectively. We observe the following:

- It was forecast by the Chase Manhattan Bank in 1972 that total 1985 U.S. energy use would be 127.4 Q; however, a figure of 98.9 Q was projected by the FEA in 1976. The difference between the two forecasts is about 30%.
- In the Chase forecast, it was optimistically predicted that nuclear energy could replace a large portion of oil in 1985. It was estimated that, in 1985, nuclear energy would contribute 35% of the total energy need. In reality, nuclear energy in 1985 might not contribute much more than it has to date, which is 5%.

Shortsightedness of U.S. Energy Policy and Planning. In 1970 in the United States there was battling among producing and consuming regions about increasing imports of low-cost foreign oil. President Nixon appointed the Cabinet Task Force on Oil Import Controls on March 26, 1969, to make an in-depth study and provide recommendations. The task force recommended an increase in oil imports from the Middle East to a maximum of 10% of U.S. oil demand (a "security ceiling") and also from Canada, Latin America, and Mexico to about 5% of the U.S. need by mid-1973. In September 1973, the Arab oil embargo started. At that time, the United States imported less than a quarter of its total oil consumption; however, today the United States imports almost 50% of its total oil.

In the 1960s and early 1970s, the manufacturing and utility sectors were required by environmental regulations to switch their combustors from burning coal to burning oil and gas. However, because of an oil shortage, they have been asked to switch back to burning coal by 1990. This shortsightedness of U.S. policy makes the forecasting of long-term industrial energy use very difficult.

Table 3-1.   A Comparison of 1985 U.S. Energy Forecasts.

Forecasts: Publication Date:	CHASE[1] 1972	FORD[2] 1974	FEA[3] PIES—1976
Total U.S. energy use (Quads)	127.4	115	98.9
Change from 1980 (%)	56.2†	41.9†	21.2†
Proportion (%) from:			
OIL	17	43.4	39.8
Domestic	8.4	24.3	26.9
Old sources		23.5	
New sources		0.8	
Imported	8.6	19.1	12.9
GAS	11	27	23.3
Domestic	9.2	23.5	22.0
Old sources	8.4	22.6	
New sources	0.8	0.9	
Imported	1.8	3.5	1.3
COAL	29	17.4	22.8
Eastern			14.5
Western			8.3
Burned			
Liquefied			
Gasified			
HYDROELECTRIC	8	2.6	3.2*
NUCLEAR	35	8.7	9.6
Nonbreeder			
Breeder			
SOLAR	N/A	**	
Solar electric			
Biomass			
Ocean			
Wind			
Solar thermal			
GEOTHERMAL	N/A	0.9**	
OTHER	0		1.3

* Including hydroelectric and geothermal.
** Including solar and geothermal.
† Exxon Company's 1980 projection used as the base year for calculating percentage change.[6]

Uncertainty in World Oil Prices and Supply.   The United States imports almost one-half of its oil, a significant portion of this imported oil coming from the Middle East and other OPEC countries. The oil market is sensitive to Middle East politics, and its prices and supply are subject to political changes. Table 3-2 shows the impact of world oil prices on projections for U.S. energy supply. Depending on the level of world oil prices, the 1990 U.S. energy supply forecast varies from 104.8 (world oil price at $16.00/barrel) to 87.9 Q (world oil price at $44.00/barrel) in 1979 dollars.

| | 1978 Actual | 1977 Annual Report | | 1990 Projections | | | | | |
| | | | | 1978 Annual Report | | | 1979 Annual Report | | |
		Series C	Series F	C Low	C Middle	C High	C Low	C Middle	C High
World Oil Price (Dollars per Barrel)	15.50	17.00	27.00	16.00	20.00	26.00	27.00	37.00	44.00
Domestic energy supply									
Oil	20.7	20.1	23.5	21.9	23.1	25.8	18.1	19.6	20.3
Gas	19.5	16.7	17.4	16.7	17.4	18.3	18.3	18.7	18.7
Coal	15.0	27.5	29.4	28.7	31.2	33.6	28.5	29.3	29.5
Nuclear	3.0	10.3	10.4	9.5	9.4	9.5	8.1	8.2	8.1
Other:	3.0	5.0	5.0	3.5	3.5	3.5	3.6	3.7	3.6
Subtotal, domestic production	61.2	79.6	85.7	80.3	84.6	90.7	76.7	79.5	80.3
Oil imports	17.1	28.8	20.5	24.5	17.0	7.8	17.0	11.7	9.5
Gas imports	0.9	2.6	2.5	2.1	2.0	0.95	1.8	0.8	0.8
Coal imports	−0.9	−2.1	−2.1	−2.1	−2.1	−2.1	−2.7	−2.7	−2.7
Subtotal, net imports	17.2	29.3	20.9	24.5	16.9	6.65	16.1	9.8	7.6
Total supply	78.4	108.9	106.6	104.8	101.5	97.3	92.8	89.3	87.9
Supply prices									
Crude oil (dollars per barrel)									
Domestic (wellhead)	9.80	15.32	22.76	15.54	18.92	24.33	26.29	35.71	43.14
Imported-landed U.S.	15.86	16.71	27.21	16.36	20.15	25.66	27.00	36.54	44.07
Average refinery acquisition cost	13.57	16.17	24.87	16.18	19.78	25.31	26.89	36.40	43.88
Natural gas (dollars per million Btu)									
marginal price Southwest	2.19	2.56	2.43	2.41	2.62	2.82	3.43	3.68	3.40
Coal (mine entrance, dollars per ton)									
High-sulfur bituminous,									
Northern Appalachia	NA	27.63	28.26	31.34	32.49	33.72	34.92	34.92	35.79
Low-sulfur subbituminous,									
Northwestern Great Plains	NA	10.20	10.80	10.40	10.40	10.40	9.40	9.40	9.40
Rate of Growth in Real GNP[a]	NA	3.70	NA	3.60	3.50	3.40	2.80	2.60	2.50

[a] Data represent compound rate growth in real GNP from the years preceding the date of the Annual Report to 1990.

Source: EIA.[7]

35

These forecasts assume that, despite oil price changes, there will be no drastic shortage such as an oil embargo.

THE WORLD AND U.S. ENERGY SUPPLY IN 1990

Because the United States imports oil and natural gas, and exports coal in the world oil markets, the world energy supply is important to the U.S. energy market. On the other hand, because the industrial sector competes for energy with the other three sectors, the U.S. energy outlook will strongly affect industrial energy availability and prices. The world and U.S. energy issues that will affect energy use in the industrial sectors are presented below.

World Energy Supply

Assuming slower world economic growth (3.5% per year) and less energy intensity, compared to the pre-energy crisis period (i.e. before 1973), the 1990 world energy supply is shown in Fig. 3–1. The 1990 world energy supply would be about 270 quadrillion Btu.[17] Oil will still contribute about one-half of the 1990 world energy need; coal will contribute 20%; and natural gas will provide 18%. This projection indicates a slight increase (2%) in the share of coal and a significant reduction (9%) in the share of oil, which is forecast to be replaced by nuclear energy and synthetic fuels. As shown in the figure, most energy needs will still be met by conventional energy sources in 1990. Synthetic fuels will contribute about 2%;* solar energy will contribute less than 0.5% in 1990. Unconventional gas sources such as tight sands, Devonian shale, and the geopressurized zone will not be available in large quantity in the early 1990s.

It is expected that the cost of energy supplies, adjusted for inflation, will rise significantly in 1990 because the most accessible and lowest-cost energy sources have been almost developed. A rising proportion of the oil and natural gas supply may come from basins in the frontier areas, along with more costly production methods. On the other hand, synthetic fuels production, available on a small scale in 1990 to supplement oil and natural gas,* will be much more costly than oil and gas from the historical sources (i.e., the Middle East). A long lead time may also be expected for any new energy technologies to be commercialized. These factors will thus cause an upward trend of energy costs in 1990.

---

* Depending on the Reagan Administration's energy policy, it may be substantially smaller than 2%.

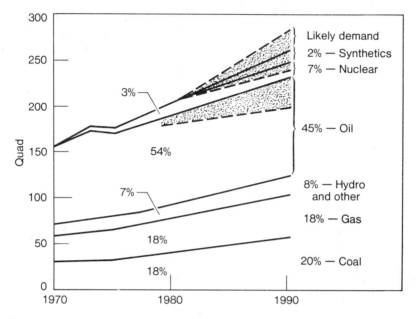

Fig. 3–1.   World energy supply projections in 1990. Source: Data obtained from Exxon.[6]

The world energy supply will probably not meet the world energy demand in 1990 if the present energy consumption pattern continues (i.e., no further substantial conservation program is implemented). This uncertainty is shown in the shaded area in Fig. 3–1. There will be a world energy shortage if any of the following three situations occurs:

- The likely demand exceeds the projected supply as shown in Fig. 3–1 (i.e., the dotted demand line which is on top of the solid supply line).
- Although the projected demand does not exceed the projected energy supply, the projected supply from nuclear energy and synthetic fuels fails to materialize.
- Even if the above two situations do not occur, if a political or military maneuver is initiated in the Middle East, oil may not be able to flow out of the producing countries in the Middle East.

World Energy Demand.   Assuming that slower economic growth (3.5% annually) and less energy intensity will persist between 1980 and 1990, and that annual energy demand growth rates in the United States, Canada, Europe, Japan, and others are 0.8, 2.4, 2.0, 2.9, and 5.8%, respectively, the total

1990 world energy demand would be about 270 quadrillion Btu.[17] This represents an increase over the 1978 level of about one-third. Energy demand will vary significantly from this projection if economic growth is more rapid than expected or if projected energy conservation does not materialize. For example, the assumption of the annual U.S. energy demand growth rate of 0.8% is extremely low, compared with the actual growth of 4.3% in 1965–1973 and that of 1.0% in 1973–1978. Continuous conservation efforts must be implemented in the United States between 1980 and 1990 in order to reduce the energy growth rate to 0.8% annually if a basic economic growth is to continue.

Any variation in energy demand growth rate will substantially change the 1990 world energy demand. For example, if economic rates vary by 0.5% per year (4% instead of the projected 3.5%), energy demand could vary by about 12 Quads in 1990. This is indicated by the top shaded area in Fig. 3–1.

Failure of Supply from Nuclear Energy or Synthetic Fuels. Nuclear energy probably will not contribute its projected share of energy needs because of the recent Three-Mile Island accident. The prospect of the nuclear industry depends entirely on the public's attitude toward nuclear safety and waste disposal, and the effort of the industry and government to reduce the public's suspicions. In addition, the commercialization of synthetic fuels may not occur in 1990 in the projected quantity (i.e., 2% of world energy needs or 2.6 million barrels/day).

Oil Disruptions in the Middle East. Any prolonged disruption of oil production, transportation, and distribution of Middle East oil will cause a world energy shortage and push world energy prices drastically upward. An oil price of $100 (in constant 1980 dollars) per barrel in 1990 is not unthinkable, given the current political instability in the Middle East.

U.S. Energy Supply

The U.S. energy supply outlook in 1990 is strongly affected by specific assumptions about future circumstances. Any significant changes from these assumptions would substantially modify the projections shown below.

Assumptions. Assumptions regarding the U.S. economy, energy policy, oil imports, world oil prices, and the environment are discussed below. These are abstracted from the EIA 1979 "Annual Report to Congress."[7]

The Economy.   There will be a slower rate of economic growth, on the average just under 2.6% annually through 1990,* compared with 5.3% over the 1965–1973 period. The reason for this slowdown is lower productivity and slower growth rates in the labor force, as well as higher energy costs (which will cause investments to be made for saving energy in order to maintain the current standard of living).

Energy Policy.   Energy laws, conservation programs, and new technology studies will encourage higher domestic energy production and lower energy demand. Major legislation includes the Natural Gas Policy Act (NGPA) and the Powerplant and Industrial Fuel Use Act (FUA). On the supply side, NGPA will phase out most price controls on natural gas at the wellhead by 1985; the Carter Administration's crude price decontrol plan would have brought domestic crude prices in parity with world prices by the end of 1981. In addition, subsidies for new technologies such as synthetic fuel production stimulate the development of new sources. On the demand side, FUA will force industrial plants and major electric utility plants to substitute coal for oil and natural gas. Conservation programs such as the Building Energy Performance Standard (BEPS) and the Auto Efficiency Standards (AES) mandate improved energy efficiency, and thus will reduce energy consumption in the residential, commercial, and transportation sectors.**

Oil Imports and World Oil Prices.   Increasing imports from OPEC nations will be required and also allowed. World oil prices will increase about 2.5% annually after an adjustment for inflation between now and 1990. Accordingly, the world oil price in 1990 will be $37/barrel in mid-1979 dollars, compared with the price of $28.90/barrel in December 1979.

The Environment.   Existing environmental regulations will be complied with, but no more severe regulations will be expected. Environmental goals will be achieved at a rate compatible with economic growth and energy development.

Projection for 1990 U.S. Energy Supply.†

As shown in Fig. 3–2, the 1990 U.S. total energy supply is projected to be about 89 quadrillion Btu.[7] In this 1990 supply, oil would still be the most important source of energy, accounting for 35% of the total U.S. 1990 energy needs. Coal would

---

* TRENDLONG 2004, recently published by Data Resource Incorporated, is used.

** Because of the Reagan Administration's energy policy, these assumptions are subject to change.

† This projection was too high. DOE has already lowered it by 5%.

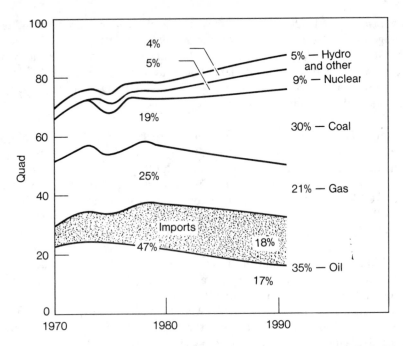

Fig. 3–2. U.S. energy supply in 1990. Source: Table 3–2, "Annual Report, Middle World Oil Price" [7]—$37/barrel with some modifications.

be second, growing from 1980's 19% to 30%. Natural gas would be third with 21%. Nuclear energy is estimated to grow from 5% in 1980 to 9% in 1990, even though significant difficulties and uncertainties are facing the nuclear industry. Much of the growth in coal supply is projected to come from western coal. The share of western coal is expected to reach 45% of total coal production in 1990, compared to 32% in 1980. Hydro, geothermal, and solar power combined are estimated to supply 4% of the total U.S. energy requirements in 1990. The growth of hydro and geothermal power is limited because of site availability. Solar applications, mostly water heating, will also be limited because of high initial costs and uncertain markets. Unconventional energy sources such as synthetic fuels and tight-sand gas will be insignificant in 1990. An optimistic projection for synthetic fuels in 1990 is between 0.6 and 3* quadrillion Btu. (This is discussed further in the section "New Technologies and Conservation" later in this chapter.)

Although oil's share of total energy supply declines from 46% in 1980

---

* Projected by Exxon,[6] these fuels include coal gas, naphtha, and liquefied petroleum gas, shale oil, and coal liquids.

to 35% in 1990, the share of oil imports is expected to increase from a 1980 level of about 45% to a 1990 level of more than 50%. The magnitude of the share depends on world oil prices and synthetic oil development. The bulk of these imports will continue to be supplied by OPEC members. The availability and price of these imports will therefore largely depend on the policies of the Middle East OPEC governments, which are politically unstable. Consequently, it is clear that U.S. energy shortages may occur at any time in the coming years if there is any disruption of the Middle East oil flows.

INDUSTRIAL ENERGY AVAILABILITY AND PRICE IN 1990

This section provides some general trends regarding industrial energy outlook by objectively observing the Energy Information Administration's and other recently published projections.[6,7] (Readers who need projections for specific industries should consult either the DRI[8] or the Wharton Energy Model,[16] or build their own industry-specific energy models.)

Assumptions

The assumptions for 1990 projections of industrial energy availability and price are the same as those listed under "U.S. Energy Supply" in the preceding section.

Overview of the Forecast

Because decisions in industry are largely based on cost factors, there will be rapid shifts from oil to alternative fuels by 1990, particularly coal consumption, which more than doubles in 1990. New industrial boilers, which produce steam by burning oil, gas, or coal, invariably will choose coal or gas. Existing oil-fired industrial boilers will either convert to coal or gas, or will be retired earlier than the end of their physical life. In the electric utility sector, existing oil-fired generating plants will be retired earlier or converted to coal as quickly as their public utility commissions permit. Oil backout legislation for the utilities to shift from oil burning to coal burning is in progress in Congress.

This conversion will be enhanced by the FUA and conservation programs. FUA forces industrial plants and electric power plants to substitute coal for oil and gas. Conservation programs such as Building Energy Performance Standards and the Voluntary Industrial Energy Conservation Targets mandate improved energy efficiency.

Even with the conversion of oil-fired plants to coal, petroleum product prices rise faster than those of almost all other fuels. The exception is natural gas, which includes a surcharge related to the price of residual fuel oil. The surcharge is a result of the incremental pricing provisions of the Natural Gas Policy Act. Electricity prices rise most slowly because of overcapacity and reduced demand due to customers' intensive conservation efforts.

One caveat must be carefully observed for the industry energy forecasts made in this section: The forecasts have not systematically addressed energy savings from likely changes in processes or the installation of more efficient combustors. These forecasts simply extrapolate the trend of the industrial energy conservation efforts occurring between 1973 and 1978. As the U.S. General Accounting Office pointed out, the reduction in energy use between 1973 and 1978 was largely the result of bookkeeping efforts such as light turn-offs and machine tune-ups. Further energy savings must be achieved through investments in process changes and large-scale retrofitting in existing facilities. Being unable to take into account the likely savings from process changes, these energy demand forecasts are likely to be greatly overestimated (i.e., the projection would be much higher than the actual 1990 observations).

The 1990 Forecast

The 1990 projections for industrial energy availability and price are shown in Fig. 3–3 and Table 3–3. Table 3–4 shows the growth rates for energy uses and prices. These projections are taken from the Energy Information Administration's 1979 Annual Report to Congress, middle world oil price $37/barrel.[7]

Industrial Energy Availability. The industry energy availability shown in Fig. 3–3 and Table 3–3 represents a highly overestimated view for future industrial energy needs. Most of the oil and a significant portion of natural gas will be replaced by coal after 1985 because of high oil prices, the PUA conversion requirement, and the NGPA gas incremental pricing provisions. This view, which is clearly shown in Fig. 3–3, is supported by the expectation that the industrial sector will respond promptly to these economic and regulatory pressures, and that coal and coal-fired equipment will be available without uncertainty about environmental regulations.

Oil consumption declines over 20% annually through 1985, from about 20% of the total in 1978 to 5% in 1985 and 1990. In particular, distillate and residual fuel oil provided nearly 12% of the 1978 total industrial energy needs; by 1990, it accounts for only about 2% of industrial fuel use.

In contrast, industrial coal consumption, which declined over the last few decades for environmental and economic reasons, increases substantially to replace oil use. By 1990, coal accounts for nearly 27% of industrial energy, compared with 15% in 1978. Most of this projected increase in coal consumption is in boilers; some of the increase is in process heat applications.

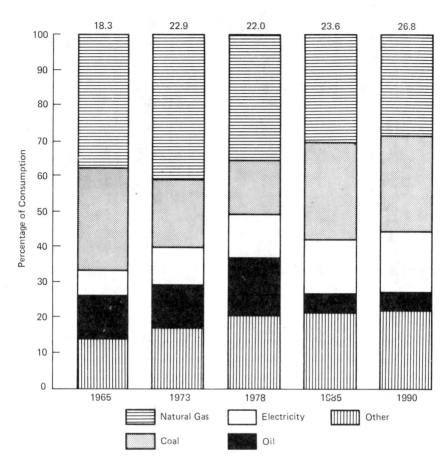

Fig. 3-3.   Industrial energy fuel shares by source: medium world oil price, 1965-1995 (quadrillion Btu per year). Source: EIA.[7]

Natural gas is used in heaters and boilers, and as feedstock and raw material. Because of the Powerplant and Industrial Fuel Use Act and the Natural Gas Policy Act, boiler natural gas use declines between 1978 and 1985. Beyond 1985, increases in the process heat use of natural gas outweigh this decline. On the other hand, the raw material and feestock use of natural gas increases throughout 1990. By 1990, natural gas accounts for about 30% of industrial energy requirements, compared with 36% in 1978.

Electricity use increases significantly between 1978 and 1990. It accounts for about 19% of the industrial market by 1990, compared with 12% in 1978 (this excludes energy loss allocated for power generation, transmission and distribution). Part of this increase is to replace oil use.

Table 3-3. Industrial Energy Consumption and Prices: History and Projections for Three Base Scenarios, 1965–1990 (Quadrillion Btu and 1979 Dollars per Million Btu).

Fuel	History[a]			Projections					
				1985			1990		
	1965	1973	1978	Low	Mid	High	Low	Mid	High
Electricity	1.50	2.30	2.70	3.60	3.60	3.60	4.50	4.60	4.60
Price	6.41	5.96	8.34	11.27	11.32	11.48	11.88	12.18	12.26
Distillate[b]	0.70	0.90	1.20	0.30	0.30	0.30	0.40	0.40	0.40
Price	2.29	2.33	3.60	5.62	6.34	7.46	5.66	7.18	8.62
Residual[b]	1.20	1.30	1.50	0.20	0.10	0.10	0.20	0.10	0.10
Price	1.24	1.72	2.49	4.65	5.49	6.50	4.70	6.22	7.42
Liquid gas	0.30	0.60	0.80	0.70	0.70	0.60	0.80	0.70	0.60
Price	1.90	2.37	3.42	6.71	7.45	9.56	6.71	8.83	10.80
Coal	5.40	4.40	3.40	6.50	6.50	6.50	7.10	7.30	7.20
Price	1.03	0.98	1.34	2.09	2.10	2.10	2.24	2.26	2.28
Natural gas[b]	6.80	9.60	7.90	7.60	7.20	7.10	8.70	7.80	7.50
Price	0.76	0.75	1.56	3.36	3.47	3.56	4.06	4.85	4.91
Other[c]	2.50	3.80	4.50	5.20	5.10	5.10	6.20	6.00	5.90
Total consumption	18.30	22.90	22.00	24.00	23.60	23.30	28.00	26.80	26.30
Average price	1.61	1.73	2.98	4.86	4.97	5.13	5.40	5.90	5.97

[a] Sources of historical data are (1) *State Energy Fuel Prices by Major Economic Sector* 1960 to 1977, Preliminary Report and Documentation, July 1979, DOE/EIA-0190 and (2) *State Energy Data Report*, Statistical Tables and Technical Documentation 1960–1978, DOE/EIA c214(78).

[b] Includes refinery consumption, excluding raw materials.

[c] Includes feedstocks, raw materials, and refinery consumption of still gas and oil.

Note: The analysis of industrial energy consumption treated the following four components separately: process heaters and small boilers, major boilers, refineries, and raw materials. The text discusses some of the different trends in these areas.

Source: EIA.[7]

Table 3–4. Industrial Energy Consumption and Prices: Compound Annual Growth Rates, Projections for Mid-World Oil Price Scenario 1965–1995 (Percent).

Fuel	History[a]		Projections	
	1965–1973	1973–1978	1978–1985	1985–1990
Electricity	6.1	3.1	4.0	4.9
Price	−0.9	7.0	4.5	1.5
Distillate	3.7	5.1	−16.2	1.6
Price	0.3	9.1	8.4	2.5
Residual	1.4	1.8	−29.8	3.3
Price	4.1	7.7	12.0	2.5
Liquid gas	9.3	5.7	−2.5	1.5
Price	2.8	7.6	11.8	3.5
Coal	−2.6	−4.7	9.7	2.1
Price	−0.7	6.6	6.6	1.5
Natural gas	4.4	−3.8	−1.3	1.5
Price	−0.2	15.8	12.1	6.9
Other	5.1	3.3	1.9	3.1
Total consumption	2.9	−0.9	1.0	2.6
Average price	0.9	11.5	7.6	3.2

[a] Sources of historical data are (1) *State Energy Fuel Prices by Major Economic Sector* 1960 to 1977, Preliminary Report and Documentation July 1979, DOE/EIA-0190, and (2) *State Energy Data Report*, Statistical Tables and Technical Documentation 1960–1978, DOE/EIA c214(78).

Note: The analysis of industrial energy consumption treated the following four components separately: process heaters and small boilers, major boilers, refineries, and raw materials. The text discusses some of the different trends in these areas.

Source: EIA.[7]

Finally, we should note that the industrial fuel mix, excluding raw materials and feedstocks, is senstive to changes in world oil prices, owing to the substitution of natural gas for oil in process heaters. However, because FUA restricts the use of gas in boilers with capacities greater than 100 million Btu per hour, coal would be the only economic boiler fuel, and thus the fuel mix for industrial boilers alone is not sensitive to changes in world oil prices.

Industrial Energy Prices. Projections for 1985 and 1990 industrial energy prices are shown in Table 3–3; the price growth rates are shown in Table 3–4. Prices for petroleum products are the highest, excluding electricity which is a converted fuel. Coal is currently the least expensive fuel; in 1978, its price per Btu was about 14% below that of natural gas and 46% below the price of residual oil. This price advantage increases over the forecast period. In 1990, coal price per Btu is only one-half of that of natural gas and less than one-third of the price of petroleum products. Even though the coal price is low, the capital costs for transporting and burning coal within environmental regulation limits are substantially higher than those of natural gas and oil. The competitiveness of coal is improved by NGPA and FUA as previously discussed.

Petroleum prices rise faster than those of all other fuels except for natural gas. According to NGPA, the natural gas price for industrial customers includes a surcharge related to the price of residual fuel oil.* The surcharge increases with the price of oil, thereby raising the price paid by industrial customers.

Electricity prices rise slowest because of customers' conservation efforts and utilities' overcapacity in the next decade. The prices of electricity grow even slower than coal in 1978–1985. The primary reason for this slow growth is the significant use of nuclear energy, which offsets a part of coal price increases.

The Role of Energy Conservation. The overall industrial energy intensity, as measured by energy consumed in Btu per constant dollar of manufac-

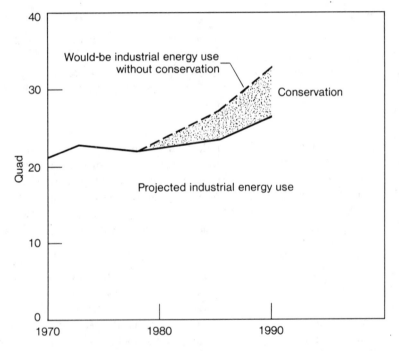

Fig. 3–4. Energy from conservation in the industrial sector (the shadow area).

* See Chapter 9 for further discussion.

turing value added, is expected to decline further between now and 1990, owing to intensive conservation efforts. Industrial energy intensity is estimated to be reduced by 14% in 1985 and 23% in 1990, compared with that of 1978.[7] In other words, industrial energy savings will be about 3.5 quadrillion Btu in 1985 and 6 quadrillion Btu in 1990 (as shown in Fig. 3–4). Savings in the industrial sector will mostly come from energy management, retrofitting of industrial facilities, and recycling.

Because savings from changes in processes have not been accounted for in this forecast (as previously discussed), projected energy savings in 1990 can even be significantly higher if the market penetration of new energy-efficient processes can be forecast and quantified. Potential energy savings from changing processes are much higher than those from retrofitting, recycling, and energy management.

NEW ENERGY TECHNOLOGIES AND CONSERVATION

In the previous sections, conventional energy source (i.e., oil, gas, coal, nuclear, and hydro) and conservation were highlighted because they will provide most of the energy needs in 1990. However, a few prominent new energy sources are likely to be available commercially in 1990 although their production would be insignificant in quantity.

Overview of New Energy Sources

A significant number of new energy technologies have been commonly mentioned as future potential energy sources. The following technologies can be commercialized by 1990 and are of importance to the industrial sector:

- Coal gasification
- Coal liquefaction
- Solvent-refined coal
- Shale oil
- Biofuels—methanol, medium- and low-Btu gas
- Fuel cell
- Geothermal energy
- Solar (heating and cooling)

Among these sources, the first five technologies can produce synthetic fuels, which most likely will be available to industrial users. Fuel cell energy has the potential for being used in sugar refineries and in pulp and paper mills in 1990 to produce electricity and steam simultaneously. However, the application of solar and geothermal energy depends on locality, and the temperature generated from this application may not be sufficient for most industrial processes although these energy sources have been available for use.

Coal Gasification. Coal gas was popular until cheaper natural gas took over. Now it is becoming important again as a pollution-free fuel. Coal gasification is a process that converts coal into high-Btu or medium-Btu gas. The basic process involves crushing coal to powder, and then heating this material under pressure of steam and oxygen to produce gas. In situ gasification is a promising method, that is, burning the coal while it remains underground to produce gas. An optimistic estimation for coal gas is 0.6 quadrillion Btu (0.6 TCF) in 1990.[18]

Coal Liquefaction. Coal liquefaction is a process that converts coal into liquid hydrocarbons and related compounds by hydrogenation. Bituminous coal is pulverized and treated with hydrogen gas under appropriate temperature and pressure to form liquid fuels. An optimistic estimation of coal liquids is 0.1 quadrillion Btu in 1990.[18]

Solvent-Refined Coal. The solvent-refined coal (SRC) process is used to purify coal. Pulverized coal is mixed with an aromatic solvent and reacted with hydrogen gas in a hot, pressurized process. Coal is thus dissolved and gives up ash and other impurities before becoming solidified again as an ash-free, concentrated fuel. SRC contains about 16,000 Btu per pound, with an ash content of 0.1% and a low sulfur content of about 0.5%. It is solid at room temperature but can be liquefied. This product will be used in industrial boilers and electric power plants. No estimation of production is made here.

Shale Oil. Shale oil has good potential for commercial production in 1990. The Green River formation in Colorado and Utah has the potential of producing oil at about 1200 thousand barrels per day for four centuries.[22] The development of this resource base will meet formidable challenges from environmentalists and local residents because of water and solid-residue issues, air-quality regulations, and worker-relocation problems. However, thanks to recent successes in developing in situ processes and a relaxation of environmental standards, shale oil production is optimistically expected to be between 1.3 and 1.9 quadrillion Btu* in 1990.

---

* Exxon projected synthetic oil (coal liquids and shale oil) to be between 1.4 and 2 quadrillion.[6] This projection is used in subtracting the coal liquid estimation of 0.1 quadrillion.

Biofuels. Biofuels are fuels derived directly or indirectly from biomass (organic plant material). The production and collection of biomass and its use, or conversion to fuels, is generally referred to as bioconversion. Biomass can be converted into gaseous or liquid fuels by thermal or biological processes. Biofuel can be widely used by current biomass producers (agricultural and forestry) and by small to medium-size industries located near available sources. The sources of biomass are forestry, agricultural, and municipal wastes and residues. The maximum economic potential for biofuels is estimated at 5% of the U.S. fossil fuel consumption.[19] Projected production of biofuels in 1990 is near zero.[18]

Fuel Cell. Fuel cell power plants convert the energy of a fuel directly to electricity by an electrochemical process, which is characterized by higher efficiency, lower environmental pollution, and lower initial capital costs than conventional power generation. The fuel cell process can be used to cogenerate electricity to save energy in sugar refineries and in pulp and paper mills. A fully matured commercial product of fuel cell power plants would probably not be available before 1990, even though initial units could be operated by 1985.[20]

Geothermal Energy. Geothermal energy consists of superheated steam and liquid water trapped in geological structures. It is regionally concentrated in the Southwest, predominantly in California, Idaho, and Utah. For industries located in these areas, geothermal energy may provide a good alternative energy source. However, there are limitations in application of geothermal energy: When geothermal fluids are extracted and used, dissolved salts and metals in the geothermal brines will corrode the well pipes and ultimately build a scale on the pipes that restricts the flow of the fluids. Identified reserves can reach 79.4 quadrillion,[21] but the actual implementation in 1990 is expected to be insignificant.

Solar Energy. Solar energy, for practical purposes, is eternal, and it can include wind, tidal, and ocean thermal energy. The technology available by 1990 for industrial use is mainly solar heating and cooling. Solar collectors normally supply water to a heating system at less than 140°F to 180°F. This temperature range being far below that of process needs, an auxiliary back-up heating system is required. The duplication of solar and conventional heating systems raises the magnitude of initial capital investments to such an economically undesirable level that solar energy may not be widely used in industries in 1990.

Importance of Conservation during the Development of New Energy Sources

Comparison of energy from industrial conservation and new technologies in 1990, as shown in Table 3–5, indicates that energy conservation is far more productive than new technology in terms of providing energy. Because new technologies may not provide needed energy, it is clear, for most industries, that the most reliable source of energy in the coming decade would be conservation. Conservation efforts include large-scale retrofitting of existing facilities and replacing current equipment with more energy-efficient processes.

Table 3–5. Comparison of Energy from Conservation and New Technology in 1990.

Technology	Energy Produced or Conserved in 1990 (10^{15} Btu)
Conservation	larger than 6
Coal gas	0.6
Coal liquids	0.1
Solvent-refined coal	NS
Shale oil	1.3–1.9
Biofuels	~0
Geothermal energy	NS
Solar energy	NS

NS: Not significant.

CONCLUSIONS

Energy projections for the world, the United States, and the industrial sector have been examined in this chapter. The world energy need may exceed the world energy supply if conservation does not take effect, if new energy sources are not timely, or if interruptions occur in the Middle East oil flows. Since the United States will import about one-half of its oil needs, its energy outlook will be subject to world energy crunches. On the other hand, the development of new energy sources in the United States is sensitive to political intentions; therefore, new energy sources may not contribute as much as expected in 1990. The only reliable solution in the coming decade for industries is conservation. Conservation here is defined as investments in large-scale retrofitting of existing facilities or replacing them with more energy-efficient processes. Conservation will pay off in the coming decade of uncertain energy supply and rocketing energy prices.

REFERENCES

1. Chase Manhattan Bank. "Outlook for Energy in the U.S. to 1985," June 1972.
2. Ford Foundation. *A Time to Choose,* final report by the Energy Policy Project of the Ford Foundation. Cambridge, Mass.: Ballinger Publishing, 1974. The case of Historical Growth Scenario with High Imports is used.
3. Federal Energy Administration (FEA). "National Energy Outlook," February 1976. The reference case of Business as Usual is used.
4. FEA Project Independent Blueprint. Natural Gas Final Task Force Report (under the direction of the Federal Power Commission), Washington, D.C., November 1974.
5. Cherniavsky, E. "Brookhaven Energy System Optimization Model," BNL #19569, Brookhaven National Laboratory, Upton, N.Y., December 1974.
6. Exxon Company, U.S.A. "Energy Outlook: 1977–1990," January 1977, and "Energy Outlook: 1980–2000," December 1979.
7. Energy Information Administration. "Energy Information Administration Annual Report to Congress," Department of Energy, 1978, 1979, 1980 (prepublication draft).
8. Data Resources, Inc. "Energy Review," Autumn 1978.
9. Wilson, G. *Workshop on Alternative Energy Strategies (WAES), Energy: Global Prospects, 1985–2000.* New York: McGraw-Hill, 1977.
10. Brookhaven National Laboratory. "Alternative Scenarios to Develop R&D Strategies," BNL report, Brookhaven National Laboratory, Upton, N.Y., 1978.
11. Dantzig, G. B. "Formulating a PILOT model for Energy in Relation to the National Economy," SOL 75–10, Stanford University, April 1975.
12. Manne, A. S. "ETA-MACRO: A Model of Energy-Economy Interactions," pp. 1–45 in C. J. Hitch (ed.), *Modeling Energy–Economy Interactions—Five Approaches.* Washington, D.C.: Resources for the Future, 1977.
13. Belden, R. D., Jr. "Report on the Forum Project: FOSSIL 1 Analysis of Future Trends in the U.S. Coal Industry," December 1977, DSD-#96: EMF Report 2, Vol. 3, September 1978.
14. Cazalet, E. G. "Generalized Equilibrium Modeling: The Methodology of the SRI-GULF Model, SRI Report," SRI International, Menlo Park, Calif., May 1977.
15. Adler, J. R., Cazalet, E. G., et al. "The DFI Energy–Economy Modeling System," Decision Focus Report (draft), Decision Focus, Inc., Palo Alto, Calif., May 1978.
16. Klein, L. and Finan, W. "The Structure of the Wharton Arrival Energy Model," presented at the EPRI Modeling Forum, October 1976.
17. Exxon Corp. "World Energy Outlook," December 1979.
18. U.S. Department of Energy. "A Comparative Assessment of Five Long Run Energy Projections," prepared by Brookhaven National Laboratory, DOE/EIA/CR–0016–02, December 1979.
19. Electric Power Research Institute (EPRI). "Biofuels, A Survey," EPRI ER–746–SR, June 1978.
20. EPRI. "Fuel Cell Power Plants for Dispersed Generation," EPRI TS-1/54321 Rev., July 1980.

21. Muffler, C. "Geothermal Resources," in D. Brobst and W. Pratt (eds.), "U.S. Mineral Resources," U.S. Geological Survey Professional Paper 820, Washington, D.C., 1973.
22. Culbertson, W. and Pitman, J. "Oil Shale," in D. Brobst and W. Pratt (eds.), "U.S. Mineral Resources," U.S. Geological Survey Professional Paper 820, Washington, D.C., 1973.

PART II
ENERGY MANAGEMENT
AND POLICY

This section presents a discussion of principles, procedures, and application of comprehensive energy conservation planning (i.e., a systems analysis approach to energy conservation problems).

4
Comprehensive Energy Conservation Planning

INTRODUCTION

This chapter presents comprehensive energy conservation planning for both industrial and national energy planners. The comprehensive energy conservation planning method is defined as a management technique that carefully evaluates overall organization costs and benefits of alternative investments in energy conservation measures, and then recommends which conservation measure or combination of measures is more favorable at what energy price. It is a systems analysis for effective energy management. The objective of energy conservation planning is to enable industrial firms or governmental bodies to achieve maximum energy efficiency at minimum cost.

Comprehensive energy conservation planning includes motivation, principles, procedure, significance, tasks required, and applications.

MOTIVATIONS FOR COMPREHENSIVE ENERGY PLANNING

Motivations for comprehensive energy planning vary from firm to firm and from institution to institution. Basically, two different groups of energy planners need this planning: planners in industrial firms and those in governmental bodies.

For profit-seeking industrial firms, a primary incentive for energy conservation planning is to manage investment carefully in order to improve profitability and reduce production costs. Major reasons for such planning can be summarized as follows:

- Continuing desire to maintain profitability when facing increasing energy costs
- Urgent desire to maintain adequate and stable energy supply for undisturbed production schedules
- Need to manage large capital required for conservation technologies that possess substantial conservation potential
- Requirement to comply with government's energy conservation targets or efficiency standards
- Need to present to consumers an image of energy efficiency

The need for effective conservation programs for industrial companies is urgent because of recent energy chaos. On the other hand, there is competition between large investments for energy conservation and investments for other, nonconservation efforts. Comprehensive conservation planning carefully examines costs and benefits of all alternative investments in energy conservation and determines at what energy price these investments should be made.

For governmental energy planners, comprehensive energy conservation planning is needed for the following reasons:

- To ensure adequate energy supply and possible energy independence
- To maintain a balance between energy development and environmental protection in implementing conservation strategies and tactics
- To satisfy various governmental goals and interest groups in implementing conservation policies

Because conservation is treated as an alternative energy source, comprehensive conservation planning is necessary to ensure effective usage of this source. In particular, conservation can be relatively cheaper and faster in producing results than new energy sources. It is regarded as a quick-fix, short-term solution for current energy problems.

However, in the process of implementing various conservation policies, conflicts will arise between the goals of government and institutions and between different affected interest groups. These conflicts call for comprehensive energy planning so that impacts of conservation policies can be foreseen, and conflicting goals and competing interest groups can reach agreements before any policy is implemented.

PRINCIPLES OF ENERGY CONSERVATION

Two principles govern energy conservation policies: maximum thermodynamic efficiency and maximum cost-effectiveness in energy use.* The first principle is to achieve maximum energy efficiency permitted by the thermodynamic laws, and the second principle is to minimize costs in achieving this efficiency. The efficiency of energy use depends on thermodynamics, and the cost of energy use responds to economics of resource allocation. Therefore, a comprehensive energy conservation plan is a guide to the most effective use of the nation's energy resources.

Maximum Thermodynamic Efficiency

Energy efficiency is measured by the first and second laws of thermodynamics. The first law indicates that energy can be neither created nor destroyed; it can only be changed from one form to another. Therefore, if we neglect the chemical conversion of energy into mass or vice versa, the net increase in the energy content of a particular system in a given time interval is equal to the energy content of the material leaving the system plus the work done on the system and the heat added to the system. Figure 4–1 shows this energy flow in a typical device.

The second law of thermodynamics states that heat always flows spontaneously and unidirectionally from a hot body to a cold one, never the reverse; the entropy of an isolated system cannot decrease. In other words, the availability of energy always decreases, never the reverse, because the entropy is zero for a perfectly ordered system (maximum energy potential for work). The entropy increases when the availability of energy for work decreases.

One can increase energy efficiency by reducing heat loss and heat discharge

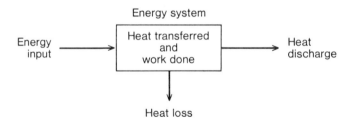

Fig. 4–1. Energy flow through a typical device.

---

* Principles of comprehensive energy conservation planning are outlined in this section and detailed in Chapter 5.

or increasing heat transfer and work rate. There are two indices for measuring energy efficiency: (1) energy quantity index (the ratio of work done plus heat transferred over energy input), also called the first-law efficiency index; and (2) energy quality index (availability of energy output over availability of energy input), also called the second-law effectiveness coefficient.

These two indices will be used to identify inefficient equipment to be replaced and compare candidate conservation technologies. (Details are discussed in Chapter 6.)

Maximum Cost-Effectiveness

Maximum cost-effectiveness of energy conservation is achieved when other sources are economically used to replace energy in production in order to maximize energy efficiency. This substitution includes using nonenergy resources to replace energy, sacrificing convenience and time to save energy, and deploying a future energy source for present use. (More details about substitution are found in Chapter 5.)

Cost-effectiveness can be measured as a ratio of benefits over costs, the ratio being used to choose the conservation alternatives that will maximize energy efficiency and simultaneously minimize the cost of production. Here, the costs and efficiency of whole organizations are emphasized, rather than those of individual energy devices, because the impact of conservation on the whole organization must be accounted for. Chapter 7 details the measurement of cost-effectiveness for various energy conservation measures.

PROCEDURE FOR COMPREHENSIVE ENERGY CONSERVATION PLANNING

The procedure of comprehensive energy conservation planning consists of five steps: (1) specify goals; (2) identify energy-inefficient facilities; (3) synthesize alternative conservation measures; (4) evaluate costs and benefits; and (5) compose an optimal combination of alternatives.

The relationship among the five steps is sequential, as shown in Fig. 4–2. The procedure of comprehensive energy conservation planning can start at any step, and its output will be the result of the previous step. For example, if we start the procedure at step 2 (identify energy-inefficient facilities), the final output of the procedure will be the result of step 1—an optimal conservation target under a given set of socioeconomic conditions (including expected future energy prices). (This example is discussed further in later sections and in Chapter 10.)

If we start the procedure at step 1 (specify conservation goal), then the output of the procedure will be the result of step 5—an optimal combination

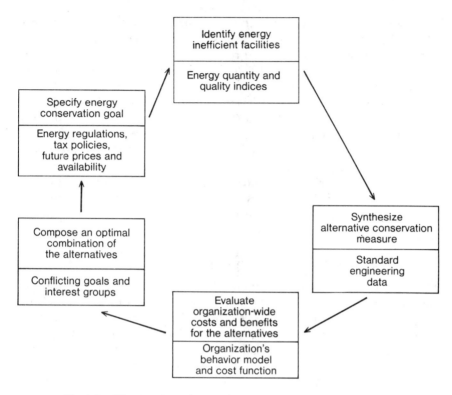

Fig. 4–2. The procedure of comprehensive energy conservation planning.

of alternative conservation measures, which will produce a compromise among the organization's goals and interest groups. (This example is detailed in Chapter 11.)

Specifying Energy Conservation Targets

The first step for comprehensive energy conservation planning is to specify the organization's conservation targets. These targets determine the level of management that will be involved and the detail of the comprehensive energy conservation planning. There are three levels of conservation efforts:

- Administrative and information programs, which are relatively easy and inexpensive to implement. These are the initial stage of conservation, the main characteristic of this stage being to replace (save) energy with time and convenience. These programs include tune-ups, light turn-offs, and small adjustments in production processes. They will save between

3 and 7% of the current energy consumption; the main responsibility of these programs falls on operators and low-level managers.

- Re-equipping, retrofitting, and recycling through small incremental investment, for an additional 5 to 10% savings. Middle-level managers will make these decisions.
- Major production process changes through large capital expenditures, for total savings of 20 to 90%, depending upon the nature of each industry and facility. These decisions will be made by top-level managers. Factors to be considered in determining conservation targets are energy regulations (e.g., mandatory energy conservation target, efficiency standards, etc.), energy tax incentives (e.g., tax credits for recycling equipment), expected future prices and availability, and future production commitments and market potential.

Identifying Energy-Inefficient Facilities

The second step is to identify the facilities and processes that waste the most energy. The first and second laws of thermodynamics are used to evaluate energy quantity and quality indices. The facilities and processes that have the lowest values of the two indices are candidates for replacement.

As discussed before, the first index is the energy efficiency index (energy quantity index), which corresponds to the first law of thermodynamics. This is used to measure the quantity of energy. The efficiency index is defined as the ratio of useful energy output (E_{use}) to total energy input (E_{in}):

$$\text{Efficiency} = E_{use}/E_{in} \qquad (4\text{--}1)$$

The useful energy output consists of work and heat transferred.

The second index is the energy effectiveness coefficient (energy quality index), which corresponds to the second law of thermodynamics. It is used to measure the quality of energy. The effectiveness coefficient is defined as the ratio of energy availability of output (A_{out}) to energy availability of input (A_{in}):

$$\text{Effectiveness} = A_{out}/A_{in} \qquad (4\text{--}2)$$

This effectiveness definition is analogous to that of efficiency, and often its value is close to that of efficiency.

The availability A_i of stream i is defined as follows:

$$A_i = M_i \Sigma_j (h_j - T_o\, s_{ij}) \qquad (4\text{--}3)$$

Where M_i is the mass flow rate, h_j is the enthalpy, T_o is the temperature of the reference atmosphere, s is the entropy, i can be output or input stream, and j is one of the components in the stream.

We use these two indices to detect maximum energy loss (i.e., the most inefficient facilities and processes) by comparing energy quantity and quality indices. The facilities with the lowest values of the two indices are chosen as candidates for replacement.

Synthesizing Alternative Conservation Measures

Synthesizing alternative conservation measures is the third step. In this step, alternative energy-saving measures (technologies and programs—e.g., recycling, retrofitting and re-equipping, and new processes) that can properly replace the inefficient facilities and processes identified in task 2, are taken to achieve the conservation goal specified in step 1. Standard engineering data concerning these alternatives will be analyzed to determine their energy savings and costs.

These measures can be classified according to three criteria: method of installation, method of heat use, and size of investment. This classification is summarized as follows:

- Method of installation
  - Recycle
  - Retrofit
  - New process
- Method of heat use
  - Waste heat recovery
  - Waste material use
  - Waste heat utilization
  - Process efficiency improvement
- Size of investment
  - Administrative and information programs to create awareness and reduce individually controlled energy use
  - Small incremental investments to recover waste heat, alter process flows, and retrofit facilities for better heat use
  - Major capital expenditure to redesign production processes over time

This step results in a set of alternative feasible measures, each having specific performance characteristics such as energy requirements, capacity, reliability, and other production requirements. It also includes a set of data on the environmental and socioeconomical impacts of each system on the institution analyzed.

Evaluating Overall Organization Costs/Benefits of Alternative Measures

The fourth step for comprehensive energy planning is to evaluate overall organization costs and benefits of alternative conservation measures. A behavior model and a cost function representing the operation of the organization under study are used to evaluate costs and benefits for the alternative measures. The purpose of this step is to take into account compound impacts interacting among elements in the organization. These impacts are a composite result of the measures and thus have not appeared in standard engineering data in step 3.

Composing an Optimal Combination of Alternative Conservation Measures

As stated in the section on the need for comprehensive energy conservation planning, there are conflicting goals (e.g., energy development versus environmental protection) and competing interest groups (e.g., industrialists versus labor unions) involved in making decisions for the investment of conservation technologies and programs. A method of ordering various goals and interest groups is needed in this step. Thus, hierarchical analysis will be used to derive scales of priorities for these activities (goals and groups). Activities involved in the decision process may be grouped on different levels that form a hierarchy according to the structure of the process, whereby lower-level elements can be compared pairwise with respect to higher-level components, and a method of weighting yields the overall priorities for the activities on the lowest level. These overall priorities indicate the order of preference among feasible policies in achieving the conservation goal.

SIGNIFICANCE OF COMPREHENSIVE ENERGY CONSERVATION PLANNING

The procedure for comprehensive energy conservation planning offers two advantages: the use of a microeconomic approach and the convenience of constant review of energy conservation policies.

Microeconomic Approach

In contrast to the econometric aggregate approach, a microeconomic approach is used here to measure costs and benefits of energy conservation measures for a given organization under study. The analysis consists of:

- Evaluation of costs and benefits of conservation measures—recycling, retrofit, and new process
- Use of a behavior simulation of the organization to assess the adoption of the alternative measures in existing facilities

The costs and benefits of a standard facility are analyzed for each of the conservation measures. Most operational changes to reduce energy consumption with little or no costs have already been made. Consequently, these alternative measures will reduce energy consumption for existing facilities and increase production costs.

As stated before, energy conservation is a matter of substitution. Future energy sources replace the present ones; time and convenience replace energy use; and nonenergy resources replace energy. Clearly, these substitutions happen in the micro-production level. However, since the 1973 Arab oil embargo, most economists as well as national policy-makers have focused on the national aggregate elasticity of GNP with respect to energy, and neglect microlevels of substitution. Econometric models such as PILOT, Kennedy-Niemeyer, Wharton, Hudson-Jorgenson, Hnyilicza, and DRI-Brookhaven were developed to estimate the elasticity of GNP based on respective aggregate substitution between energy and other production factors. Both the Kennedy-Niemeyer and the PILOT models, which assume limited substitution, display very small aggregate elasticities (between 0 and 0.1). The remaining models, which include higher substitution possibilities, tend toward large, long-run elasticities (between 0.3 and 0.5).[1]

Evidently, the substitution between energy and other production factors in microlevels is the key in determining the elasticity of GNP. Few studies regarding this substitution have been undertaken. Because each industry is unique in terms of production and conservation technologies, aggregate substitution for the industrial sector may be inappropriate. This method thus evaluates the possibilities of energy substitution for individual companies in industry.

Convenience for Constant Review of Conservation Policies

The procedure of comprehensive energy conservation planning is purposely set up for convenience, allowing constant review of energy conservation policies. The reason is simply that the procedure is so systematized that it is applicable for any organization such as a plant, an industrial company, a local government, or the federal government. Because the five steps of the procedure can be formulated through models or mathematical equations, the procedure can be easily converted into a computer model and maintained with minimum effort. This conversion enables energy planners of organizations to review and plan their energy conservation efforts periodically with little difficulty and at minimum cost.

TASKS REQUIRED FOR COMPREHENSIVE ENERGY CONSERVATION PLANNING

Major tasks for comprehensive energy conservation planning are as follows:*

- Define the organization under study. This task consists of specifying the framework, boundary, and operations of the organization.
- Prepare models to represent the function of the organization defined in task 1. Models here are defined as sets of cause-effect relationships (including rules, formula, methods). This task consists of model construction and exercise.
- Collect accountable data. This task includes data acquisition and data transformation.
- Verify/validate models.

The relationship of these tasks is shown in Fig. 4–3.

Define the Organization

The organization under study is defined through specification of its boundary, framework, and operations. The framework sets the performance criteria and outlines the basic rules of its progress. The boundary defines the organization's domain and environment, and the operations define its hierarchy and structure of activities. Task 1 will enable energy conservation planners to understand exactly the objective and functions of the organization they are examining.

Prepare Models to Represent the Organization's Functions

After the organization has been defined, models (sets of cause–effect relationship) can be constructed to simulate the organization's functions. These models include the following relationships for the five steps outlined in the procedure of comprehensive energy conservation planning as shown in Fig. 4–2:

- Rules for specifying the organization's conservation targets (step 1)

---

* The tasks and models required for comprehensive energy conservation planning are outlined here. They are presented in detail in Chapter 8.

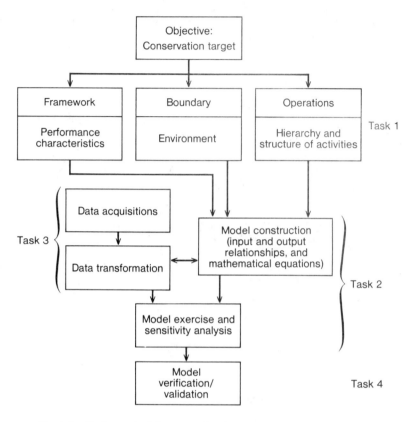

Fig. 4–3. Tasks required for comprehensive energy conservation planning.

- Energy quantity and quality indices to identify energy-inefficient facilities (step 2)
- Formulas to compute energy savings and incurred costs for energy conservation technologies (step 3)
- Behavior (or production) simulation models and cost functions to represent the input–output relationship for the organization (step 4)
- Hierarchical analysis methods to establish priorities for goals and interest groups (step 5)

This task also requires model exercises to test the sensitivity of models.

Collect Accountable Data

In this task, data that were specified in the previous task for the preparation of models (including rules, formulas, and methods) are acquired. In addition, data transformation may be needed to estimate parameters and coefficients in the models for the previous task.

Verify/Validate Models

Model verification and validation have a similar purpose; both are used to make sure the model constructed represents the functions of the organization under study. Model verification is defined as a method to ensure accuracy of forecast and computer codes of models. Model validation is more broadly defined as a method to ensure the correctness of underlying assumptions, theories, forecasts, and computer codes.

APPLICATION OF THE COMPREHENSIVE ENERGY CONSERVATION PLANNING METHOD

In this section, we consider potential users, ways of application, and how to use the method. Primary copper production is the example chosen to illustrate the use of the method.

Potential Users

This method is intended to be used by energy planners of industrial firms, government bodies, and large commercial organizations, who have incentives to plan and implement energy conservation policies carefully. Obviously, the potential users must have fairly large organizations, be concerned about their high energy costs, and consider their organizations' various goals and affected interest groups.

Ways of Application

The five steps in the procedure of comprehensive energy conservation planning represent five ways of application, if any sequential four of the five steps are performed. For example, if steps 2 through 5 (identify inefficient facilities, synthesize alternative measures, evaluate costs and benefit, and compose an optimal combination of the alternatives) are performed, then an optimal conservation target under an expected energy price results. In other words, the purpose of that exercise is to determine an optimal conservation target under a predicted energy price. Similar exercises can be performed for the remaining four steps. Thus, the uses of the comprehensive energy conservation method can be summarized as follows:

- Determine optimal conservation targets under predicted energy prices or determine energy prices required for a given level of conservation efforts
- Indicate marginal energy-inefficient facilities and processes to be replaced under given energy prices
- Determine optimal introduction of energy savings technologies and programs under given energy prices
- Compute the organization-wide costs and benefits of a combination of energy conservation technologies and programs
- Calculate optimal combinations of conservation technologies and programs that will effect compromise among goals and interest groups in achieving a given energy conservation target. (See Chapter 11.)

The first use is presented as an example in the following section and discussed in detail in Chapter 10.

An Example of How to Use the Method: Primary Copper Production

Primary copper production is analyzed to demonstrate how to use the procedure for comprehensive energy conservation planning. The reason for this choice is that primary copper production is one of the most energy-inefficient processes in the United States.*

In this example, the first of the previous five steps is described briefly. (A detailed discussion is given in Chapters 8 and 10.)

Description of Primary Copper Production. Primary copper production consists of four processes: mining, beneficiation, smelting, and refining. The energy need in each process is shown in Table 4–1. The values in this table do not include the energy consumed for transporting ores, materials, and products-in-process. The procedure is explained below step by step to demonstrate how it is applied.

*Identify Energy-Inefficient Facilities in Primary Copper Production** (step 2 in the procedure). The energy quantity and quality indices of individual production processes are shown in Table 4–1. The company analyzed is assumed to consist of integrated plants to take hot materials directly from previous processes; the resulting efficiencies for each process are much higher than the overall efficiency. Examining these indices shows that the most

---

* For details, see Chapter 6.

Table 4–1. Energy Consumption and Efficiencies of Copper Production Processes.

Copper Production Process	Energy Consumption (10^6 Btu/short ton)	Quality Index Efficiency (%)	Quality Index Effectiveness (%)
Primary processes			
Mining	15.12	0.0*	0.0*
Beneficiation	35.87	64.1	76.4
Smelting	35.84	44.0	48.0
Refining	12.00	46.0	53.0
Total	98.83	3.5	9.2
Secondary process	10.3	N/A	N/A

* No energy was considered as input in the mining process, because no heating devices except mechanical forces are used in the mining process.

Sources: U.S. Bureau of Mines[2] and Battelle Columbus Laboratories.[3]

conservation potential exists in mining and smelting processes. Using scrap will save almost 90% of energy used to produce copper from ores (shown in Table 4–1).

*Synthesize Alternative Conservation Measures** (step 3 in the procedure). Three promising options† in mining and smelting processes, which can conserve energy without reducing production, are: using scrap instead of ores; replacing reverberating smelting with flash smelting; and retrofitting tonnage oxygen plants in reverberatory smelting processes. The costs and major benefits of the three conservation measures are shown in Table 4–2.

Table 4–2. Costs and Benefits of the Three Energy Conservation Measures.

Conservation Measure	Cost Increase	Benefits Energy Savings	Pollution Reduction
Using copper scrap	depends on facilities	up to 90%	up to 90%
Flash smelting	7%	60%	10–15%
Retrofitting tonnage oxygen	$1.2/ton refined copper	7%	less than 7%

* For details, see Chapter 7.

† Leaching and hydrometallurgy processes, which may save a large proportion of the energy use for milling ores, are still at the experimental stage and therefore not discussed here.

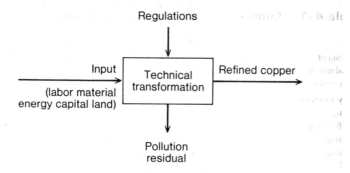

Fig. 4–4. A basic copper production process.

Evaluate Organization-Wide Costs and Benefits† (step 4 in the procedure). Cost functions and production (behavior) models are used to evaluate organization-wide impacts of alternative combinations of conservation measures. The purpose of cost functions and production models is to simulate a copper production process whose basic format is shown in Fig. 4–4.

A cost function can be formulated as follows:

$$\text{cost/benefit} = f(\text{production quantity, price of production factors}) \qquad (4\text{–}4)$$

where f is a symbol of function. Equation 4–4 shows that cost/benefit is a function of production quantity and prices or production factors. A production (behavior) model can be expressed as follows:

1. quantity = g_1 (capacity, utilization, energy conservation requirements, sales commitments, transportation, pollution controls)

2. energy consumption reduction (%)
   = g_2 (technologies, age of facilities, location, etc.)

$$(4\text{–}5)$$

Equations 4–4 and 4–5 are used to evaluate organization-wide costs and benefits for feasible combinations of conservation technologies and programs. Feasible combinations of alternative conservation measures will be obtained

---

† For details, see Chapter 8.

from Equation Set 4–5, and then input to Equation 4–4 to yield their overall costs/benefits.

Compose an Optimal Combination of the Alternative Energy Conservatiᵒn Measures (step 5 in the Procedure). If every interest group agrees that the industry's goal is cost minimization, then the optimal combinations of the alternative conservation measures under various conservation goals are shown in Table 4–3. This is a simplified case, which does not need to use a hierarchical analysis to set priorities for various goals and interest groups. Chapter 11 will show a case that does need hierarchical analysis.

Determine an Optimal Conservation Target (step 1 in the procedure). The result of this step is the output of the procedure in this example (i.e., to compute an optimal conservation target). To determine an optimal target, we need to know expected energy prices. Therefore, the question is rearranged as "At what energy price can energy conservation be optimal?" and "Within which range of energy conservation targets?"

Data concerning energy saved (∇E) and cost increase (ΔC) for the producer in Table 4–3 can be used to estimate equations governing the relationship between the two variables for the three companies as follows:

$$\Delta C = 1.41\, \nabla E \qquad\qquad (4\text{–}6)$$

where ΔC is in thousands of 1974 dollars incurred in the installation of energy conservation measures, and ∇E is energy saved in 10^9 Btu.

If prices for capital, labor, and materials are maintained at the 1974 level, the required average price increase (ΔP) to offset conservation costs can be expressed as follows:

$$\Delta P = \frac{\Delta C}{\nabla E} \qquad\qquad (4\text{–}7)$$

Equation 4–6 is substituted in Equation 4–7, and the average energy price increase required for reducing energy consumption to levels within 15% of the 1974 level is computed as 93% of the 1974 price. (The average 1974 energy price was estimated at $1.504/million Btu.)

Therefore, when the energy price by primary producer A is increased by 93% of the 1974 price, assuming constant nonenergy prices, then it is optimal for producer A to specify a conservation target, within 15% of the 1974 energy consumption level.

Table 4-3. Economics of Energy Conservation for Primary Copper Producer A.

Energy Saved (∇E)		Conservation Methods Implemented†	Economic Impacts		
10^9 Btu	% of Original Consumption	(Production in 10^3 Short Tons)	Cost Increase, ΔC ($\$10^6$)	% of Original Cost	Average Cost Increase ($\Delta C/\nabla E$: $\$10^6$ Btu)
1.376	5	(A) 7.2 (C) 292.5 (B) 189.4	2.544	.51	1.85
2.753	10	(A) 7.2 (C) 188.9 (B) 293.0	4.033	.81	1.46
4.129	15	(A) 7.2 (C) 85.3 (B) 396.6	5.523	1.12	1.34

† (A)—scrap; (B)—flash smelting; and (C)—tonnage oxygen.

SUMMARY

In this chapter, a method for comprehensive energy conservation has been discussed. The method indicates how to determine a conservation target, identify energy-inefficient facilities and processes, synthesize alternative conservation measures, evaluate costs and benefits for the alternative, and compose an optimal combination of the alternatives. In addition, needs, principles, procedures, tasks, and applications for comprehensive energy conservation planning have been examined. The example of primary copper production, one of the most energy-inefficient industries, has been analyzed.

REFERENCES

1. Energy Modeling Forum. *Energy and the Economy,* Vol. 1. September 1977 (Stanford University).
2. Rosenkranz, R. "Energy Consumption in Domestic Primary Copper Production," Bureau of Mines Information Circular, 8698.
3. Battelle Columbus Laboratories. "Evaluation of the Theoretical Potential for Energy Conservation in Seven Basic Industries," final report to Federal Energy Administration, National Technical Information Service, July 1975.

5

Principles of Energy Conservation

INTRODUCTION

In Chapters 2 and 3, the importance of conservation was discussed; in Chapter 4, the procedures of comprehensive energy conservation planning were outlined. This chapter presents principles of energy conservation.

The economics of energy conservation is a complex problem. It involves not only the relationships in production and consumption at a given time in a generation, but also the relationships between generations over time. The problem is further compounded by the characteristics of uncertainty of supply, expectation of generations, and international politics mostly in the Middle East. Many problems in the economics of energy conservation can be solved through empirical studies. However, decisions regarding energy conservation policies are so urgent that they cannot wait for an empirical solution that requires a long data period and extensive time for research. A sound set of fundamental principles and, hence, an appropriate analytical framework are necessary for both policy making and empirical research. This chapter provides a basis for energy conservation planning and research.

DEFINITION OF ENERGY CONSERVATION

Like capital, labor, and material, energy is one of the production factors used to produce final goods and services. In economic terms, energy is a demand-derived good. That is, energy is an intermediate good whose demand depends on the demand of the final (end-use) goods and services it produces.

For example, gasoline is not the final good or service; transportation is the final service. The demand for gasoline is determined by the demand for transportation.

Energy conservation can be defined as the substitution of energy with capital, labor, or material and time. This definition also covers the substitution of scarce types of energy (e.g., oil) with abundant types of energy (e.g., coal) or the substitution of energy with convenience. For example, people will turn lights off when they are not in their offices, or will reduce driving by carpooling.

We call this first definition production factor substitution. As shown in Fig. 5–1a, PP is an assumed production combination possibility curve on which any combinations of energy and nonenergy factors (e.g., capital) will produce an equal amount of goods and services. If point B is chosen instead of A on the PP curve, less energy is used, but more nonenergy production factors are needed. This shifting theoretically should produce the same amount of goods and services. However, in many cases, energy used to be wasted. For example, we did not turn lights off and did not tune up engines often. Therefore, savings of energy actually will help produce more goods and services because the energy saved from wasting can be used to produce. This is one of the greatest benefits of conserving energy.

Consumers need to be encouraged not only to conserve energy, but also to conserve end-use products and service by using less of them or choosing more energy-efficient ones. We also have to educate producers to invest in energy-efficient technologies for producing goods and services.

On the other hand, primary energy resources (oil, natural gas, coal, and

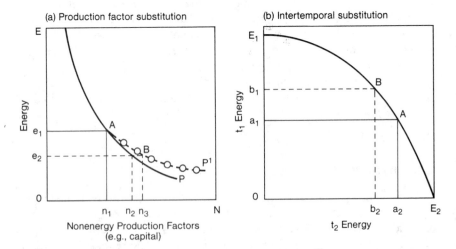

Fig. 5–1.   Definition of energy conservation: production factor and intertemporal substitution.

uranium) are stock-derived products; that is, they yield services that can be stored and then used in other times or other generations. These resources are limited in quantity and depletable over time. Consequently, we are competing with future generations in using up primary energy resources.

Energy conservation can also be defined as the substitution of this generation's energy with that of future generations. We call it "intertemporal substitution." As shown in Fig. 5–1b, E_1E_2 is the total resource curve on which any combinations of this generation's production (t_1) and the future generation's (t_2) production will deplete the total resource base. For example, point A on the E_1E_2 curve indicates that we would produce a_1 amount of energy in this generation and a_2 amount in the future generation. However, we may choose the production combination of B instead of A on the E_1E_2 curve. More energy would therefore be produced in this generation and less in the future generation.

In this sense of energy conservation, we need to consider the following two points before demanding more conservation efforts:

- Whether the depletion of the given resource can be offset by new explorations and findings. That is, is the resource really limited and hard to replace in a short time?
- The intertemporal welfare of consuming this type of energy. That is, is the marginal utility of consuming this energy in this generation larger than the utility of consuming it in the next generation?

Because future utility of consuming a given type of energy is discounted by a rate, for example, 10% in computing total intertemporal welfare (like computing intertemporal revenues), and because often there will be substitutes invented in future generations, if we assume the utility of consuming a unit of the given energy is the same for the present and future generations, it is always more beneficial to consume it in the present generation. The temptations and rationale to consume more in this generation can therefore be justified. Consequently, it is difficult to persuade whoever perceives energy conservation in this sense to conserve energy. Therefore, many industrialists and politicians insist that there is no need for energy conservation.

Nonetheless, there is a good argument for conservation even under the second definition: If there is an ongoing shortage or a prominent potential for shortage of a particular type of energy, then conservation, even rationing, becomes necessary.

PRINCIPLES OF ENERGY CONSERVATION

As noted in the last chapter, the two principles governing energy conservation policies are maximum thermodynamic efficiency and maximum cost-effectiveness in energy use. The first and second laws of thermodynamics measure the efficiency of energy use; allocation of available production factors determines the cost-effectiveness of conservation. These two principles are discussed in this section. Thermodynamics and energy efficiency are detailed in the next chapter.

Maximum Thermodynamic Efficiency in Energy Use

Generally when energy is used for industrial processes, heat is obtained through energy devices such as boilers from primary energy sources, then transferred to work through heat exchangers, and finally discharged into the environment (as shown in Fig. 4–1 in Chapter 4).

Three important factors thus need to be considered:

● How much heat is available?
● How good is the heat available (i.e., how high is its temperature)?
● How well is the heat transferred (i.e., what is the heat loss in transfer)?

First, industrial processes require an adequate quantity of heat available for production. The quantity of heat available for work can be measured by the application of the first law of thermodynamics. Second, the heat available must be of adequate quality for the purpose of work; for example, heat available at 200°F cannot be used directly to heat steam to 300°F. Heat quality (availability) can be measured by applying the second law of thermodynamics. Finally, the heat must be transferred from the source to the work place where it is used for production. This is the problem in heat transfer.

Maximum thermodynamic efficiency of energy use is defined as maximum work done in production by using a given amount of primary energy input, as defined in the following simple form:

Maximize: Work = (Energy input)
$$- \text{(Energy loss in transfer)} - \text{(Energy discharge)} \quad (5\text{--}1)$$

The terms in this equation will be explained later. Note that, in Equation 5–1, the conversion of energy into mass of energy systems, or vice versa, is neglected. For example, in a power generator, if water of 100 Btu heat content and 3000 Btu heat enter a boiler, the generator discharges steam of 2000 Btu heat content through turbines, and loses 900 Btu into the atmosphere through surface losses and blowdown. The work done in the boiler would be 200 Btu in the present situation (3000 + 100 − 2000 − 900 = 200 Btu) if we assume that there is no chemical reaction to the boiler. The thermodynamic efficiency of this generator is substantially low; however, we can maximize the efficiency (work) by the following conservation measures:

- Using condensers to recapture the heat discharged as many times as possible
- Reducing heat loss with better heat exchangers
- Replacing the boiler generator with a fuel cell power generator

Each of the above three measures will increase thermodynamic efficiency. The magnitude of improvements varies over implementation of these measures as does the implementation cost. Before the cost-effectiveness of the above conservation options is discussed, the terms in Equation 5–1 are detailed.

Work. Work is an interaction between a system (e.g., boiler) and its surroundings (e.g., turbines). It is done by a system on its surroundings if the sole external effect of the interaction could be the movement of a desired mass in the surroundings. In most cases, work is done because of a force on a moving boundary of a system. Its magnitude is equal to the product of the force and the displacement of its point of application in the direction of the force. In the case of power generation, the force is the heated steam from the boiler, and the work done is the turning of turbines. The magnitude of work done is measured by the number and force of turns made by the turbines.

Energy Input. Energy enters most process equipment (energy systems) as chemical energy in the form of fossil fuels, as sensible enthalpy in fluid streams, as latent heat in vapor streams, or as electrical energy. To measure energy input, we need to meter the quantity of fluid flowing or the electrical current. However, we may not need to submeter every flow continuously because temporary installations can provide sufficient input data. In the case of furnaces and boilers that use pressure ratio combustion controls, measurements of the control flow meters are sufficient. For electrical energy flows, ammeters or kilowatt-hour meters can measure the current entering the system. The chemical composition of the stream sometimes affects energy input. For air, water, and other pure substances, no test is required, but for fossil fuels, the composition must be determined by chemical analysis or be secured from fuel suppliers.

Energy Loss. Energy is lost to the ambient environment from the process equipment through radiative and convective heat transfer. Radiative heat loss takes place by means of light and other electromagnetic radiation, whereas convective heat transfer occurs when the surface of hot material is displaced by cool gas. Infrared measurement techniques can accurately measure radiative heat loss, but it is difficult to measure convective heat transfer exactly owing to nonuniformity of the temperature distribution over the surface of a process unit.

Heat Transfer. There are three major mechanisms of heat transfer: conduction, convection, and radiation. In most cases, heat is transferred through heat exchangers for purposes of energy conservation. Conduction is affected by the conductivity of the material through which the heat is being transferred, and by the area, temperature, and direction of the heat flow. Convection is the transfer of heat by the flow of a material. It depends on physical properties, area, and fluid temperature. Radiation is the transfer of heat energy by electromagnetic means between two materials whose surfaces "see" each other, and is generally governed by absolute temperature, areas, and positions of the two surfaces.

Energy Discharge. Energy discharge depends on composition, discharge rate, and temperature of each outflow from the process unit. The composition of exhaust products can be determined by Orsat analysis, chromatographic

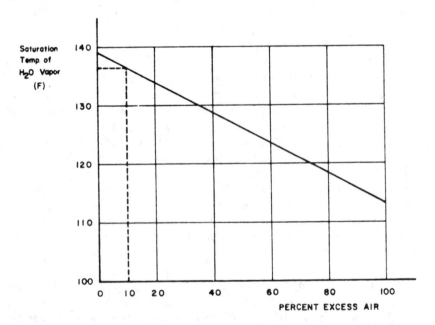

Fig. 5–2. Natural gas combustion chart. Source: National Bureau of Standards.[1]

testing, or determination of the volumetric fraction of oxygen or carbon dioxide.

Figure 5–2 is an example of a chart that can be used for determining the quantity of excess air in the combustible mixture. Incomplete combustion is the major source of energy discharge.

As noted, energy discharge depends on the speed of the exhaust flow. The discharge rate and temperature of outflows can be measured with simple engineering devices.

Maximum Cost-Effectiveness in Energy Use

Maximum cost-effectiveness in energy use is achieved if the production factor or intertemporal substitution is made to maximize energy efficiency at the least cost. From this definition, maximum cost-effectiveness in energy use is determined by two cost components: (1) conservation costs (costs for implementing energy conservation efforts in order to save energy) and (2) energy costs. The tradeoff of the two cost components is shown in Fig. 5–3 and explained below.

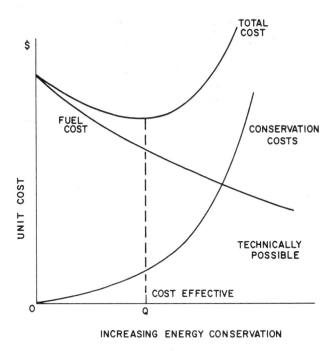

Fig. 5–3. Unit maximum cost-effective conservation is achieved in the neighborhood of point Q if constant energy prices are assumed.

Conservation Costs. As shown in Fig. 5–3, unit conservation costs increase as the degree of conservation increases. This is understandable because the least expensive conservation methods are the first to be implemented. When more conservation efforts are needed to achieve higher thermodynamic efficiency in energy use, more substitution must be made. That is, more substitutes, such as capital, material, labor, and time for energy, are needed; and they become more expensive to get. To achieve technically maximum efficiency, the cost for conservation efforts would be economically undesirable, as shown in the far-right side of the figure.

Energy Costs. Unit energy costs will decline as more conservation measures are implemented to reduce energy use, if constant energy prices are assumed. The reason for this decline is that less energy is required in each unit of production when energy use becomes more thermodynamically efficient.

Maximum Cost-Effectiveness. Examination of total cost for energy and conservation indicates that maximum cost-effective conservation is achieved in the neighborhood of point Q of Fig. 5–3, if constant energy prices are assumed. In this neighborhood, maximum thermodynamic energy use is achieved at the least cost. Clearly, the level of maximum cost-effective conservation is substantially lower than that of technically maximum thermodynamically efficient conservation when one takes into account the use of all production factors (capital, labor, energy, materials, and time).

Cost-effectiveness in energy use can be measured by a ratio of benefits over costs due to conservation. Details of the measurement are discussed in Chapter 7.

ECONOMICS OF ENERGY CONSERVATION POLICY

Discussion of principles of energy conservation is inevitably followed by the question of the economics of conservation policy: What would be the relation between costs and conservation policy in industrial production? The economics of conservation policy in industrial production depends on the price structure of production factors, the availability of conservation technologies, and the government's fiscal and regulatory policies. Given the present price structure of production factors, conservation technologies, and governmental fiscal and regulatory policies, a typical curve is presented in Fig. 5–4 to show the relationships between production cost increases and energy savings per unit production over the base year for various conservation policies.

Each of the four segments, A, B, C, and D, in the curve has its own significance. In segment A, the initial efforts to conserve energy would result

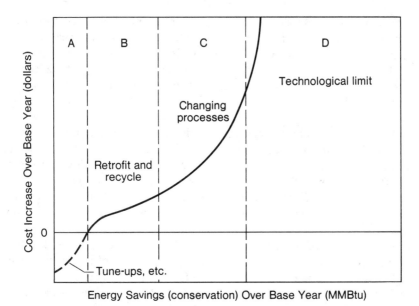

Fig. 5–4. Curve of economics of conservation policy with the current price structure.

in cost savings in an *economic* sense even under the present price structure for production factors, because in this segment time and convenience are used to replace energy, which is considered to have been *wasted* previously. (Time and convenience had been preferable to energy when energy prices were low.) Because time and convenience are not counted as cost components when used to replace energy, production costs would decline slightly. Examples of conservation measures in these stages are light turn-offs, engine tune-ups, and frequent oiling. The following three important factors associated with this stage of conservation should be noted:

- The difficulty of conserving energy is minimal, and relatively inexpensive procedures are required.
- The amount of energy saved in this stage is not substantial because, if it were significant, the procedure would already have been taken to reduce production costs.
- Average energy savings are potentially between 3 and 7%.

The second stage of conservation efforts is shown in segment B in Fig. 5–4. In this stage, because producers expect moderate future energy price increases and feel some pressure from government conservation programs, they start to retrofit existing facilities and recycle scraps to conserve energy.

Labor and materials with a small proportion of capital are used to replace energy in this stage of conservation. Examples of conservation measures are: implementing better insulation, installing devices for waste heat recovery (e.g., heat pumps), recycling scrap materials, and establishing simple energy management programs. The significance of this stage of conservation is summarized as follows:

- Under the present price structure, production costs due to these conservation efforts may increase slightly, but, with respect to expected moderate energy price increases in the near future and government's conservation programs, these efforts would be economical in the midrun.
- Producers expect at least moderate increases in relative energy price in the near future, and they feel the pressure of government's energy conservation programs.
- Conservation policy mostly consists of retrofitting, recycling, and installing easy energy-management plans. These measures often use materials, labor, and a small proportion of capital to replace energy.
- Because these conservation efforts are small-scale and not well coordinated among production divisions, this stage of conservation may suffer a diseconomy of scale.
- Average energy savings are potentially between 5 and 10%.

In some energy-intensive industries, the fear of rapid energy price increases, energy supply uncertainty, and enormous governmental pressures for conservation may force producers to consider further conservation efforts to reduce energy consumption and costs. This stage of conservation is shown as segment C in Fig. 5–4. In this stage, in order to replace energy, large capital investment is required in technologies that will change production processes. Examination of overall production plans is made to ensure that minimum production cost will be achieved in expectation of rapid future price increases. Investment decisions are usually made by top-level management because of the large size of capital expenditures, and energy savings potential is substantial. The characteristics of this stage of conservation efforts are summarized as follows:

- Energy-intensive producers are deeply affected by rising energy prices relative to nonenergy prices and by an uncertain energy supply; their productivity is declining, and their ability to compete with foreign producers is falling. In addition, the government is pushing for more conservation because of rapidly increasing dependence on unstable foreign oil supplies, which may endanger U.S. national security and economic growth.
- The producers will reexamine the energy efficiency of their production

processes, and will replace inefficient ones with new processes and/or substantially modify the existing facilities with computerization of their energy management systems.

- Because of rescheduling whole production processes and computerizing energy management systems, the producers will have a better opportunity to enjoy an economy of scale and reduce their production costs in the face of rising energy prices.
- Because of the large size of capital investment and extensive impacts on production due to major changes of processes, investment decisions are usually made by top-level management.
- Average energy savings potential is between 20 and 90%, depending upon the particular industry.

Since there is a technological limit on energy savings with current technologies, energy conservation efforts beyond this limit will definitely raise conservation cost to an unacceptable level, as shown in segment D in Fig. 5–4. However, when expected energy prices keep rising, new conservation technologies will be developed. These technologies will increase energy savings and reduce conservation costs.

Current Conservation Efforts in the Industrial Sector

The U.S. General Accounting Office[2] and Department of Energy[3] have both indicated that recent improvements in industrial energy efficiency have been achieved primarily through easily implemented administrative and small investment programs (conservation stages A and B as shown in Fig. 5–4). The level of improvements shows that further conservation results will depend principally upon larger capital investment (stage C in Fig. 5–4). For this reason, continued acceleration of the energy-conservation trend is unlikely if no further governmental actions, rapid price increases, or energy shortages are expected in the near future.

In the remaining section, we will focus on the relationships between conservation and expected energy prices. However, these relationships can also be analyzed to obtain relationships between conservation and government's actions.

Conservation and Expected Future Prices

If future energy prices are expected to increase, there will generally be more effort to conserve energy. For industrial energy conservation, it is the magnitude of expected future energy prices that makes the difference in corporation decision-making because the corporation decision-makers are profit-maximizers. In the course of seeking maximum profits, they will plan future investments so as to minimize future energy use and costs if energy prices are expected to increase substantially in the future.

In contrast to this, recent rising oil prices and the federal government's voluntary industrial energy conservation program have not substantially improved industrial energy use.[2] Although total industrial energy consumption was reduced by 4.2% from 1972 to 1976, most of this reduction resulted from a slowing down in industrial production rather than conservation.

These contradictions can be explained as follows:

- Energy prices historically have been behind the trend of general inflation. As discussed in Table 2-2 (Chapter 2), because the 1974–1975 price increases did not significantly increase the percentage of production costs for energy, producers had not invested more in conserving energy.
- Addition of environmental-control equipment has substantially increased energy consumption in many energy-intensive basic industries such as copper and steel. For example, owing to pollution controls, total energy consumed by the primary copper companies reporting increased from 166×10^{12} Btu in 1972 to 176×10^{12} Btu in 1976 (a 6% increase), despite a drop in refined copper production from 1.95 to 1.80 million short tons.[4]

The fact that inexpensive energy prices discourage energy conservation can be further confirmed by the U.S. General Accounting Office (GAO) report.[2] In the report, GAO stated that industries with high energy consumption generally have the knowledge and technical expertise to conserve energy; however, U.S. oil and gas price regulation has made energy-saving investments less attractive than they otherwise might have been. The question thus arises: At what energy price will conservation be economically attractive in the industrial sector?

The answer to this question can be found in Fig. 5–5, which is derived from Fig. 5–4. The curve in the figure governs the relationship between energy savings (ΔE) and cost incremental (ΔC) over the base year. The expected unit energy price increase in the future (Δp) required to make the energy-saving investments attractive would at least be

$$\Delta p \geqslant \frac{\Delta C}{\Delta E} \qquad (5\text{--}2)$$

That is, the future price increase would at least be able to compensate costs incurred due to conservation.

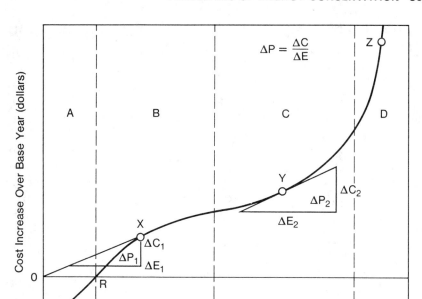

Fig. 5–5. Required energy price increases for further conservation with the current price structure.

Price increases required for the four stages of conservation efforts discussed in Fig. 5–4 are described as follows:

- A different magnitude of price increase expectation is needed for each stage of conservation.
- In the initial stage of conservation, only relatively moderate energy price increases are required to justify efforts for retrofitting, recycling, and simple energy management planning. Section RX on the curve in Fig. 5–5 indicates this stage.
- Expectations of substantial energy price increases are needed for industries to justify large capital investment to change production processes, computerize energy management systems, and modify major facilities. Section XY on the curve in Fig. 5–5 indicates this requirement.
- Beyond point Y on the curve in Fig. 5–5, further conservation requires expectations of very rapid price increases because incremental energy savings become extremely expensive owing to the technological limit of current conservation methods.

Economics of Energy Conservation Technological Developments

When expected future energy price increases are high, and conservation efforts under current technologies are costly, advanced energy-saving technologies will be developed and marketed. These technologies will be able to save more energy at lower costs than the current ones. As shown in Fig. 5–6, under the current conservation technology mix T_1, the minimum price increase required to save q_1 (amount of energy over the base year) is tan α_1,* under current conservation (which is identical to the curve in Fig. 5–5). If there are better conservation technology mixes developed, such as T_2 and T_3, more energy savings (q_2 and q_3) can be reached even with the same price increase requirement. In other words, if more advanced conservation technology mixes (T_2, T_3) are developed, lower energy price increase requirements (tan α_2 and tan α_3) and higher conservation possibility (q_2' and q_3') can be achieved. Availability of conservation technologies is therefore the key to industrial conservation. This point is further explained in Chart B in Fig. 5–6, which shows that more advanced energy conservation technologies reduce the minimum expected future energy price requirement for further conservation.

However, Chart B in Fig. 5–6 also shows that there are limits to development of energy conservation technologies. Energy conservation technological advances become uneconomical beyond point Z as shown in Chart B; because too much resource has been used to develop new conservation technologies, the return of the resource declines. In addition, technically, there is a minimum energy requirement for any given production process. The closer the conservation technologies come to this minimum, the more difficult and expensive it becomes to further reduce energy consumption. Accordingly, Chart B shows that, under the current economic structure, optimal conservation technological advance is near point Z. If our conservation technologies have not been developed to this point, we must spend more money on research and development for more advanced conservation technologies. On the other hand, if the conservation technological developments have exceeded the optimal point, we must invest this capital in other technologies.

OPTIMUM ENERGY CONSERVATION

Energy conservation is a matter of substitution—substitution for other production factors or substitution for future energy. Substitution forces reallocation of economic resources. In discussing optimum energy conservation, we thus

---

* The minimum price increase required is expressed as $\Delta C_1/\Delta E_1$ (shown in Fig. 5–5, which is tan α in Fig. 5–6).

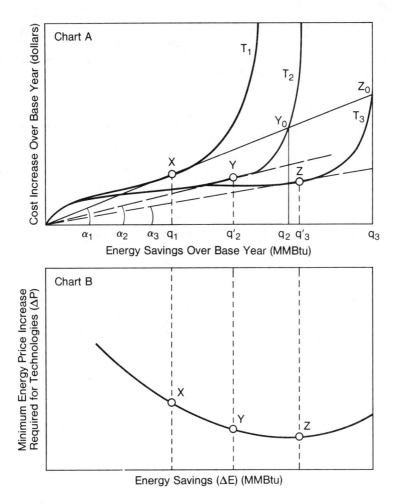

Fig. 5–6. Economics of conservation technological developments with the current economic structure.

have to deal with social welfare because industrial energy conservation affects the whole society.

Social welfare will increase during this reallocation if either (1) total satisfaction of some individuals (generations) increases without decreasing the total utility of other individuals (generations), or (2) the increment in utility to the recipient group is greater than the decrement in utility to the sacrificing group. This is called the principle of compensation, in modern welfare economic terms. However, when the principle is applied in the energy conservation case, it may not guarantee that, although some individuals (generations) will be better off, no individual (generation) will be worse off and that, accordingly, a better interpersonal (intergeneration) welfare will be achieved.* Two reasons for this failure are as follows:

- Energy is an essential substance for many individuals to maintain their standards of living. For example, gasoline is indispensable for commuters if there is no public transportation.
- Compensation to a present generation for a greater level of energy conservation can be made only through reward of other energy resources. The intergenerational social welfare may not increase if coal is used up in one generation and petroleum is largely conserved until a later generation because, in the later generation, petroleum may become of no value and coal may become of great value to the society.

What, then, is the optimum level of energy conservation for an intergeneration society? The optimum level of energy conservation is what each society in each succeeding generation thinks it to be. To be specific, at least three important conditions must be attained between individuals before aggregate intergeneration welfare can be maximized:

- In production, the last unit (marginal rate) of substitution between a given source of energy and any other production factors must be equal. For example, it is not a sufficient condition that coal be used up and petroleum and natural gas be largely conserved, or that labor be so overused in order to conserve energy that the society's standard of living declines.
- In consumption, the last unit (marginal rate) of substitution between a given source of energy and any other goods and services must be equal for every pair of individuals. It is also not a sufficient condition that an old man in Maine die of coldness due to extensive conservation of heating oil and a Californian drive his Lincoln Continental.
- In governmental policy-making, the last unit of benefit over cost in conservation must be equal among all fiscal and regulatory energy policies. The last unit of energy savings over costs of each government action or piece of legislation must be equal. For example, the last unit of energy savings over costs from an energy tax credit law must be equal to that from an industrial voluntary conservation program or the natural gas price decontrol.

---

* E. Hardy and O. Scoville expressed the same opinion for stock resources such as oil.[5]

SUMMARY

Principles of energy conservation and economics of conservation policy have been presented in this chapter. Energy conservation consists of production factor substitution and intertemporal substitution. Some may consider that, because energy conservation is an intertemporal substitution and more resources will be found in the future, conservation is unnecessary. On the contrary, because of dependence on foreign oil and the associated uncertainty of its supply sources, active conservation or even rationing is needed if the maintenance of stable economic growth is desirable.

The two principles of energy conservation are maximum thermodynamic efficiency and maximum cost-effectiveness. To achieve maximum thermodynamic efficiency, nonenergy productive factors (e.g., capital, labor, and time) are used to substitute for energy. To achieve maximum cost-effectiveness, inexpensive nonenergy factors, energy types, or future energy sources are substituted for expensive types of energy, such as today's oil.

The economics of conservation policy thus depends on the collective value judgment of resources by each generation in the society. In general, the initial phase of conservation will increase productivity, and further conservation efforts will improve market competitiveness for U.S. producers, if energy prices are expected to increase substantially in the near future.

However, conservation is a collective effort for all elements of the society. To obtain an optimum energy conservation policy, producers, consumers, and government bodies must work together to ensure that the last unit of substitution for energy is equal among all other production factors or future energy sources and that the last unit of governmental actions results in equal cost-benefit among all legislation and programs.

REFERENCES

1. National Bureau of Standards. "Waste Heat Management Guidebook," prepared for Federal Energy Administration, NBS Handbook 121, Feb. 1977.
2. U.S. General Accounting Office. "Report to the Congress," EMD-78-38, June 1978.
3. U.S. Department of Energy. "Industrial Energy Conservation Strategic Plan," prepared by Office of Conservation and Solar Applications, Division of Industrial Energy Conservation, July 1978.
4. U.S. Department of Commerce and U.S. Department of Energy. "Voluntary Industrial Energy Conservation," Progress Report 5, DOE/CS-0035, reprinted March 1978.
5. Heady, E. and Scoville, O. "Principles of Conservation Economics and Policy," Bureau of Agricultural Economics, U.S. Department of Agriculture, Research Bulletin 382, July 1951.

6
Thermodynamics and Energy Efficiency

INTRODUCTION

In this chapter, we evaluate the energy efficiency of national energy end-uses, energy-intensive industries, and selected energy devices by applying the first and second laws of thermodynamics. Energy efficiencies of the four copper production processes (mining, flotation, smelting, and refining) are calculated as an example of identifying energy-inefficient facilities that are suitable targets for conservation. Limitations on high energy efficiency are subsequently examined.

There are many terms for energy efficiency such as "energy effectiveness," "the first-law efficiency," and "the second-law efficiency." In this discussion we will use energy efficiency as a general term referring to the state of energy use in achievement of a given work.

This chapter consists of four major sections: (1) thermodynamics and definition of energy efficiency; (2) energy efficiency for major national energy end-uses, energy-intensive industries, and energy devices; (3) identifying energy-inefficient processes for improvement—the example of copper production; and (4) limitations on achieving high energy efficiency.

THERMODYNAMICS AND DEFINITIONS OF ENERGY EFFICIENCY

Energy efficiency is governed by the first and second laws of thermodynamics. The first law measures the quantity of energy, and the second law measures its quality (the availability of energy).

Quantity of Energy

The first law establishes a standard energy balance, defined as the ratio of useful output energy (E_{use}) to total input energy (E_{in}), as follows:

$$\eta = E_{use}/E_{in} \qquad (6\text{-}1)$$

Here, η is called the *first-law efficiency index*.

The first law states that energy can be neither created nor destroyed; it can only be changed from one form into another. Therefore, if we neglect the chemical conversion of energy into mass or vice versa, the net increase in the energy content of a particular system in a given time interval is equal to the energy content of the material leaving the system plus the work done on the system and the heat added to the system. Accordingly, the numerator in Equation 6–1 will be the sum of work and heat transferred, and the denominator will be energy input.

The theoretical maximum value of the first law efficiency index can be either greater than 1.0 (e.g., heat pumps), less than 1.0, or equal to 1.0. If the value is less than or equal to 1.0, then the index is conventionally called a coefficient of performance (COP); if the value is greater than 1.0, the index is called an efficiency.

The first law efficiency index has two characteristics: (1) its maximum values depend on the type of energy system and on surrounding temperature, and may be greater than, less than, or equal to 1.0; (2) it does not measure energy quality (availability).

Quality of Energy (Availability)

The second law of thermodynamics governs the availability of each material stream through the process being studied. It states that spontaneous heat flow is always unidirectionally from a hot body to a cold one, never the reverse, and that the entropy of an isolated system increases in such a process. In other words, the availability of energy always decreases, never the reverse, because the entropy is zero for maximum energy potential; the entropy increases when the availability of energy for work decreases.

Entropy is a function of the disorder (randomness of energy distribution) within a system. The entropy is zero for a perfectly ordered system and increases with increasing disorder (i.e., with greater randomness). Entropy change is a measure of energy degradation or disordering during a process. Work is always done at the expense of some increase in the randomness (degradation of order in the total universe).* Randomness reduces the strength of energy that theoretically can be transformed into work.

---

* Universe, as used here, means the system plus all surrounding matter.

Availability is therefore defined as the maximum work that can be obtained from a material or material stream by bringing the material to complete equilibrium with the environment by reversible processes.

A reversible process is a theoretical process for which all changes within a system and its surroundings can be "undone," and the universe returned to its initial condition. This concept is important because entropy is defined as

$$ds = \frac{dQ}{T}\bigg]_{\text{reversible}} \qquad (6\text{--}2)$$

where ds is an increment of entropy and dQ is the infinitesimal heat transfer at absolute temperature T. Only in a reversible process is the entropy of the universe held constant (the entropy increase in one part of the universe being equal to the entropy decrease in another part). In all irreversible processes, the entropy of the finite universe increases. This leads to corollaries of the second law that state that the minimum energy is achieved in the performance of a process if the process takes place reversibly, and therefore maximum work is produced in reversible work-producing processes.

The availability of a stream (A) is defined as follows:

$$A = M \cdot \Sigma_j (h_j - T_0 \cdot s_j) \qquad (6\text{--}3)$$

where M is the mass flow rate, h_j is the enthalpy, T_0 is the temperature of the reference atmosphere, s is the entropy, and j is a component. The summation is done over all components j in each stream. The enthalpy, h_j, depends only on the component and stream temperature, but the entropy generally depends upon the concentration of each component in the stream as well as the temperature.

Thus, availability is affected not only by pressure and temperature differences, but also by differences in chemical potential and electrical potential between the material and the environment. Unlike energy, availability can be destroyed; indeed, in any real process it is destroyed. Availability is destroyed when work is not obtained from a process in which there is a decrease of some potential difference (e.g., pressure, temperature, etc.).

The quality of energy in a device can be measured with a ratio of availability of stream output (A_{out}) to availability of stream input (A_{in}) as shown in Equation 6–2.* We name this ratio the *second law effectiveness coefficient* (ϵ):

---

* There are other definitions for effectiveness. For example, Hall et al. also defined it as the ratio of the increase in availability of the process to the decrease in availability for the process, or as the ratio of total output availability to the total input energy.[2]

$$\epsilon = A_{out}/A_{in} \tag{6-4}$$

The maximum value of ϵ is 1.0 in all cases.

For single-source, single-output devices, a more convenient measurement of second-law effectiveness coefficients may be expressed as follows:

$$\epsilon = \frac{B_{min}}{B_{actual}} \tag{6-5}$$

where B_{min} is the minimum amount of available energy required to deliver a given amount of work, and B_{actual} is the actual amount of available energy used. The denominator is the same as that in the first-law efficiency index (Equation 6-1). The numerator is the theoretical minimum permitted by the first and second laws. It is a task minimum, not a device minimum.

If the output of an energy device is heat or work (but not both), Q, then Equation 6-4 can be transformed:

$$\epsilon = \frac{\dfrac{B_{min}}{Q}}{\dfrac{B_{actual}}{Q}} = \frac{\dfrac{1}{\eta_{max}}}{\dfrac{1}{\eta}} = \frac{\eta}{\eta_{max}} \tag{6-6}$$

where η is first-law efficiency, and η_{max} is the maximum efficiency permitted by the second law of thermodynamics.*

The second-law effectiveness coefficient (ϵ) provides immediate insight into the quality of performance of any single-source, single-output energy system relative to its ideal performance. It measures theoretical potential for improvement and pinpoints sources of unwarranted fuel waste.

In summary, second-law effectiveness is analogous to efficiency and often is close to efficiency, as is the case with power plants. However, under certain circumstances, the two can differ substantially. For example, an electric hot water heater providing hot water at 110°F (43°C) from cold water of 32°F (0°C) may have a second-law effectiveness of $\epsilon = 0.041$ and first-law efficiency of $\eta = 0.9$ (see the next section for detailed computation). The reason for this difference is that the effectiveness coefficient is computed by dividing by the maximum efficiency of the heater ($\eta_{max} = 7.35$).**

In the case of single-source, single-output devices, ϵ and η are proportional as shown in Equation 6-6, differing by the scaling factor η_{max}, which is determined by the materials available. In other words, for processes such as power plants in which $\eta_{max} = 1.0$, the effectiveness coefficient and the

---

* For more theoretical discussion, see references 3–6, 8, 9 and 16.
** See under "Efficiency of Energy Devices," in the subsection on the electric hot water heater.

efficiency index are almost identical. Since the second-law effectiveness coefficient more accurately indicates the potential for energy savings, the maximum of ϵ becomes a technical goal to be considered in conservation policy. However, the constraints of economics, time, environment, and heat requirement affect pursuit of this goal. We will discuss these limitations in later sections of this chapter.

ENERGY EFFICIENCY FOR MAJOR NATIONAL ENERGY END-USES, ENERGY-INTENSIVE INDUSTRIES, AND ENERGY DEVICES

In this section, we evaluate energy efficiencies for major national energy end-uses, seven energy-intensive industries, and commonly used energy devices, by using the efficiency equations developed in the previous section.

Effectiveness of Major National Energy End-Uses

It is possible but difficult in practice to determine the energy effectiveness coefficients of national energy end-uses.* Since the national energy end-uses consist of multi-source, multi-output systems, theoretically Equation 6–4 should be used to compute their energy effectiveness coefficients. However, the American Institute of Physics[12] used Equation 6–5 to roughly estimate the effectiveness coefficients for energy end-uses as shown in Table 6–1.

The table indicates both the relatively large energy use for space and water heating and automobile transportation, and the poor thermodynamic efficiency of energy use for these purposes. Even though industrial energy uses have higher-than-average fuel efficiency, there is room for improvement because their energy effectiveness coefficients are relatively low, in the range of 0.25–0.30. Since about 40% of energy used nationally is used in the industrial sector, the potential benefit of industrial energy conservation is quite significant. Among industrial energy uses, process steam and direct heat are the most promising candidates for waste heat recovery technologies.

Efficiency of Seven Energy-Intensive Industries

In this section, Equations 6–1 and 6–4 will be used to evaluate the energy efficiency and effectiveness of specified production process units in seven energy-intensive basic industries:[2] steel, copper, aluminum, container glass, rubber, paper, and plastics.

---

* First-law efficiency indices are not computed because of difficulties in estimating input energy for each national energy end-use category, which consists of numerous energy users under various operating conditions.

Table 6-1. Effectiveness of Major National Energy End-uses.

	Fraction of Total Fuel Consumption (1968)	Estimated Overall Energy Effectiveness (ϵ)
Space heating	18	.06
Water heating	4	.03
Air conditioning	2.5	.05
Refrigeration	2	.04
Industrial uses		
Process steam	17	.25
Direct heat	11	.30
Electric drive	8	.30
Electrolytic processes	1.2	—
Transportation		
Automobile	13	.10
Truck	5	.10

Source: Data selected from AIP[12] and SRI.[11]

A process unit is defined as any process, subprocess, or combination of processes up to an entire plant or industry. For example, in the steel industry specific furnace types were analyzed as process units (blast furnace, basic oxygen furnace, open hearth); in the paper industry several identifiable but closely coupled processes were analyzed as a unit (the kraft pulping process as analyzed included not only kraft pulping but also the recovery systems for black and green liquors); and in the rubber industry the batch reaction for polymerization was analyzed as two units (preheating and reaction). Thus, a process unit can be as large or as small as the investigator wants, or needs, it to be.

The most important step in setting up a process unit analysis is deciding on specific boundaries for the process unit. The analytical results, especially the process efficiency and effectivenesses, are quite sensitive to the detailed specification of process and its stream entry and exit conditions. All the material streams into, or out of, the process must be identified, and their pressure, temperature, flow rate, and composition must be determined. The flow streams are specified by their composition, pressure, temperature, and mass flow rate. Composition specification includes identification of the chemical species, mole fraction or number of moles, molecular mass, and partial pressure in the environment.

The flow streams are designated as inflow or outflow by positive or negative values respectively for the mass flow rate. Each outflow stream, and each outflow of heat may be designated as a useful output, and thus be used in the calculation of efficiency. All work outputs are considered useful.

In determining the first-law efficiency of a given process unit, heat balance is checked.* Since the first law of thermodynamics requires equality between the energy inputted to the system and the energy (and work) outputted from the system, an iterative procedure is used to successively balance the heat input and output flows. The iterative process ends when a maximum number of iterations is reached (usually eight), or when the heat flows are balanced to within 0.5% of the total input energy.

When the heat flows have been satisfactorily balanced, the second-law effectiveness of this process is evaluated on the basis of energy availability. The availability of each flow stream is calculated from the enthalpy, entropy, and pressure of the stream. Heat losses are assumed to be destructions of availability. Work inputs and outputs are considered to be fully equivalent to availability inputs and outputs. Since the second law of thermodynamics is an inequality, there is no way to balance the availability flows for a process in the way that energy flows must balance to satisfy the first law.

As shown in Equations 6–3 and 6–4, calculating the energy and availability of a flow stream requires evaluating the enthalpy and entropy of the components of the flow stream. Entropy (s) is calculated by using the following equation:

$$s = \frac{H - G}{T} \qquad (6\text{--}7)$$

where G is the adjusted standard tabulated value of Gibbs free energy of formulation, and T is the absolute temperature. Enthalpy is expressed in Btu/lb mole and entropy in Btu/lb mole K at a given temperature (Kelvin).

Resultant Energy Efficiency. The energy efficiency indices and effectiveness coefficients of seven energy-intensive industries were analyzed by the above methods[2] and summarized in Table 6–2. Except for the rubber and selected plastics industries, these energy-intensive industries, which consume a significant portion of U.S. energy used in the industrial sector, do not effectively use their energy sources. Among them, copper is the most energy-inefficient industry, owing to the low-grade ore used (0.5 to 1.2% copper content), and will thus serve as an example throughout the book.

Each of these industries consists of several production processes, each of which has its own energy efficiency. The difference between these efficiencies is often substantial. For example, the energy efficiency indices of the Bayer and anode baking processes in aluminum production are 0.02 and 0.55, respec-

---

* Methods and results of Battelle Columbus Laboratories report are used.[2]

Efficiency (η) and Effectiveness (ϵ) of Processes

Industry	Energy Use, % of Total U.S. Industrial (1976)	Overall Industry η	Overall Industry ϵ	Process	η	ϵ	Process	η	ϵ	Process	η	ϵ	Process	η	ϵ
Steel	12.59	.42	.40	(Coking)	.956	.92	(Blast furnace)	.709	.759	(Basic oxygen furnace)	.894	.866	(Soaking pit)	.85	.88
Copper	0.29	.035	.092	(Mining)	0	0	(Concentration)	.64	.76	(Smelting)	.44	.48	(Refining)	.46	.53
Aluminum	2.48	.296	.326	(Bayer process)	.02	.25	(Alumina calciner)	.21	.47	(Anode baking process)	.55	.60	(Hall process)	.44	.52
Glass	1.14	.239	.221	(Melting)	.718	.535	(Melting and regenerating)	.335	.257						
Rubber	—	.924	.962	(High-temperature production)	.987	.98	(Monomer recovery)	.97	.99	(Drying)	.97	.99	(Low-temperature production)	.984	.989
Paper	5.66	.443*	.445*	(Ground-wood pulping and bleaching)	.824	.848	(Kraft pulping)	.78	.62	(Three-stage kraft bleaching)	.838	.99	(Five-stage kraft bleaching)	.784	.995
Plastics	1.42	NA**		(Low-density polyethylene)	.835	.931	(High-density polyethylene)	.751	.879	(Polystyrene)	.827	.979	(Polyvinyl chloride)	.655	.903

* Estimated by calculating ϵ and η of the combined kraft pulping and kraft bleaching processes due to varieties of paper products. The actual ϵ and η should be lower.

** We studied four selected plastics industries, low-density polyethylene (LDPE), high-density polyethylene (HDPE), polyvinyl chloride (PVC), and polystyrene (PS). These four classes account for 57% of the total U.S. plastic production. Owing to their difference, no overall efficiency was computed.

Source: Prepared from data reported by Battelle Columbus Labs.[2]

tively, an approximately 25-fold difference. Therefore, after having identified the inefficient industries, we need to identify the most promising processes for the application of energy conservation technologies.

Efficiency of Energy Devices

Common energy devices for systems such as electric motors, air conditioners, refrigerators, power plants, and heaters mostly use single sources of energy and produce a single form of output (either work or heat). Therefore, Equation 6–6 is used to evaluate their energy effectiveness. Table 6–3 summarizes the first-law efficiency and second-law effectiveness values for these devices. The energy efficiency of an electric hot water heater and power generation from fossil fuels are calculated below as examples.

Electric Hot Water Heater. If 90% of the electric energy supplied to the water tank transforms the electricity into heat and circulates the hot water to the industrial compound, then the efficiency index (η_{ew}) is 0.90. Electrical energy is high-quality energy (high availability) produced from lower-quality energy, (approximately 33% of the energy in a fuel can be converted into electricity in an electrical generation plant). Thus, if the system is considered to include the generating plant as well as the water heater, the result is an efficiency index ($\eta_{overall}$) of 0.30. This is significantly lower than most oil- or gas-fired hot water heaters, which have efficiency indices of approximately 0.6–0.7.

The laws of thermodynamics show that to deliver an amount of heat Q to heat water, the minimum amount of available energy required is:

$$B_{min} = Q\left(1 - \frac{T_0}{T}\right) \tag{6–8}$$

where T is the temperature at which the heat is delivered, and T_0 is the ambient temperature. B_{min} is the same as the amount of work required to drive an ideal, reversible heat pump delivering heat Q at temperature T.

When a fuel is burned, the available energy is very close to the energy released in combustion, ΔH, and thus the maximum efficiency index for the heating of water is:

$$\eta_{max} = \frac{Q}{\Delta H_{min}} = \frac{Q}{B_{min}} = \frac{1}{1 - \dfrac{T_0}{T}} \tag{6–9}$$

Devices.

Type of Device or System	Numerator in η	Denominator in η	η†	ϵ‡	Standard nomenclature for η
1. Electric motor	Mechanical work output (W_{out})	Electric work input (W_{in})	W_{out}/W_{in}	η	Efficiency
2. Heat pump, electric	Heat Q_2 added to a warm reservoir at $T_2{}^*$	W_{in}	Q_2/W_{in}	$\eta\left(1-\dfrac{T_0}{T_2}\right)$	Coefficient of performance (COP)
3. Air conditioner or refrigerator, electric	Heat Q_3 removed from cool reservoir at T_3	W_{in}	Q_3/W_{in}	$\eta\left(\dfrac{T_0}{T_3}-1\right)$	COP
4. Refrigerator, absorptive**	Q_3	Heat Q_1 from hot reservoir at T_1	Q_3/Q_1	$\eta\dfrac{(T_0/T_3)-1}{1-(T_0/T_1)}$	COP
5. Heat engine	W_{out}	Q_1	W_{out}/Q_1	$\eta\dfrac{T_1}{T_1-T_0}$	Thermal efficiency
6. Power plant	W_{out}	Fuel-heat of combustion $|\Delta H|$	$W_{out}/|\Delta H|$	η	Efficiency
7. Geothermal plant	W_{out}	Q_1	W_{out}/Q_1	$\dfrac{\eta}{1-(T_0/T_1)}$	Efficiency
8. Solar hot water heater	Q_2	Q_1	Q_2/Q_1	$\eta\dfrac{1-(T_0/T_2)}{1-(T_0/T_1)}$	COP or efficiency
9. Heat-powered heating device***	Q_2	Q_1	Q_2/Q_1	$\eta\dfrac{1-(T_0/T_2)}{1-(T_0/T_1)}$	COP or efficiency

* T_1 (hot) > T_2 (warm) > T_0 (ambient) > T_3 (cool).

** Any heat-power device for cooling.

† $\eta = E_{out}/E_{in}$ (Equation 6-1).

‡ $\epsilon = \eta/\eta_{max}$ (Equation 6-6) where η_{max} is the maximum first-law efficiency index permitted by the second law of thermodynamics.

*** A furnace is a special case; for it $\eta_{max} = 1$.

Source: Data collected from AIP.[12]

If this water heater provides hot water at 110°F ($T = 316$ K) from cold water of 32°F ($T_0 = 273$ K), then we obtain $\eta_{max} = 7.35$. Thus, the heater, cited above with $\eta_{ew} = 0.9$ and $\eta_{overall} = 0.3$, has thermodynamic effectiveness coefficients ($\epsilon = \eta/\eta_{max}$) of

$$\epsilon_{ew} = \frac{\eta_{ew}}{\eta_{max}} = \frac{0.9}{7.35} = 0.122 \qquad (6\text{-}10)$$

and

$$\epsilon_{overall} = \frac{\eta_{overall}}{\eta_{max}} = \frac{0.3}{7.35} = 0.041$$

This efficiency can be improved if economic heat pumps or waste heat recovery equipment are installed. The coefficient of performance (η_{hp}) of the heat pump in this case can reach 2.43.*

Power Generation from Fossil Fuels. Normally η for power plants is subject to a theoretical maximum value, the Carnot efficiency, that is less than 0.87. When a fuel is burned, the available energy is very close to the heat of combustion, and $\epsilon = \eta$ (row 6 in Table 6–3); η is in the range 0.3 to 0.4 for most operating power plants. The Carnot efficiency is 0.87 for a flame temperature of 2000°C and 0.64 for a peak steam temperature of 550°C. Owing to the unavailability of high-temperature resistant material, it is impractical to reach an efficiency of 0.87. However, for conversion of chemicals into electricity, the fuel cell, with its optimal efficiency (ϵ_{max}) of 1.0 is the ideal device.[7]

IDENTIFYING ENERGY-INEFFICIENT PROCESSES FOR IMPROVEMENT—THE EXAMPLE OF COPPER PRODUCTION

Production Process Description**

Mining sulfide ores, often containing significant amounts of iron as well as copper, provides the major source of copper in the United States. About 80% of the domestic primary copper supply is produced from open pit mining of ores of this type containing from 0.5 to 1.2% copper. These ores are concentrated by flotation, smelted to separate copper from the gangue material, and electrolytically refined to produce a high-purity grade of copper. Sulfide ores containing lesser amounts of copper, oxide ores, and mixed oxide–

---

* The COP of heat pumps in this case can be 2.7. Since heat pumps are used to heat hot water, this COP must be multiplied by the water heater's efficiency index ($\eta_{ew} = 0.90$ as computed before). Therefore, we obtain $\eta_{hp} = 2.7 \times 0.9 = 2.43$.

** Most of this description is abstracted from reference 2.

sulfide ores are initially processed by leaching followed by cementation, electrowinning, or ion exchange. The copper-rich product of the leaching operation normally enters the conventional primary production process during either smelting or refining. Other sources of copper include native copper, high-grade sulfide ores (3 to 10% copper) and by-product ores separated during the processing of other metals. Processing of these ores is similar to that of concentrating grade sulfide ores.

A fairly good estimate of energy consumption in the production of primary refined copper probably can be obtained by assuming that all copper ores are processed according to the conventional beneficiation-smelting-refining route used to process concentrating grade ores.

A flow sheet for primary refined copper production from a concentrating grade sulfide ore, including an estimated materials balance, is given in Fig. 6–1. In constructing this flow sheet, a copper ore containing 0.76% copper was assumed. This ore grade represents the weighted average copper content of ores described in a detailed survey of flotation processing reported in 1970.[14] This same study suggested that the recovery efficiency of the flotation process averaged 82.5%. Metallurgical efficiencies during smelting and refining were estimated as 98 and 99%, respectively.

Large amounts of scrap, about 30% of the total scrap recycled each year based on contained copper content, are processed with primary copper. Scrap is charged into the converter in the smelter or into the anode furnace in the refinery. The remaining scrap material is processed by secondary producers or brass mills to produce both refined and alloy grades of copper using a variety of smelting and refining techniques.

Beneficiation. Copper ores of the sulfide type contain from 0.6 to 2% copper in a variety of mineral forms. These ores also contain significant amounts of valuable by-product elements, such as molybdenum, gold, silver, selenium, and tellurium. Beneficiation can be accomplished by flotation, taking advantage of the tendency of nonwetted particles to cling to and rise with air bubbles passed through a water suspension containing the particles. For flotation to be effective, it is first necessary to mechanically separate the desirable minerals from the surrounding gangue minerals. This is done by crushing and grinding, a particle size of about 35 mesh permitting good separation. A typical flow sheet for the concentration of copper from a concentrating grade sulfide ore is shown in Fig. 6–2.

Crushing and Grinding. Copper ore as mined may exist in fairly massive chunks. This ore is crushed in several stages using jaw or cone crushers to produce a particle size of from 1 to 2 inches in diameter. Grinding is accomplished wet in rod and ball mills to achieve a particle size of about 35 mesh. Regrinding is used at an intermediate stage to achieve greater copper recovery.

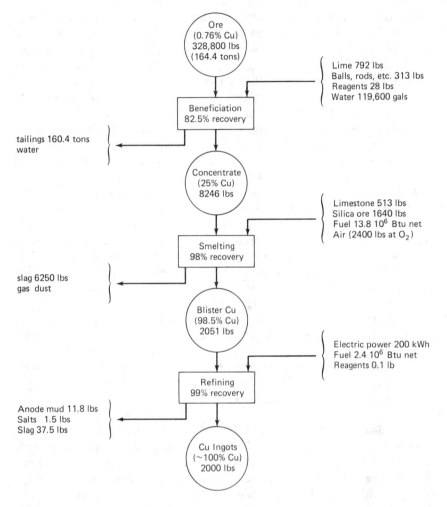

Fig. 6–1. Materials balance for primary refined copper production. Source: Battelle Columbus Labs.[2]

Flotation. Flotation is accomplished in rectangular tanks agitated to maintain the ground ore in suspension and to produce a steady supply of air bubbles to affect flotation. The copper-bearing particles form a froth on the surface of the tank, which is removed by scrapers. The wetted gangue-mineral particles settle to the bottom of the cell and are removed. Approximately 10 pounds of chemicals are required in treating the amount of ore necessary to produce 1 ton of refined copper. Lime is also required to control pH, approximately

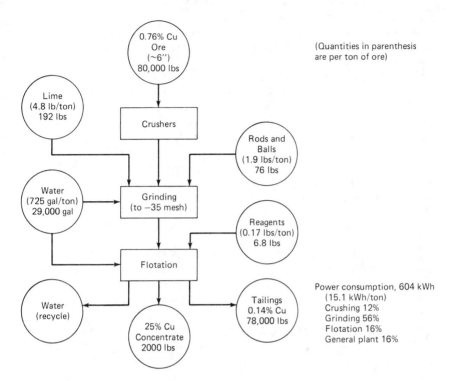

Fig. 6–2. Materials balance for flotation beneficiation of a sulfide copper ore. Source: Battelle Columbus Labs.[2]

1200 pounds per ton of refined copper produced. The system shown in Fig. 6–3 is representative of current flotation practice.

Smelting. Smelting operations are conducted to produce from concentrated copper sulfide ore an impure form of copper, referred to as blister copper, containing about 98.5% copper. Three basic operations are involved: roasting, in which the sulfur content of the concentrate is reduced and water is removed; matte smelting, in which the concentrate is melted with an appropriate flux to permit separation of much of the gangue material from a copper-iron-sulfur liquid referred to as matte; and conversion, in which iron and sulfur are removed from the matte by oxidation. A flow sheet for the smelting process is shown in Fig. 6–4.*

---

* Roasting was omitted from the process sequence shown in Fig. 6–4 for reasons discussed in the following section.

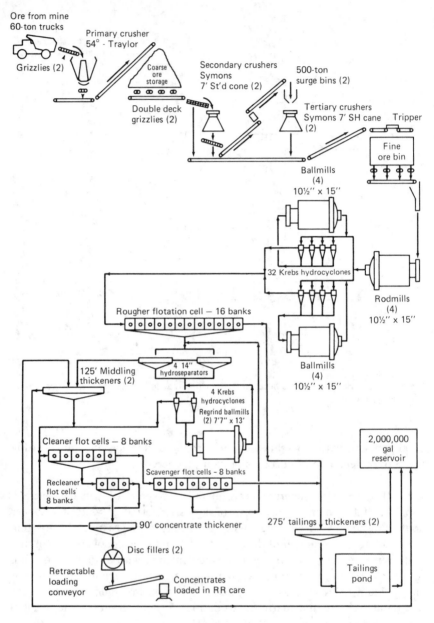

Ore from mine
60-ton trucks

Primary crusher
54° - Traylor

Grizzlies (2)

Coarse ore storage

Secondary crushers Symons 7' St'd cone (2)

500-ton surge bins (2)

Double deck grizzlies (2)

Tertiary crushers Symons 7' SH cane (2)   Tripper

Fine ore bin

Ballmills (4) 10½" x 15"

32 Krebs hydrocyclones

Rougher flotation cell — 16 banks

Rodmills (4) 10½" x 15"

125' Middling thickeners (2)

4 14" hydroseparators

Ballmills (4) 10½" x 15"

4 Krebs hydrocyclones

Regrind ballmills (2) 7'7" x 13'

Cleaner flot cells — 8 banks

2,000,000 gal reservoir

Recleaner flot cells 8 banks

Scavenger flot cells - 8 banks

90' concentrate thickener

275' tailings thickeners (2)

Disc fillers (2)

Retractable loading conveyor

Concentrates loaded in R R care

Tailings pond

Fig. 6–3.   Flow sheet of typical flotation process. Source: Bureau of Mines.[14]

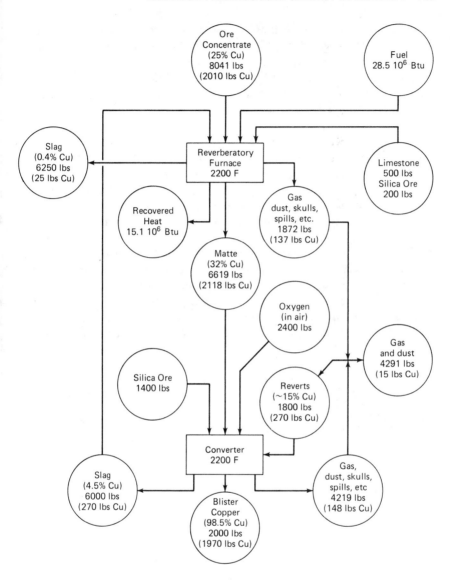

Fig. 6–4. Flow sheet and materials balance for copper smelting. Source: Battelle Columbus Labs.[2]

Roasting. Roasting is conducted by gradually heating the concentrate while it is exposed to air to a temperature of about 1650°F. Although still used at several smelters, roasting is gradually being eliminated in favor of direct smelting of wet concentrates.

Matte Smelting. Matte smelting is performed in large, batch-operated reverberatory furnaces heated to about 2200°F. The furnace is charged with roasted
or dried concentrate, additional copper-containing material such as high grade
ore, scrap, converter slag, etc., and fluxing additions. Fluxes are added as
necessary depending upon the composition of the furnace charge, to produce
a slag composition favoring maximum separation of gangue material from
copper.

As the charge is melted, a majority of the gangue material dissolves in
the flux, forming a silicate slag that separates from and floats on top of the
matte. The slag will contain less than 1% copper and is discarded, while
the matte is transferred as a liquid to the converter. From 10 to 20% of
the sulfur content of the charge is converted to SO_2 during matte smelting
and is discharged with the flue gases, which may contain as much as 1.0%
SO_2.

Heat input during smelting is required to melt the charge components;
little heat is generated by chemical reaction during melting as presently conducted. Approximately 15% of the heat input is discarded with the slag,
and 5% is contained in the matte which is sent to the converter. Another
25% is lost by radiation, conduction, etc. Of the remaining 55%, discharged
in the flue gases, about half is recovered by heat exchangers in the exhaust
gas system.

Conversion. Conversion consists of blowing air through the matte to oxidize
iron and sulfur, leaving a relatively impure form of copper. Conversion is
autogenous, and is conducted in refractory-line vessels with tuyeres along
the bottom surface for admitting air.

Refining. Blister copper contains small amounts of elements such as arsenic,
bismuth, and iron that are detrimental to desired properties, as well as amounts
of gold and silver which are well worth recovery. Purity is upgraded by
refining operations. Typically, an initial fire refining operation to remove
the bulk of the base metal impurities followed by electrolytic refining to
remove the remainder of the base metals and to salvage the precious metals
is used to refine copper. A flow sheet including an approximate materials
balance is shown in Fig. 6–5.

Fire Refining (Anode Furnace). Most copper smelters have fire refining operations on site, the molten blister copper being transferred directly to a refining
furnace following which anodes are cast. Electrolytic refineries will also have
fire refining capability to permit processing of in-plant and purchased scrap.
Refining is conducted in small reverberatory furnaces operated at about
2200°F. The molten metal is first oxided by blowing air into the melt at a

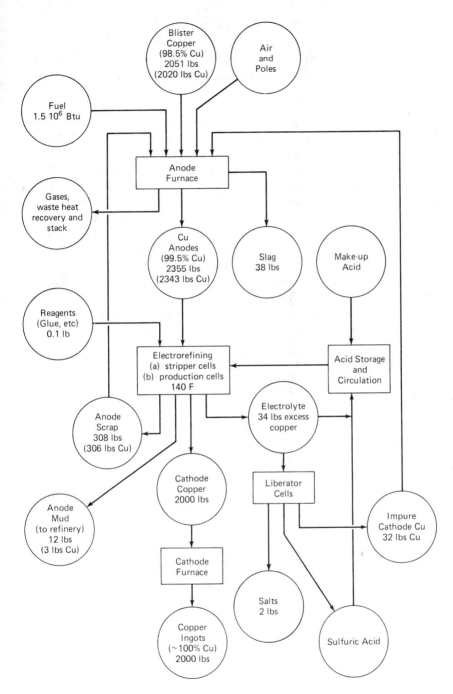

Fig. 6–5. Flow sheet and materials balance for electrolyte refining of copper. Source: Battelle Columbus Labs.[2]

low rate or by agitating the melt surface to promote atmospheric contact. Impurities are removed either as a slag formed on the melt surface or in the flue gas if volatile.

Electrolytic Refining. Copper is electrolytically refined in an aqueous copper sulfate electrolyte containing about 200 grams per liter of free H_2SO_4 and 40 grams per liter of copper. Copper is deposited in stripper cells on rolled copper cathodes and periodically stripped off to provide starter sheets. These sheets supply cathodes for the production cells where the bulk of the refining is conducted. Precious metals do not enter the solution, and are deposited at the bottom of the tank as a slime (anode mud) which is removed for separate processing. Base metal impurities are removed from the electrolyte as necessary by special electrolyte purification treatments.

A well-designed electrolyte refining plant requires about 200 kWh of electrical energy per ton of copper produced. Auxiliary energy consumption is required to heat and purify the electrolyte.

Following refining, the production cell cathode copper is melted in small reverberatory or electrolytic furnaces and cast into useful shapes.

Secondary Processing. A number of brass mills, primary producers, foundries, ingot makers, and secondary smelters process copper scrap, most of them producing brass or bronze ingot. A few secondary smelters produce either fire-refined or electrolytically refined copper. Over 30% of the scrap is processed by electrolytic refining, mostly by the primary producers. Over one-half of the remainder is processed directly by brass mills.

Scrap processing by secondary smelters depends upon the grade of scrap (purity, form, etc.), and the scrap may enter the secondary process at a variety of different points. Many of the processing steps parallel those used in the production of primary metal, such as fire refining and electrolytic refining processes.

Prerefining Processes. A variety of shredding, crushing, and briquetting treatments are used to size scrap into forms suitable for charging into the refining furnace. Specialized processes are also used to remove organic coatings (burning), and to remove insulating material (stripping).

Smelting Processes. Low-grade scrap, typically containing from 10 to 35% copper and consisting of iron scrap, process slags and fluxes, mixed sweepings, etc., is usually processed in a blast furnace. Coke, limestone, and mill scale are charged into the furnace with the low-grade scrap. Iron silicate slags are formed which dissolve the bulk of the impurity elements while a relatively impure form of copper, referred to as black copper, is formed containing the bulk of the copper present in the charge. Black copper will contain from 70 to 85% copper.

Refining of Alloy Scrap. Well-identified alloy-grade scrap can be remelted and fire-refined to produce brass and bronze ingot. This type of scrap processing is done by brass mills as well as by secondary processers. Significant amounts of scrap are processed in this manner.

Energy-Inefficient Processes in Copper Production

Having described the copper production processes, we now examine their energy-efficiency. As shown in Table 6–2, copper is the most inefficient industry among the seven energy-intensive industries studied. The average energy consumption per short ton of primary copper production is shown in detail in Table 6–4. If we consider only the energy consumed in the processes involved in heat transfer, not that used in crushing, grinding, and transporting ores, materials, and products-in-process, the energy and material balance is shown in Fig. 6–1. This set of energy consumption data is used to measure energy efficiencies of primary copper production processes.

Efficiency of Primary Copper Production. As may be seen in Table 6–4, overall energy efficiency for the whole primary copper production process is very low, below 10%.

Detailed energy efficiencies of individual production processes are shown in Table 6–4. Because we assume the majority of companies use integrated plants that take hot materials directly from previous processes, the efficiencies

Table 6–4. Energy Consumption and Efficiency of Copper Production Processes.

Primary Production Process	Average Energy Use* (10^6 Btu/Short Ton)	Efficiency Index** η	Effectiveness Coefficient** ϵ
Mining	15.12	0	0
Concentration (beneficiation)	35.87	.64	.76
Smelting	35.84	.44	.48
Refining	12.00	.46	.53
Total	98.83	.035	.092
Secondary production process	10.3	N/A	N/A

*Bureau of Mines.[1,15]
**Table 6–2.

for each process are much higher than the overall efficiency. Examining these indices closely, we find that the most conservation potential exists in the mining and smelting processes. Unfortunately, leaching and hydrometallurgy, which may save a large proportion of the energy used for milling ores, are still limited in practice. Several energy conservation measures may apply to the smelting process; their costs and benefits will be examined in the next chapter.

Efficiency of Secondary Copper Production. The efficiency of secondary copper production depends on the quality of copper scrap. On the one hand, if the grade of the scrap is low, and it is used in smelters, then the savings of energy is only the amount used in mining and flotation; the efficiency is not improved substantially from that of primary copper production. On the other hand, if the scrap is of high grade and is used in refineries, its energy savings can be up to 90% of the energy used for primary copper production.

LIMITATIONS ON ACHIEVING HIGH ENERGY EFFICIENCY

The energy-efficiency approach provides a valid technological evaluation of how effectively energy resources are used. However, the laws of thermodynamics do not consider economic, environmental, temporal, and nonenergy resource factors. The concern for effective energy utilization stems from just these factors, namely, a shortage of fuel supplies, resultant high energy prices, and limits on the capacity of the environment to absorb the wastes. If there were no limits on the amount of energy that could eventually be supplied at an affordable price, and if the energy could be used by society without associated damage to the environment, there would be no need to consider conservation or energy efficiency.

In addition to these nontechnical factors, there are two convenience constraints (time allowed for performance of the task and size of equipment) and two physical constraints (friction for control and the temperature limits of given materials or fuels). These constraints technically prevent devices from reaching their theoretical maxima of energy efficiency.*

Time for performance is the most crucial constraint. To perform any task with the minimum amount of available energy it is necessary that the task be carried out reversibly, that is, "infinitely slowly." Any task performed in a "finite" time interval will have an internal irreversibility that increases in some unspecified way the faster the task is performed. In developed industrial society, time sometimes is more valuable than energy, and mass produc-

---

* Many of these viewpoints were discussed in references 13, 17.

tion requires rapid task performance. Therefore, a task performed rapidly but less energy-efficiently may be more useful to the society than a slower, more efficient one. This is particularly apparent in industrial production and air transportation.

The size of equipment required in industrial production is important because of its cost and required installation area. For any process involving heat flow, a thermodynamic efficiency approaching 100% would require that the temperature gradient across any boundary where heat is transferred approach zero. Under such conditions the heat flux would approach zero. A corollary indicates that high thermodynamic efficiency requires the use of very large heat exchangers. The materials embodied in heat exchangers require energy, as well as other resources, for their fabrication. Clearly, making very large devices may be more wasteful than sacrificing a little in the thermodynamic efficiency of the device. For example, in the operation of electric storage batteries, reversibility requires a very low current drain per unit area of plates. To draw currents of a useful size, large batteries with extensive plate area must be used. Nearly reversible operations would require enormous batteries, too large to use in electric cars.

Friction is often needed to accomplish a given task although it reduces thermodynamic efficiency. For example, without friction between tires and the surface of the road, it is impossible to control automobiles. However, because much of the available work is required to overcome friction, the availability of energy decreases, with a lessening of thermodynamic efficiency.

The temperature limits of the materials from which energy devices are made also prohibit maximum thermodynamic energy efficiency. The development of materials capable of withstanding the high stresses presented by high temperatures has encountered formidable problems. Although most fuels can be burned at higher temperatures than the materials surrounding the combustion products can withstand, the temperature is maintained at a "practical" level by controlling the combustion fuel-air mixture. In principle, the temperature can be reduced in this way to within a small increment above the ambient temperature and yet give the same total energy release. Since the available energy is directly dependent on the absolute temperature, higher temperatures have theoretical advantages in maximizing energy efficiency. However, real materials must be used, and these have temperature limits that result in lowered thermodynamic efficiency.

The above constraints on thermodynamic efficiency are mostly those that stem from conscious decisions about the way we build a machine or carry out a process or task. Inefficiency may also result from our failure to comprehend a problem fully, and even from legal and institutional constraints.

Engineers have always designed processes to achieve the desired goals at minimum total cost, because this is what society demands of them. Often this has meant that less than maximum efficiency was attained, as designs were optimized to use less of the more expensive resources—scarce commodities, expensive labor, complex controls—in performing the given task. Energy resources have traditionally been given relatively low priority, because they were abundant, and inexpensive. Today, with much higher energy costs, the engineer is finding that designs to save fuel are becoming more consistent with economic goals.

In the age of inexpensive fuel, equipment was designed to use much high-quality energy for tasks that really required no more than low-quality energy. United States industry generally has not bothered with recovering and utilizing low-quality "waste" heat because its value is often very low.* Where the quantity and quality of the released heat have been high, as in hot gases from a blast furnace, the heat has often been used for preheating incoming air and fuel. In other cases the heat has been released to the atmosphere or a nearby water body. It can be expected that the rising cost of fuel will make it economical for industry to recover an even larger fraction of heretofore rejected heat. All such regenerative use and recovery of previously wasted heat will result in improved use of available energy and, hence, improved thermodynamic efficiency.

Thermodynamic efficiency can also be limited by restrictions imposed by society. Certain social decisions have adversely affected efficiency, sometimes drastically. For example, installation of SO_2 scrubbers reduces the net thermal efficiency (effectiveness coefficient) of a coal-fired steam electric generating station by approximately 4%. (Overall heat rate is increased by about 10%.)**

SUMMARY

In this chapter, we have applied the first and second laws of thermodynamics to evaluating energy efficiency of the nation's major energy end-uses, seven energy-intensive industries, and selected energy devices. The two laws offer two methods, the energy efficiency index and the energy effectiveness coefficient, for assigning quantitative values of how efficiently energy resources are used. The two methods generally produce identical values if used for processes for which their absolute maximum values are 1.0 and differ if the energy efficiency index is greater than 1.0.

Thermodynamics can be used to identify inefficient industries, inefficient production processes in these industries, and inefficient devices within these processes. For example, the copper industry is the most inefficient of the

---

* If electric power price is set at $4.00/10^6$ Btu, and steam production cost at $1.00/10^6$ Btu for 1,250 psig, the waste heat recovered is worth $.4/10^6$ Btu for 50 psig, and $.1/10^6$ Btu for 0 psig.

** Estimated by T. T. Frankenberg of American Electric Company, Canton, Ohio.[17]

seven industries studied, and the mining and smelting processes of copper production are the most likely candidates for improvement. The laws of thermodynamics provide valid methods of measuring energy efficiency, but do not consider economic, environmental, and other nonenergy factors. An overall consideration is needed when one uses these efficiency indices.

REFERENCES

1. Rosendranz, R. "Energy Consumption in Domestic Primary Copper Production," U.S. Bureau of Mines Information Circular 8698. Washington, D.C.: U.S. Government Printing Office, 1976.
2. Battelle Columbus Laboratories. "Evaluation of the Theoretical Potential for Energy Conservation in Seven Basic Industries," final report to Federal Energy Administration. Springfield, Va.: National Technical Information Service, July 1975.
3. Berg, C. A. "Energy Conservation through Effective Utilization," *Science,* Vol. 181, 128, July 13, 1973.
4. Kauzman, W. *Thermodynamics and Statistics.* New York: W. A. Benjamin, 1967.
5. Keenan, J. H. *Thermodynamics.* New York: Wiley, 1948.
6. Van Wylen, G. J. *Thermodynamics.* New York: Wiley, 1948.
7. Vielstich, W. *Fuel Cells.* New York: Wiley-Interscience, 1970.
8. NATO. *Technology of Efficient Energy Utilization,* E. G. Kovach (ed.), report of a NATO Science Committee Conference held at Les Arcs, France, 8–12 October 1973 (Scientific Affairs Division, NATO, 1110 Brussels, Belgium, 1973). [Also New York: Pergamon Press, 1975.]
9. ASHRAE. *Systems.* New York: American Society of Heating, Refrigerating and Air-Conditioning Engineers, 1973.
10. Akerman, J. R. "Automotive Air Conditioning Systems with Absorption Refrigeration" (SAE no. 710037), *SAE Transactions,* Vol. 80, 132, 1971.
11. Stanford Research Institute. "Patterns of Energy Consumption in the United States," prepared by SRI in 1970 using 1960–1968 data. Washington, D.C.: U.S. Office of Science and Technology, 1972.
12. American Institute of Physics. *Efficient Use of Energy* (the APS Studies on the Technical Aspects of the More Efficient Use of Energy), AIP Conference Proceedings No. 25, New York, 1975.
13. Mitre Corporation. "An Agenda for Research and Development on End Use Energy Conservation," Report 6577, Bedford, Mass., 1973.
14. Morning, J. and Greenspoon, G. "Flotation Trends," *Minerals Yearbook 1970,* Vol. 1, Bureau of Mines, U.S. Department of the Interior, 1972, page 80.
15. McMahon, A. "Copper, A Materials Survey," Bureau of Mines Information Circular 8225, U.S. Department of the Interior, 1965.
16. Balzhiser, R. and Samuels, M. *Engineering Thermodynamics.* Englewood Cliffs, N.J.: Prentice-Hall, 1977.
17. Rotty, R. and Vanartsd, E. "Thermodynamics and Energy Policy," ORAU/IEA 77–18, Oak Ridge, Tenn.: Institute of Energy Analysis, Oak Ridge Associated Universities, July 1978.

7
Economics of Alternative Energy Conservation Measures

INTRODUCTION

This chapter describes a typical firm's motivation for considering conservation measures and its methods of evaluating the costs, benefits, and gives an example of evaluating the three alternative conservation measures: recycling, retrofitting, and changing processes.

The investment evaluation methods discussed in this chapter assume that one prime objective of an industrial firm is either cost minimization or profit maximization. It is therefore implied that the expected cost reduction or profit increase resulting from an investment in conservation is a critical factor in determining if a firm will implement the conservation measures. Depending on the size and importance of the investment, the firm may use either comprehensive or quick evaluation methods to make investment decisions. The comprehensive method (annual conservation cost/saving method) takes into account depreciation and tax credits, but the quick methods (e.g., payback period) do not consider these factors.

For evaluation purposes, the costs and benefits of conservation measures are quantified in terms of monetary value. Unquantifiable costs and benefits are also considered with care when applying the evaluation methods. If the objective of a firm is something other than cost minimization or revenue maximization (e.g., minimizing risk, creating a desirable public image, or maximizing employment), then these evaluation methods may need modification before they are applied.

This chapter is divided into six sections: (1) motivation of implementing conservation measures, (2) annual conservation cost/saving method, (3) evaluating costs and benefits of conservation measures, (4) some quick evaluation methods, (5) special factors to be considered in evaluating conservation investments, and (6) summary.

MOTIVATION OF IMPLEMENTING CONSERVATION MEASURES

A profit-seeking firm may have any of three likely motivations to invest in conservation measures. First, it may expect the resulting benefits to exceed investment cost with regard to future rising fuel prices and uncertain fuel supplies. Second, it may be required to implement the measures because of either mandatory or voluntary conservation legislation. Third, it may implement conservation measures such as changing processes, which often produce fewer emissions, because they will enable the firm to meet new mandatory stringent pollution standards.

The costs and benefits of implementing conservation measures vary from technology to technology and from firm to firm. (See Tables 7–1 and 7–2.)

It is important that only the costs and benefits that are attributable to an investment be included in the analysis of that investment. For example, if a plant is required by mandate to add a pollution control apparatus, the decision to add a waste heat recovery system to the pollution control system should not be influenced by the costs of the pollution control system. As a further example, costs of equipment replacement or repair not required by the addition of the waste heat recovery system should not be incorporated

Table 7–1. Possible Benefits from Implementing Conservation Measures.

Fuel savings
Reduced size, hence lower capital cost, of heating or cooling equipment
Reduced maintenance costs for existing equipment
Reduced costs of production labor
Pollution abatement
Improved product
Revenue from sales of recovered heat or energy

Note: Not all these benefits will necessarily result from investment in conservation technologies; in fact, fuel savings may be the only benefit in many applications. However, examples of the other kinds of benefits shown were found in existing applications.

Source: NBS.[17]

Table 7-2. Potential Costs in Investing in Conservation Measures.*

Type of Costs	Examples of Costs
1. Pre-engineering and planning costs	Engineering consultant's fee; in-house manpower and materials to determine type, size, and location of the heat exchanger.
2. Acquisition costs of heat recovery equipment	Purchase and installation costs of a recuperator.
3. Acquisition costs of necessary additions to existing equipment	Purchase and installation costs of new controls, burners, stack dampers, and fans to protect the furnace and recuperator from higher temperatures entering the furnace due to preheating of combustion air.
4. Replacement costs	Cost of replacing the inner shell of the recuperator in n years, less net of the salvage value of the existing shell.
5. Costs of modification and repair of existing equipment	Cost of repairing furnace doors to overcome greater heat loss resulting from increased pressure due to preheating of combustion air.
6. Space costs	Cost of useful floor space occupied by waste heat steam generator; cost of useful overhead space occupied by evaporator.
7. Costs of production downtime during installation	Loss of output for a week, less net of the associated savings in operating costs.
8. Costs of adjustments (debugging)	Lower production; labor costs of debugging.
9. Maintenance costs of new equipment	Costs of servicing the heat exchanger.
10. Property and/or equipment taxes of heat recovery equipment	Additional property tax incurred on capitalized value of recuperator.
11. Change in insurance or hazards costs	Higher insurance rates due to greater fire risks; increased cost of accidents due to more hot spots within a tighter space.
12. Costs of demolishing existing facilities	Loss of salvage value; labor and materials for demolishing.

* In addition, attention should be given to the length of intended use, expected lives of related equipment, and the flexibility of alternative equipment in future modification and expansion.

Source: Partial information obtained from NBS.[17]

116

into the waste heat evaluation, even though such replacement or repair may be undertaken simultaneously for convenience.

ANNUAL CONSERVATION COST/SAVING METHOD

The annual conservation cost/saving method may be used to evaluate a conservation measure. This method takes into account depreciation and tax credits, and computes the annual incremental costs (or benefits) and annual energy savings of incorporating a conservation measure into the original production process. It computes the initial capital requirement of a conservation measure, deducts tax credits, and annualizes the capital cost over the useful lifetime of the conservation measure. Then the annual operation and maintenance (O&M) costs of the measure are computed. Comparing the sum of annualized capital and O&M costs with the cost of the original production process yields the annual cost difference of the conservation measure. The energy savings of the measure are computed by comparing the energy use of the measure with that of the original process. The benefits of reduced pollution resulting from implementation of the conservation measure are also considered an element of savings.

Computing O&M costs and energy savings is relatively straightforward. However, the annualized capital cost is significantly affected by depreciation treatment, taxes, and tax credits. These elements change cash flows and the basis for capital computation, and the significance of their impacts on annual capital cost varies.

Depreciation allowances are not cash flows per se, but they affect after-tax income. The method of depreciation selected has significant impacts on annualized capital costs. The U.S. Supreme Court has ruled that the depreciation allowance represents reduction in value through wear and tear of depreciable assets. The allowance is the sum that should be set aside for the taxable years in order that the total allowances set aside throughout the assets' useful life will, with salvage value, equal the original cost. The depreciation allowance is purely for income tax purposes, because the federal income tax is based on net income (which is equal to gross income minus expenses and depreciation). A large depreciation allowance in a profitable year decreases net taxable income and, hence, income tax, making more funds available for new investment. Because of the time value of money, it is generally desirable to take larger depreciation allowances in the early years of an asset's life as these allowances can be reinvested to produce income.

In addition, property taxes, tax credits for investments, and income taxes all affect a firm's net profit. Tax dollars are cash flows and therefore require the same attention as expenses for wages, energy, materials, capital, and land. Using the most favorable depreciation and tax methods allowed by law can result in a significant gain in after-tax income.

Methods of Depreciation

The components used to determine the basis for the depreciation allowance are the investment cost, salvage value, and useful life of the asset. Therefore, depreciation can only be taken on properties that are used in the taxpayer's trade or business or held for the production of income. It cannot be applied to inventories or land. The investment cost includes the cost of the property, capital addition to the property, and installation costs. The salvage value is an estimate of the market value at the end of the property's useful life. The useful life is the period in which the taxpayer intends to use the property. This life may have little relationship to the inherent physical life of the asset.

Straight-line, declining balance, and sum of the years-digits are depreciation methods commonly accepted by the Treasury Department. We discuss them below.

Straight-Line Depreciation. The straight-line depreciation is a uniform write-off of an asset over its useful life. The formula is as follows:

$$D_t = \frac{P - S}{n} \qquad (7\text{-}1)$$

where D_t = depreciation allowance at the end of taxable year t, P = acquired cost, S = salvage value, and n = useful life.

For example, if a waste heat economizer is purchased at a price of $10,000, is installed at a cost of $1,000, and has 10 years of useful life and $1,000 of salvage value, the yearly straight-line depreciation allowance is calculated as follows:

$$P = \$10,000 + \$1,000 = \$11,000$$
$$S = \$1,000$$
$$n = 10$$

$$D_t = \frac{P - S}{n} = \frac{\$11,000 - \$1,000}{10} = \$1,000 \qquad (7\text{-}1')$$

The straight-line method is the most commonly used method of calculating depreciation, and must be used in all cases where the taxpayer has not adopted a different method accepted or approved by the Internal Revenue Service. One may switch to this method in the latter years of the life of an asset initially depreciated under an accelerated method. We will therefore use this method to evaluate the costs and benefits of new process, retrofit, and recycling conservation technologies in the following sections.

Declining Balance Depreciation. The declining balance method is known for its accelerated write-off of assets. It provides high depreciation allowances in the early years and low allowances throughout the rest of an asset's useful life. The depreciation rate permissible under law can be as high as twice the straight-line rate. When it is twice the straight-line rate, as it most frequently is, this method of depreciation is known as the double decline method.

Under this method, the depreciation allowance at the end of the tth year is a constant fraction (d) of the book value at the end of the previous year (B_{t-1}). That is:

$$D_t = d \cdot B_{t-1} \qquad (7\text{--}2)$$

The book value at the end of the tth year is:

$$B_{t-1} = P(1 - d)^{t-1} \qquad (7\text{--}3)$$

where P is the original book value as defined in Equation 7–1. Substituting Equation 7–3 into Equation 7–2, we obtain the following equation for the declining balance depreciation:

$$D_t = dP(1 - d)^{t-1} \qquad (7\text{--}4)$$
$$\text{but} \quad d \leqslant 2/n$$

where $1/n$ is the straight-line depreciation rate. If the double declining balance depreciation method $(d = 2/n)$ is used to compute the allowances for the equipment discussed in Equation 7–1, the first depreciation allowance would be:

$$D_1 = 2/10 \cdot \$11{,}000 = \$2{,}200 \qquad (7\text{--}4')$$

This allowance is higher than the previous straight-line depreciation by \$1,200 (\$2,200 versus \$1,000).

The above calculation of declining balance depreciation does not take into account the estimated salvage value. The book value in the last year of an asset need not be equal to the resale value. If there is a discrepancy, compensating adjustments for capital or income must be made at the time of asset disposal. Also, in any year, the depreciation allowance must not exceed the book value of an asset.

If the salvage value needs to be considered, then the last year's book value must be equal to the salvage value:

$$B_n = S = P(1-d)^n$$

consequently:

$$d = 1 - n\sqrt{\frac{s}{p}} \qquad (7\text{-}5)$$

$$\text{and} \quad d \leqslant 2/n$$

Using the same example, we calculate d according to Equation 7–5:

$$d = 1 - 10\sqrt{\frac{1}{11}} = .2132, \qquad (7\text{-}5')$$

a value that exceeds the maximum allowable rate, $2/n = .2$. Thus, the double declining balance rate (.2) should be used to calculate depreciation rather than the result of Equation 7–5'.

The Internal Revenue Service allows switching from the double declining balance method to straight-line depreciation for the purpose of deferring taxes until later years. Also, declining balance depreciation in excess of 1.5 times the straight-line rate may be used only for tangible property or depreciable realty having a useful life of three years or more where the original use (any use) began with the taxpayer. Used tangible property and realty may be limited to 1.5 or 1.25 times the straight-line rate.

Sum of the Years-Digits Depreciation.   The sum of the years-digits depreciation method, like the declining balance method, is an accelerated write-off method. The name "sum of the years-digits" comes from the fact that the sum of an asset's useful life n

$$1 + 2 + 3 + \ldots + n = \frac{n(n+1)}{2}$$

is used as the denominator of Equation 7–6 for the calculation of depreciation allowances. The tth year's depreciation allowance (D_t) is:

$$D_t = \frac{n - (t-1)}{n(n+1)/2} \cdot (P - S) \qquad (7\text{-}6)$$

where n, P, and S as defined for Equation 7–1. This equation indicates that the depreciation is the highest in the beginning year of the asset's life depreciation ($t = 1$) and tapers off gradually.

Using the same example, the first year's depreciation is calculated as follows:

$$D_1 = \frac{10}{10(10 + 1)/2} (\$11,000 - \$1,000) = \frac{2}{11} \cdot \$10,000$$
$$= \$1818.18 \qquad (7\text{-}6')$$

Comparing results of the first year depreciation allowance of the three methods (Equations 7–1', 7–4', and 7–6'), we find that the double declining balance method gives the highest initial depreciation, then comes the sum of the years'-digits method, and straight-line depreciation gives the least.

Sinking Fund Depreciation and Other Methods. Sinking fund depreciation is rarely used even though it is still referred to on the Professional Engineering registration and certification examinations. The method is a decelerated depreciation method because it assumes an asset depreciates at an increasing rate. The tth year's depreciation allowance is considered to be the sum of a sinking fund plus the interest earned on the account. The sinking fund pays interest at a rate of $i\%$, and after n years, it will have a balance equal to the total amount to be depreciated, (original book value minus salvage value).

Some other recognized depreciation methods are the units of production, operating day, and income forecasting methods. The major limitation for these methods is that total allowances at the end of each year do not exceed, during the first two-thirds of an asset's useful life, the total allowance that would result if the declining balance method were used.

The units of production method allows equal depreciation per unit of output. The tth year's allowance is based on the ratio of the tth year production (U_t) to total production (U) over an asset's useful life:

$$D_t = (P - S) \frac{U_t}{U}$$

The operating day method, similar to the previous method, is based on the ratio of days used during the years (Q_t) to total days expected in a useful life (Q):

$$D_t = (P - S) \frac{Q_t}{Q}$$

The income forecasting method is based on the ratio of year t rental income (R_t) to the forecast total useful life income (R):

$$D_t = (P - S)\frac{R_t}{R}$$

The above depreciation methods must be used consistently over an asset's useful life and are subject to IRS approval. They may be used to determine the depreciation allowance whenever the declining balance method is applicable.

Taxes and Tax Credits

Taxes and tax credits affect annualized capital and O&M costs. Taxes include federal income tax, state income tax, excise tax, and property tax. Tax credits consist of general investment credits and energy tax credits.

The federal corporate income tax rate, although not constant throughout its history, has hovered around 50% in recent decades. Effective in 1975, the corporate tax rate included a normal tax of 20% on the first $25,000 of taxable income plus 22% on taxable income over 25% and a surtax of 26% on taxable income over $50,000. State income tax, local property tax, and excise tax rates vary from location to location.

The general investment tax credit was established by the 1962 Revenue Act at 7%, suspended briefly between 1966 and 1967, and terminated in 1969. The credit was restored to 7% by the 1971 Revenue Act, only to be temporarily increased by the 1975 Tax Reduction Act to 10% for qualified property acquired and placed in service after January 21, 1975 and before January 1, 1977. It was recently raised to 10% again by windfall profits tax legislation. The exact deduction for tax credits also depends on the taxable income available to cover the credits and the useful life of assets. Qualified property must be intended for use over at least seven years to qualify for the full percentage credit.* Table 7–3 shows the effect of an asset's useful life.

The energy tax credits have been revised by the windfall profits tax legislation. These credits are provided for investments in various renewable energy sources (e.g., biomass, solar, and wind). Table 7–4 shows the rates and cutoff date of these credits. Other energy taxes provided for by the National Energy Act are discussed in Chapter 9.

Energy tax credits are currently computed on the full percentage basis; however, a company cannot take the full credit, only $25,000 plus 70% of

---

* Subject to changes by new income tax laws.

Table 7-3. Effect of an Asset's Useful Life on General Investment Tax Credit.

Declared Useful Life*	Investment Tax Credit Rate
Less than three years	None
Three to four years	1/3 of 10%*
Five to six years	2/3 of 10%
Seven or more years	10%

* The useful life and the percentage may be changed over time by the Congress.

the remaining tax liability for tax year. The unused energy credit, like the investment credit, can be carried back three years and forward seven years. There is no 70% limit for the carrying, as there is in computing the investment credit. Although the government pays for part of an investment in energy equipment through credits, the usual depreciation write-off is based on 100% of the investment.

EVALUATING COSTS AND BENEFITS OF CONSERVATION MEASURES

One way to conserve energy is to import materials from abroad instead of producing them domestically. However, the analysis in this section is based on the expectation that industries will save energy without reducing their production, owing to the high deficit in balance of trade, and other employment considerations.

Three effective alternative options for most industries, which are expected to conserve energy without reducing production, are (in order of conservation potential):

- Recycling (using scrap)
- Changing processes (from existing to more energy-efficient ones)
- Retrofitting

For the purpose of demonstrating how to evaluate the energy savings and costs of the above three measures, U.S. copper production is analyzed. The copper industry is chosen for analysis because of its great potential for energy conservation (see Chapter 6).

Recycling (Using Copper Scrap Instead of Ores)

Recycling is the most efficient way to conserve energy. The U.S. Bureau of Mines estimated that there were more than 11 million short tons of scrap

Table 7-4. Energy Tax Credits Provided by the Windfall Profits Tax Legislation.

Investment	Tax Credit Rate	Cutoff Date
Biomass equipment to convert waste into fuel	More liberal provisions and the cutoff date extended three years. Credit remains 10%	1985
Cogeneration equipment	A new 10% credit	1982
Coal gasification equipment	A new 10% credit	1982
Geothermal and ocean thermal equipment	Credit raised to 15% from 10% and extended to ocean thermal	1985
Intercity buses	A new 10% credit	1985
Small-scale hydroelectric facilities	A new 11% credit for power plants of less than 125 megawatts of capacity	1985
Solar or wind equipment	Credit raised to 15% from 10%	1985
Production of shale oil and similar products	A new $3 per barrel credit to apply when domestic crude oil is priced below $23.50	1989
Production of engine-fuel alcohol	A new 30¢ to 40¢ per gallon credit to apply in some cases in lieu of a tax exemption	1992

Source: IRS[18] and *Business Week*.[19]

copper unrecovered in 1970.[4] In 1972 the primary producers consumed 30% of total unalloyed copper and copper base scrap in the United States, of which 45% was low-grade scrap and 55% was high-grade (No. 1 and 2) copper scrap. As shown in Fig. 7-1, primary producers feed low-grade scrap in smelting processes to replace part of the ore concentrate, and use high-grade scrap in refining processes as a substitute for blister copper.* In general the low-grade scrap varies in copper content, containing 25 to 40% or even less,** but the high-grade scrap may contain 94 to 99% copper. There is no difference between the products from scrap and those from original ores. The operations of smelting and refining are generally identical for scrap and ores.[1,2]

Since the scrap is introduced at different points in the production processes depending on its grade (see Fig. 7-1), calculating the energy savings associated with the various substitutions requires different methods. The high-grade scrap that substitutes for blister copper conserves almost as much energy as would have been required to produce the blister. If a company has m mines and mills, n smelters, and p refineries, then the energy saving per ton of high-grade scrap used at the kth refinery (∇e_{ijk}) is equal to the sum of the energy requirement per ton of refined copper produced at the ith mine (e_{iM}), the ith mill (e_{iF}), and the jth smelter (e_{jS}), expressed in the following equation:

$$\nabla e_{ijk} = e_{iM} + e_{iF} + e_{jS} \qquad (7\text{-}7)$$

The energy savings from feeding low-grade scrap (∇e_{ij}) to the jth smelter can be expressed as:

$$\nabla e_{ij} = e_{iM} + e_{iF} \qquad (7\text{-}8)$$

Additionally, recycling reduces the mining of ores and smelting of blister copper, and therefore reduces air pollution, the amount of land that is disturbed, and the amount of waste water.

The unit cost (Δc_{ijk}) increase from the use of high-grade scrap at the kth refinery is equal to the difference between the weighted average price in dollars per ton of copper content of the scrap purchased (Y_2) and the sum of unit cost of production at the ith mine (c_{iM}), i^{th} mill (c_{iF}), and j^{th} smelter (c_{jS}), all expressed in dollars per ton of refined copper:

$$\Delta c_{ijk} = Y_2 - (c_{iM} + c_{iF} + c_{jS}) \qquad (7\text{-}9)$$

---

* Some small amount of intermediate-grade copper scrap is used as input of the converter; however, the quantity is not significant.

** Private contact with Prince Charles Scrap Metal, Philadelphia, September, 1977.

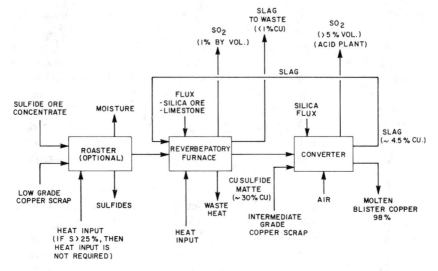

CHART A: SMELTING PROCESS

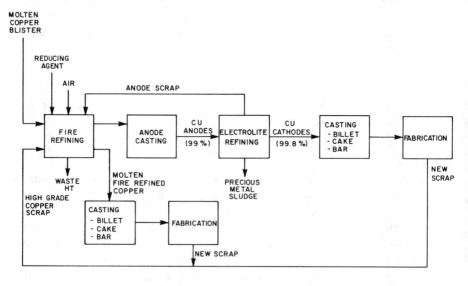

CHART B: REFINING PROCESS

Fig. 7–1.   Scrap use in primary producers. Source: Hu.[20]

Prices for No. 2 heavy (high grade) and No. 1 composition (red brass, low grade as shown in Equation 7–10) copper scrap are published daily in the *American Metals Market.* They can be used for reference.

The unit cost (Δc_{ij}) increase from feeding low-grade scrap to substitute ore concentrates at the jth smelter may be expressed in the following manner:

$$\Delta c_{ij} = Y_{2s} - (c_{iM} + c_{iF}) \qquad (7\text{--}10)$$

where Y_{2s} is the weighted average price of the low-grade scrap purchased (No. 1 composition). Observation of Table 7–5 and similar data[4] show that Y_{2s} is related to Y_{29} (producer price for refined copper which can be estimated from copper econometric models) and Y_2 (price of No. 2 heavy scrap). This relationship can be expressed as:

$$Y_{2s} = .45 Y_2 + .20 Y_{29}$$
$$(7.06) \quad (3.55)$$

standard error of estimation (s) = 7% of mean value

Statistically, the high magnitude of values of Student tests (shown in parentheses under each coefficient of the explanatory variables) strongly suggests

Table 7–5.   Trend of Annual Average Copper Prices.
(1958–1974)
($¢$/lb of 1974 dollars)

Year Y_R	Price for No. 1 Composition Y_{2s}	Price for No. 2 Heavy Copper Scrap Y_2	Producer Price of Refined Copper Y_{29}
1958	25.47	28.11	41.62
1959	29.89	35.64	49.78
1960	28.20	33.41	51.12
1961	31.60	34.57	47.98
1962	32.20	34.29	49.12
1963	33.32	35.26	49.12
1964	39.49	40.80	50.71
1965	43.66	53.49	54.86
1966	50.68	67.78	55.45
1967	44.95	49.26	57.38
1968	39.51	47.03	60.69
1969	51.62	59.16	66.22
1970	48.60	52.34	77.34
1971	38.38	35.01	65.97
1972	36.61	47.88	62.74
1973	47.96	58.57	69.07
1974	43.75	54.87	76.64

that prices for low-grade scrap are significantly affected by the price of high-grade scrap and producer copper prices. The low value of standard error indicates a good correlation.

Historically, the scrap market has been competitive, and we may expect that any increases in scrap demand will result in increases in scrap prices.

Changing Processes (Replacing Existing Furnaces with Flash Smelters)

Changing processes (replacing existing processes with new more energy-efficient processes) requires substantial capital investment. It deserves careful analysis. This subsection gives an example of how to evaluate the costs and benefits of replacing existing reverberatory smelting furnaces with more energy-efficient flash smelting furnaces.

Description of the Two Alternative Processes. Copper may be smelted in three ways: (1) by using fossil fuel combustion in a reverberatory furnace, (2) by using electrical power in an electric furnace, and (3) by using the heat of reaction from the oxidation of iron sulfide to iron oxide in a flash furnace. Currently copper smelting employs mostly reverberatory furnaces. Table 7–6 shows that 14 out of the 15 existing U.S. copper smelters are reverberatory furnaces, and the remaining one is an electric furnace.

These facilities are more than 20 years old and not energy-efficient because they were designed in the era of cheap energy. Kellog and Henderson examined energy requirements for various sulfide smelting processes and concluded that large energy savings can be achieved by using flash smelters instead of reverberatory furnaces, which do not use oxygen-enriched air.[16] We shall discuss the details of the two smelting processes below.

Figure 7–2 depicts schematically a typical reverberatory smelting furnace. In a reverberatory furnace, a fossil fuel such as oil or natural gas is burned above the copper concentrates (products-in-process from flotation). The furnace is a long rectangular structure with an arched roof and burners at one end. Flames from the burners may extend half the length of the furnace. Part of the heat in the combustion gas radiates directly to the charge lying on the hearth below, while a substantial part radiates to the furnace roofs and walls and is reflected down to the charge.

In addition to smelting the copper concentrates, a major function of a conventional reverberatory furnace is to recover copper—both chemically and mechanically—from slag produced in the copper converters. Molten converter slag is returned to the furnace, and copper sulfide matter and copper that is mechanically entrained in this slag settle out by gravity.

Flash smelting is a process in which copper sulfide ore concentrates are

Table 7-6. Existing U.S. Copper Smelters.

| Company Name/Location | Age of Plant | | Materials | | Smelting Furnaces | |
	First Year	Last Modification	Conc. (T/D)	Blister CU (T/D)	No. of Oper/ Stby.	Gas Stream Control Equipment
1. ASARCO/Tacoma, Wash.	1890	1973 (Liquid SO$_2$ plant for converter under const.)	1200	300 Electrolytically refined Cu)	1/1	Reverb gas combined with roaster gases, then to 2 ESP's in series (98.4%).
2. ASARCO/ Hayden, Arizona	1912	1971 (New converter acid plant)	2000	366	2/0	Reverb gas through W.H.B.'s and water spray chamber, then joined with roaster gases. Combined gases to ESP (98.3%) and 300-ft stack.
3. ASARCO/ El Paso, Texas	1905	1973 (New converter acid plant, under const.)	700	260 (Anode Cu)	1/0	Reverb gas pass through W.H.B. and settling flue, then join roaster gases, then to spray chamber, ESP (98.6%), and combined with treated converter gases.
4. Phelps Dodge/ Douglas, Ariz.	1910	1971 (New ESP on converters)	2260	365 (Anode Cu)	3/0	All reverb gases treated in W.H.B., then join roaster gases, and pass out 544-ft stack.
5. Phelps Dodge/ Morenci, Ariz.	1942	1964 (Roaster acid plant)	2113	470 (Anode Cu)	4/0	Reverb gases treated in two ESP's in parallel (78.5%).
6. Phelps Dodge/ Ajo, Ariz.	1950	1972 (DMA & acid plants)	680	197 (Anode cu)	1/0	Reverb gases join converter gases, then to ESP and 360-ft stack.
7. Magra/ San Manuel, Ariz.	1956	—	1700	310 (Anode Cu)	2/0	Reverb gases treated in W.H.B. and ESP (89%), then out 515-ft stack.
8. Kennecott/ Hurley, N.M.	1939	1971 (New fourth converter added)	767	234	1/1	Reverb gases treated by W.H.B. balloon flue, ESP (95%), and 500-ft stack.
9. Kennecott/ McGill, Nev.	1907	—	750	185	2/0	Reverb gases treated by W.H.B., settling flue, ESP (70–85%), then out 300-ft stack.

Table 7–6. (Cont.)

Company Name/Location	Age of Plant First Year	Age of Plant Last Modification	Materials Conc. (T/D)	Materials Blister CU (T/D)	Smelting Furnaces No. of Oper/ Stby.	Smelting Furnaces Gas Stream Control Equipment
10. Kennecott/ Hayden, Ariz.	1958	1968 (Fluid bed roaster and acid plant)	1050	220	1/0	Reverb gases treated by W.H.B., mixed with 50% of converter gases, passed through ESP (95%) and 600-ft stack.
11. Kennecott/ Garfield, Utah	1907	1968 (Removed roasters and converted to green-feed reverbs)	2200	750	3/0	Reverb gases treated by W.H.B. and ESP (50%), then out two 410 ft. stacks.
12. Anaconda/ Anaconda, Montana	1906	1973 (New acid plant)	1710	500 (Anode Cu)	3/1	Reverb gases pass thru water spray chamber and flue, then join converter gases.
13. White Pine/ White Pine, Michigan	1955	—	700	220	1/1	Reverb gases treated by W.H.B. and two ESP's in series, join converter gases, and pass out 500 ft. stack.
14. Cities Service/ Copperhill, Tennessee	1845 (Inter- mit- tent	1972 (SO₂ treat- ment of elec- tric furnace reverb gases)	300	50	1/0	Electric furnace gases mix with other plant gases, then to acid plants.
15. Inspiration/ Miami, Ariz.	1915	1972 (Elect. fur- nace, syph. conv., acid plant under const.)	840	300	1/0	Reverb gases pass thru W.H.B.'s, flue, and 275 ft. stack.

Source: EPA [5]

smelted by burning a portion of the iron and sulfur contained in the concentrates while they are suspended in an oxidizing environment. As such, the process is quite similar to the combustion of pulverized coal. The concentrates and fluxes are injected with preheated air, oxygen-enriched air, or even pure oxygen, into a furnace of special design. Smelting temperatures are attained

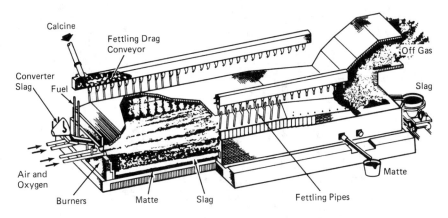

Fig. 7–2. Reverberatory smelting furnace. Source: EPA.[5]

as a result of the heat released by the rapid, flash combustion of iron and sulfur.

Two companies have developed flash smelting technology: Outokumpu Oy in Finland and International Nickel Co. (INCO) in Canada. Both offer their technology for license through either their own offices or various contractors. The major differences between these technologies are in the design of the flash smelting furnace and the oxidizing environment in the furnace. Outokumpu uses preheated air, or oxygen-enriched air, as the oxidizing medium, whereas INCO uses pure oxygen. Most of the flash smelting furnaces presently operating in the world are of the Outokumpu design.[5]

By mid-1973, 13 copper smelting installations in the world operated flash smelting furnaces ranging in capacity from 300 tons/day to 1500 tons/day of copper concentrates. One installation is of INCO design; the remaining 12 are of Outokumpu design. Schematic representations of the Outokumpu and INCO flash smelting furnaces are shown in Fig. 7–3 and 7–4, respectively.

Flash smelting furnaces can be designed to operate autogenously. Under these conditions no external source of fuel or energy is required. The heat released by the flash smelting reaction is sufficient to smelt the furnace charge. Thus, flash smelting requires a lower energy input per pound of copper produced than either reverberatory or electric smelting.

Flash furnaces typically produce high-grade mattes containing 45 to 65% copper. This is higher than the grade of matte produced at most domestic smelters, which typically contains 30 to 40% copper. High-grade mattes result in reduced secondary copper scrap processing capacity. Consequently, the adoption of flash smelting by the domestic industry would reduce, to

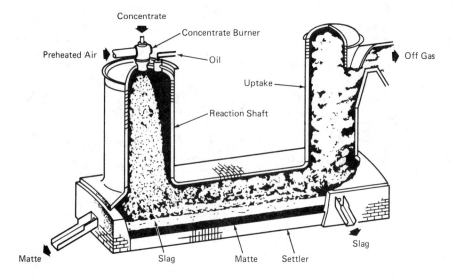

Fig. 7–3. Outokumpu flash smelting furnace. Source: EPA.[5]

some extent, the ability of the industry to reprocess copper scrap. It is likely, however, that this limitation associated with flash smelting would encourage realignment of the domestic smelting industry, rather than limit the growth of the industry or prohibit the application of flash smelting.[5] A reduction in the capacity of the domestic primary smelting industry to recover copper from copper scrap would encourage the expansion of the domestic secondary smelting industry in most cases, rather than limit the recovery of copper from scrap. Both these eventualities—adoption of flash smelting and expansion of secondary smelting—would help energy conservation in the industry.

Additionally, flash furnaces have an advantage over reverberatory or electric furnace smelting in terms of the ease with which emissions of sulfur oxides can be controlled. The concentration of sulfur dioxide in the furnace off-gases is high, normally in the range of 10 to 14%, and the process is steady-state with respect to the flash combustion reactions. The control of emissions is straightforward because the major portion of the sulfur oxides resulting from the smelting of copper concentrates is steadily discharged in the off-gases from the flash furnace, rather than in the off-gases of low concentrations or with large fluctuations.

Benefits and Costs. Energy savings of the flash smelter are shown in Table 7–7. The annual savings are about 1.4 trillion Btu in the form of gas

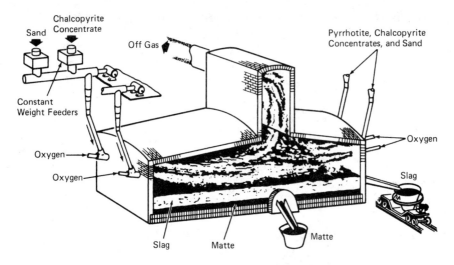

Fig. 7–4. INCO flash smelting furnace. Source: EPA.[5]

and electricity. This is approximately 60% of the reverberatory smelter's energy consumption (2.36 trillion Btu/year). Since it produces high-grade mattes, the flash smelter reduces air pollution. Ease of emission control is also one of the benefits (as noted).

Engineering data concerning energy, labor, material, and capital requirements for a standard flash smelter and a standard reverberatory smelter are shown in Tables 7–8 and 7–9, respectively. Comparing the costs in the tables, we find that the flash smelter is 62% cheaper in direct operation costs, and 9.8% more costly in capital investments than the reverberatory smelter.

Table 7–7. Energy Requirement for Smelter Technologies of 86,000 Tons/Year.

Item	Flash Smelter Quantity	Billion Btu	Green Charge Reverb. Smelter Quantity	Billion Btu
Fuel	587.5 MMcf	605.7125	2287.5 MMcf	2358.4125
Power	36.6 MMkWh	374.7474	—	—
Steam credit	(3.5 MMkWh)*	(35.8365)	—	—
Total		944.6234		2358.4125
Percentage		40		100

* This assumes the efficiency of heat transformation to be only one-third in producing electricity.
Source: Computed from EPA.[5]

Table 7–8. Direct Costs for Smelter Technologies of 86,000 Tons/Year (1973 Dollars).

Item	Flash Smelter ($1,000's)	Green Charge Reverb. Smelter ($1,000's)	Unit Cost
Supervision	132	168	$12,000/yr
Operating labor	1001	889	$3.75/hr
Fuel	235	915	40¢/Mcf
Power	366	—	1¢/kWh
Flux	456	1630	$15/ton
Maintenance labor	443	375	$3.75/hr
Supplies	455	465	—
Copper losses (slag)	688	1380	50¢/lb
Steam credit	(105)	—	1¢/kWh
Total (actual)	3671	5822	
(Unit base)	2.1¢/lb	3.4¢/lb	

Source: EPA.[5]

For the purpose of calculating the cost increase of replacing existing reverberatory smelters with flash smelters, we assume the following data:

- The useful life of smelters is 20 years.
- The straight-line method is used to compute depreciation expense.
- The real interest rate is 5% (over inflation).
- The tax credit is 10%, the company has sufficient profit to enjoy credits, and the credit is distributed evenly over the 20-year life cycle.
- The salvage value of the existing plant is 30% of the replacement cost (expected remaining life 6 years, out of 20-year life cycle*), and 20% of the replacement cost is needed for demolition. These two items are amortized over 20-year life of the new flash smelter.

The breakdown of annual costs for the two smelters is therefore computed and shown in Table 7–10. As may be noted, the new flash smelter is 18% cheaper than the new reverberatory smelter with regard to annual costs. However, if an existing reverberatory furnace is replaced with a flash smelter, the annualized costs of demolishing the existing facility and compensating for its remaining value make flash smelting 7% more costly than the reverbera-

---

* Although most smelters are about 30 years old, they have been continuously remodeled and updated. Therefore, we use a more conservative estimation.

Table 7-9. Elements of Capital Costs for Smelter Technologies (1000's in 1973 Dollars).

Total Costs	Flash Smelter	Green Charge Reverb. Smelter	Comments
A. Concentrate and flux handling	1,600	1,600	—
B. Dryer and air preheater	1,280	—	—
C. Furnace and waste heat boiler	5,900	4,600	—
D. Flue gas and dust handling	1,900–3,400	3,000–5,300	—
E. Converter aisle	5,000	5,000	—
F. Anode casting facility	1,200	1,200	—
G. Flue reactor	—	—	—
H. Power plant	2,700	4,000	—
I. Slag plant	2,700	—	—
J. Miscellaneous facilities	1,300	1,300	—
K. Site clearance	1,100	1,100	—
L. Electric wiring	2,100–3,400	2,100–3,400	—
M. Buildings and plant facilities	1,100–2,100	1,100–2,100	—
Total direct costs	30,000	27,300	—
Construction, supervision, and equipment	4,500	4,100	15% of direct costs
Engineering and home office	4,200	3,800	14% of direct costs
Indirect costs	10,500	9,500	Start-up contingency, administrative, expenses, fees (35% of directs)
Total capital	49,100	44,700	

Source: EPA.[5]

tory smelter. This fact helps explain the General Accounting Office's finding that most energy conservation activities involved only operational changes.[12]

However, if energy prices increase more rapidly than the prices of material, labor, and capital, then the flash smelter may become attractive. It may also become economical if pollution controls on copper production get tighter, since the flash smelter produces high-grade mattes, allows easier installation of emission control equipment, and therefore reduces air pollution.

Retrofitting (Installing Tonnage Oxygen Plants in Reverberatory Smelters)

Retrofitting generally cannot save substantial energy use but requires less substantial investments than changing processes. When evaluating retrofits, we must include only those direct costs and benefits that result from the retrofitting. As an example, the installation of tonnage oxygen plants in reverberating smelters is discussed below.

Table 7–10.   Annual Costs of Replacing Existing Reverberatory Smelters with Flash Smelters (Thousand 1973 Dollars).[a]

Item	Flash Smelter	Reverb. Smelter
Capital cost allocated[b]	5,892	5,364
Allocated tax credit[c]	491	447
Cost for demolishing existing facilities and compensation for its remaining value[d]	2,682	
Operation cost[e]	3,671	5,822
Total	11,486	10,739
Cost increase (%)	7	0

[a] Annual costs for smelter technologies of 86,000 tons/year.

[b] 12% of the capital cost of the new facility (7% for interest and 5% for depreciation expense, based on a straight-line depreciation calculation and a 20-year life). Total capital costs for flash and reverberatory smelters are $49.1 million and $44.7 million, respectively. See Table 7–9.

[c] Assumes that the tax credit is 10%, the company has sufficient profit to cover credits, and the tax credit is distributed evenly over the 20-year cycle.

[d] Assumes 20% for demolition and 30% for salvage value (expected remaining life 6 years, out of 20-year life cycle).

[e] 1974 prices for energy and materials were used and then converted to 1973 dollars for consistency with capital cost figures. Data from Table 7–8.

Description of the Retrofit Process.   The use of tonnage oxygen to enrich combustion air is rapidly gaining widespread use, particularly in the secondary smelting of nonferrous metals. Essentially all the oxygen in the combustion air is used up in a reverberatory smelter. Since air is 80% nitrogen, the capacity of the furnace is limited by the nitrogen in the combustion air supply. It has been calculated that 4 tons of nitrogen carry away enough heat to smelt a ton of charge.[6] Therefore, replacing part (or all) of this ambient air with oxygen will lower the total volume of nonoxygen gases introduced into the reverberatory furnace and will permit combustion of a larger quantity of fuel, producing higher flame temperatures and increasing the smelting throughput rate.[7]

Because the volume of off-gases per unit of charge to the furnace decreases, the concentration of sulfur dioxide can increase approximately to the range of 3½ to 5%,[8-10] and becomes easier to control.

Oxygen enrichment is currently employed by the International Nickel Co. Ltd. (INCO) for smelting in the nickel reverberatory furnaces[7] at its Sudbury, Ontario facilities in Canada.* This operation, however, is on an intermittent basis using surplus dump oxygen.

---

* Until recently, oxygen enrichment of the combustion air in a reverberatory smelting furnace was also practiced at the Onahama Smelting and Refining Co. copper smelter at Onahama, Japan on a small scale.[11]

Benefits and Costs. In an evaluation study, INCO Ltd., in Canada, equipped one of its coal-burner reverberatory furnaces with four water-cooled oxygen "lances," one below each of the coal-burners, directed away from the furnace side-wells. The improved fuel efficiency shows that one ton of oxygen is almost equivalent to 0.95 ton of powdered coal, and that the reverberatory smelter equipped with tonnage oxygen plants conserves about 7% of the energy required in the reverberatory smelter without tonnage oxygen plants (see Table 7–11).

To estimate the economic impacts of this conservation measure, 1974 prices for energy, labor, materials, and capital are used. Installing the reverberatory smelting processes with tonnage oxygen plants under 1974 price conditions will cost $1.42/ton of copper produced more than those without plants (see Table 7–12). This again explains the GAO findings.

SOME QUICK EVALUATION METHODS

In this section, we discuss six quick methods for evaluating small investments: the payback period, return on investment, discounted benefit/cost ratio, net present value, net annual value, and opportunity interest rate methods. Some of these methods do not consider one or several of such factors as time value of cash (interest), depreciation, taxes, and tax credits.

Table 7–11. Energy Savings from Installing Tonnage Oxygen in Existing Reverberatory Smelters (Net Copper Production = 765 Tons/Day). Improved Fuel Efficiency With Oxygen Enrichment.

Inputs	Test Furnace With Oxygen		Standard Furnace	
	Quantity (ton/day)[b]	Energy Equivalent (10^9 Btu)	Quantity (ton/day)[b]	Energy Equivalent (10^9 Btu)
Coal[a]	213	5.1120	279.6	6.71
Oxygen	80	0.257	0	—
Total		5.3577		6.71
Daily energy savings[c]		1.3527		0
Daily energy use in smelting		18.1281		19.4808
Daily savings as a percentage (%)		(7.0)		(0)

[a] Not all smelters use coal, but smelters using other energy forms are analyzed in a way similar to that shown here.

[b] From A. A. Mathews Inc.[17]

[c] Daily savings = $(6.7104 - 5.3500) \, 10^9$ Btu.

Table 7-12.   Costs of Installing Tonnage Oxygen in Existing Reverberatory Smelters with a Net Copper Production of 765 Tons/Day (1974 Dollars).

Additional capital cost required ($10³)[a]		
Oxygen plant	2,178.6	
Ladies (four with 30-ton capacity)	77.0	
Converters (two of dimensions 13 ft x 13 ft. Includes gun and tuyeres, gates, control and instrumentation for tuyeres, and motor)	1,018.6	
Total	3,274.2	
Preparation costs and taxes (20% of $3,274,200.00)	654.84	
Total capital cost increase	3,929.04	
A. Annual capital cost ($10³)[b]		471.485
Annual operation cost increase		
Labor cost for converters	101.86	
Oxygen production cost[c]	185.64	
B. Total operation cost increase ($10³)		287.5
C. Credit from the reduction of powdered coal ($10³)[d]		332.866
D. Tax credit (10%) ($10³)[e]		39.290
Average cost increase per ton of copper produced ($)[f]		1.42

[a] From Bennett et al.[2]

[b] 12% of $3,929,040.00. Assumes 7% for interest and 5% for depreciation expense, based on straight-line depreciation method and 20-year life.

[c] Assumes ratios of costs of each input are same as for 200 tons per day. These ratios are then inflated to 1974 dollars. Data from Carrillo et al.[4] Costs per ton oxygen are: power, $3.50; labor, $1.50; miscellaneous, $1.50. Total cost is therefore $6.50 per ton of oxygen produced. Annual cost is based on 80 tons/day for 357 days.

[d] Credit per ton = $14.00. Annual credit = $14.00 × (279.6 − 213) × 357.

[e] Assumes that the tax credit is 10%, that the company has sufficient profit to cover credits, and that the tax credit is distributed evenly over the 20-year life cycle.

[f] (A + B − C − D)/(765 × 357).

Payback Period Method

A payback period is sometimes called a payout, payoff, recovery, or break-even period. This method determines the number of years required to recover investing capital by resultant benefits. Costs and benefits are often calculated before taxes and not discounted for interest. The payback period is defined as follows:

$$\text{Payback period (PP)} = \frac{\text{First costs}}{\text{Yearly benefits} - \text{Yearly costs}} \qquad (7\text{-}11)$$

Investment costs are defined as first costs, often neglecting salvage value. Benefits are usually defined as the resulting net change in incoming cash flow or the reduction in net outgoing cash flow. Yearly benefits and costs either are assumed constant from year to year or are averaged.

For example, if a furnace recuperator costs $10,000 to purchase and install, $300/year on average to operate and maintain, and by preheating combustion air is expected to save an average of 2000 Mcf of burner gas yearly at $.70/ Mcf, the payback period is:

$$PP = \frac{\$10,000}{\$.7 \times 2000 - \$300} = 9.1 \text{ years}$$

The above calculation indicates that if the recuperator can last longer than 9 years 1½ months, it is economically justified.

The payback method is simple and useful for speculative investors who desire rapid recovery of the initial investment; it also provides good indicators for assets whose expected life is so highly uncertain that the break-even life of the investment is a major criterion for decision making. However, neglecting the cash flows beyond the payback period and failing to discount different years' proceeds may distort the real financial value of the assets under consideration. These shortcomings can be overcome simply by using average present values of yearly costs and benefits in the denominator of Equation 7–11. The method of calculating present values is explained in the discussion of benefit/cost ratio (see below).

Return on Investment Method

The return on investment (ROI) or return on assets method computes average annual net benefits (after depreciation) as a percentage of the original book value of the investment. This method also does not discount benefits for interest (i.e., disregards the timing of cash flows). The formula is shown as follows:

$$\text{Return on investment (ROI)} = \frac{\text{Average annual net benefits}}{\text{Original book value}} \times 100 \quad (7\text{-}12)$$

If a waste heat economizer costs $15,000, is expected to last for 10 years, requires $200 for yearly operation, maintenance, and repair, and will save $5,000 in fuel oil annually, the return on investment based on straight-line depreciation is calculated as follows:

$$\text{Annual depreciation, straight-line method} = \frac{\$15,000}{10} = \$1500$$

$$\text{ROI} = \frac{\$5000 - (\$1500 + \$200)}{\$15,000} \times 100$$

$$= 0.22 \times 100 = 22\%$$

The return on investment method, like the payback method, is simple to use and is commonly accepted. However, the results can be biased owing to the timing of cash flows and treatments of depreciation. Average present values of yearly benefits and costs can be used in the numerator in Equation 7–12 to improve timing accuracy of cash flows. In addition, a consistent depreciation method should be used to evaluate returns on alternative investments.

Discounted Benefit/Cost Ratio Method

The discounted benefit/cost (B/C) ratio method expresses benefits as a proportion of costs, where benefits and costs are discounted to either a present value or an annual value equivalent. The formula is as follows:

$$B/C = \sum_{j=1}^{n} \left[\frac{S_j + R_j}{(1 + i)^j} \right] \bigg/ \sum_{j=1}^{n} \left[\frac{I_j - V_j + M_j}{(1 + i)^j} \right]$$

where B/C = benefit/cost ratio, n = number of time intervals over which the investment is analyzed, S_j = energy cost savings in year j, R_j = revenue received in year j, I_j = investment costs in year j, V_j = salvage value in year j, M_j = maintenance and repair costs in year j, i = interest rate, and $1/(1 + i)^j$ = present value discount formula.

The ratio $(S_j + R_j)/(1 + i)^j$ is the present value of the benefits (energy savings and revenues) received in year j. The benefits are discounted by interest rate i over j years to become those of the current year.

The B/C ratio takes into account timing of cash flows. This ratio must be greater than 1.0 in order for an investment to be worthwhile.

The use of the B/C ratio method to compare alternative investments is illustrated in Table 7–13. We assume the investment choice to be between Plan X, the addition of a large recuperator design to preheat combustion air to a high temperature, and Plan Y, the addition of a smaller recuperator capable of a lesser amount of preheating but sized to retrofit the existing equipment. To simplify the calculation, the expected costs and benefits are assumed to be those shown in the same table. The long-term interest rate

Table 7-13. Comparison of Various Evaluation Methods.

Investment Alternative	First Cost	Expected Economic Life	Annual Fuel Savings	B/C Ratio	Evaluation Methods*		Opportunity Interest Rate
					Net Present Value†	Net Annual Value	
Plan X	$20,000	6 years	$7,000	1.32	$10,819	$1,715	26.5%
Plan Y	$12,000	10 years	$5,000	2.09	$13,094	$2,610	40.0%

* Based on a discount rate of 15% and no salvage value for the equipment.
† Plan X's life is prorated to 10 years.

is 15%, and the equipment has no salvage value. The B/C ratio computed for Plan X is 1.32, and for Plan Y 2.09. Other things being equal, Plan Y is more attractive; the additional fuel savings realized by the larger recuperator do not compensate for its additional costs.

The B/C ratio is simple and easy to understand but is influenced by whether an item appears in the numerator or in the denominator (i.e., as a disbenefit or cost). Therefore, consistent classification of cost and benefit items is needed for all alternative investments. In addition, because the benefit/cost ratio for the overall investment declines as the investment is expanded toward the most efficient level, a smaller, less efficiently sized project may have a higher ratio for the total investment than a larger more efficiently sized project. This problem can be avoided by applying the benefit/cost ratio method to evaluate the efficiency of increments of an investment, rather than the total investment.[14]

Net Present Value Method

The net present value (or net benefits) method is very similar to the B/C ratio method. Instead of computing the ratio, it calculates the difference between the present value of the benefits and the costs resulting from alternative investments. The formula is as follows:

$$NPV = \sum_{j=1}^{n} \frac{(S_j + R_j) - (I_j - V_j + M_j)}{(1 + i)^j} \qquad (7\text{--}13)$$

where NPV = net present value benefits, n = number of time intervals over which the investment is analyzed, S_j = energy cost savings in year j, R_j = revenue from sale of excess energy received in year j, I_j = investment costs in year j, V_j = salvage value in year j, M_j = maintenance and repair costs in year j, and $1/(1 + i)^j$ = present value discount formula.

The acceptance criteria of a project, as evaluated with the net present value method, are that (1) only those investments having positive net benefits will be accepted (unless the project is mandatory), and (2) when selecting among mutually exclusive investments, the one with the highest positive net benefits will be chosen (or the one with the lowest negative net benefits if none of the alternatives has positive net benefits and the project is mandatory).

In using the net present value method to evaluate alternative investments, a common time period must be observed in order to obtain unbiased costs and benefits. Therefore, if two projects have different life cycles, costs and benefits must be prorated over the same period of time.

Using the same example as that shown in Table 7–13, the net present values of Plans X and Y are $10,819 and $13,094, respectively (shown in Table 7–13). Therefore Plan Y is chosen.

The net present value method has the advantage of measuring the net effect of an investment over its life, taking into account the opportunity cost of capital. However, it has two disadvantages. One is that it computes only net benefits and does not distinguish between a project that has large benefits and costs and one that has smaller benefits and costs, as long as the two projects result in equal benefits. This problem can be avoided if benefit/cost ratios (previously discussed) are also computed for evaluation. Another possible disadvantage of this method is that it is very sensitive to the discount rate, which is the firm's opportunity cost of capital. If the appropriate discount rate is uncertain, various discount rates should be used to evaluate investments to test for sensitivity of the outcome to the choice of rates.

Net Annual Value Method

Net annual value (or net annual benefits) is the net present value times the capital recovery factor as shown below:

$$A = (\text{Net present value of an investment*}) \cdot (\text{Capital recovery factor**})$$

$$= \left(\sum_{j=1}^{n} \frac{(S_j + R_j) - (I_j - V_j + M_j)}{(1 + i)^j} \right) \cdot \left(\frac{i(1 + i)^n}{(1 + i)^n - 1} \right)$$

The net annual value method is used to evaluate alternative investments when they have different life expectations and when it is difficult to prorate a common life cycle in applying the net present value method.

If we use the same example shown in Table 7–13, the net annual values for Plans X and Y are $1,715 and $2,610, respectively (shown in Table 7–13). Plan Y is clearly more attractive economically.

The concept of an equivalent annual net benefit (net annual value) may be easier to understand than the concept of a present equivalent of all cash flow over the period of analysis (net present value). However, this method fails to distinguish between projects of unequal magnitudes yielding equal net benefits. B/C ratios should be used to supplement this method.

Opportunity Interest Rate Method

This method is also called the internal rate of return method. It calculates the opportunity interest rate (discount rate) an investment is expected to yield so that total discounted benefit and costs are break-even. Different life

---

* See Equation 7–13.
** For derivation of the capital recovery factor, see Appendix A.

expectations of alternative investment will not affect the results. The criterion for selection among alternatives is to choose the investment with the highest rate of return. The concept of the opportunity cost of capital is used because the capital can be invested elsewhere to generate interest.

The formula is expressed as follows: Find discount rate i, such that

$$\sum_{j=1}^{n}\left[\frac{S_j + R_j}{(1 + i)^j}\right] \simeq \sum_{j=1}^{n}\left[\frac{I_j - V_j + M_j}{(1 + i)^j}\right]$$

where n = number of time intervals over which the investment is analyzed, S_j = energy cost savings in year j, R_j = revenue received in year j, I_j = investment costs in year j, V_j = salvage value in year j, and M_j = Maintenance and repair costs in year j.

The discount rate is usually calculated by a process of trial and error; that is, the net cash flow is computed for various discount rates until its value is reduced to zero.

The example in Table 7–13 is used again. The opportunity interest rates for Plans X and Y are 26.5% and 40.0%, respectively (see Table 7–13). It is clear that Plan Y is preferable.

The opportunity interest rate is easy to understand in concept and leads to conclusions consistent with the other methods. However, it is more cumbersome to calculate and also has the disadvantage of overlooking larger-scale projects. This problem and its solution are similar to those discussed for the B/C ratio method.

SPECIAL FACTORS TO BE CONSIDERED IN EVALUATING CONSERVATION INVESTMENTS

In addition to previous analyses, two special factors must be considered when evaluating conservation investments: uncertainty and inflation in costs and benefits of investing technologies. In the previous illustrations of evaluation, future prices of benefits and production factors (e.g., equipment and energy) were assumed to be known. However, uncertainty and inflation may distort the previous analyses because they alter the course of future prices and benefits.

Uncertainty

Uncertainty can be defined broadly as the disparity between the predicted and the actual. It encompasses two specific concepts, *risk,* in which the probability of occurrence of an event can be predicted; and *uncertainty,* in which the expected chance of occurrence of an event cannot be predicted. The uncertainty can occur in estimates for costs, benefits, energy savings, economic lives, and discount rates of conservation investments.

Probability analysis can generally be used to deal with the risk portion of the uncertainty, and sensitivity and break-even analyses can be used to ensure that critical variables are taken into account in the evaluation of investments.

Probability analysis involves using expected values of costs or benefits rather than simply their "point" estimates. The expected value of costs or benefits can be expressed in simple form as:

$$E = \sum_{j=1}^{N} [P(X_j) \cdot V_j]$$

where $P(X_j)$ = probability that the outcome will be X_j, V_j = cost (or benefit) of outcomes X_j, E = expected value, and N = number of outcomes.

For example, the cost of installing a heat pump may be roughly estimated at $2,000, but the following might be a more accurate statement of costs:

Possible Situation	Probabilities	Cost if Situation Develops	Expected Cost
No difficulties in installation	0.20	$1,500	$ 300
No serious difficulty	0.70	3,000	2,100
Serious difficulty	0.10	7,500	750
			$3,150

Thus a more accurate estimation of cost in this situation would be $3,150.

Probability analysis quantitatively incorporates uncertainty into the investment evaluation. However, it requires more information than previously described evaluation methods (i.e., probabilities of outcomes and costs or benefits of each outcome). Depending on the need for decision-making, one can use more sophisticated methods for probability analysis.[13]

Sensitivity analysis can be used to assess the consequences of assuming alternative values for the significant variables in the analysis. By determining the effect on the outcome of potential variation in a factor, the analyst identifies the degree of importance of that estimate or assumption and can then seek more information about it if desired. For example, the profitability of a furnace recuperator might be tested for sensitivity to the expected utilization rate of the furnace.

Break-even analysis, a third technique for dealing with uncertainty, focuses on a single key variable that is regarded as uncertain, and calculates the minimum (or maximum) value of the variable required to achieve a given outcome. For example, other factors being known, one might solve for the rate of escalation in fuel prices required for an investment in a heat exchanger to break even. The opportunity cost interest rate and payback period methods described above are some variations of break-even analysis.

Inflation

In the previous investment analysis, we assumed that all costs and revenues inflate at the same general rate, and that they therefore remain constant in real terms. With this assumption, renewal costs and other future expenses and benefits are evaluated at present prices. This assumption has helped simplify the analysis and in many cases results in reasonable evaluation.

However, when income tax effects are considered, the real after-tax return to the firm can be distorted by inflation even with the assumption that future receipts and expenditures will respond fully and evenly to inflation.

Inflation will tend to have a detrimental effect on an investment financed principally by equity funds (i.e., nonborrowed funds) because: (1) tax deductions for depreciation are unresponsive to inflation; (2) terminal value of equipment is responsive to inflation and will be reflected in the capital gains tax; (3) tax deductions for fixed interest on the borrowed portion of capital are unresponsive to inflation, such that the present value of the deductions diminishes over time; and (4) inflation in receipts tends to move a firm into higher tax brackets.

A simple way to deal with inflation is to adjust costs and benefits with an inflation rate (α) as follows:

$$NR = \sum_{i=1}^{n} \frac{R_i - C_i}{(1 + r)^i (1 + \alpha)^i} \qquad (7\text{--}14)$$

where NR = net return, n = years of an investment's useful life, R_i = revenue in the i_{th} period, C_i = costs in the i_{th} period, r = discounted (interest) rate, and α = inflation rate.

If some prices are projected to increase at a rate different from the general inflation rate, then different price indices should be applied for the corresponding costs and benefits in Equation 7–14.

SUMMARY

In this chapter we have presented one comprehensive method and six quick methods for evaluating energy conservation investments, evaluated three promising conservation technologies for the copper production as an illustration, and discussed two special factors (inflation and uncertainty) to be also considered in evaluation.

By using the comprehensive method (annual conservation cost-saving method) we concluded that under the 1974 price structure of copper production factors, the three conservation methods (recycling, retrofitting, and changing to new processes) may not be economically attractive because implementation of these measures increases production costs. This explains the U.S. General Accounting Office's finding that most industrial energy conservation activities involved only operational changes.[12]

REFERENCES

1. Bonczar, E., and Tilton, J. "An Economic Analysis of the Determinants of Metal Recycling in the U.S.: A Case Study of Secondary Copper," final report to the U.S. Bureau of Mines, May 1975.
2. Bennett, H. et al. "An Economic Appraisal of the Supply of Copper for Primary Domestic Sources," Bureau of Mines Information Circular 8598. Washington, D.C.: U.S. Government Printing Office, 1973.
3. Battelle Columbus Laboratories. "Evaluation of the Theoretical Potential for Energy Conservation in Seven Basic Industries," final report to the Federal Energy Administration, National Technical Information Service, 1975.
4. Carrillo, F. et al., "Recovery of Secondary Copper and Zinc in the United States," Bureau of Mines Information Circular IC 8622. Washington, D.C.: U.S. Government Printing Office, 1974.
5. Office of Air and Waste Management (EPA). *Background Information for New Source Performance Standards: Primary Copper, Zinc, and Lead Smelters,* Vol. 1. Research Triangle Park, N.C.: U.S. Environmental Protection Agency, 1974.
6. Private communication, J. M. Henderson (ASARCO) letter to D. F. Walters (EPA), Nov. 24, 1972.
7. Saddington, R., Curlook, W., and Queneau, P. "Use of Tonnage Oxygen by the International Nickel Co.," pp. 261–269 in *Pyrometallurgical Processes in Nonferrous Metallurgy.* New York: Gordon and Breach, 1967.
8. Senrau, K., "Control of Sulfur Oxide Emissions from Primary Copper, Lead and Zinc Smelters–A Critical Review," *Journal of the Air Pollution Control Association,* Vol. 21, No. 4, 185–194, April 1971.
9. Kupryakov, Yu P. et al., "Operation of Reverberatory Furnaces on Air-Oxygen Blasts," *The Soviet Journal of Nonferrous Metals;* English translation: Vol. 10, No. 2, 13–16, February 1969; Russian edition: Vol. 42, No. 2, 14.
10. Smith, P., Bailey, D., and Doane, R. "Minerals Processing: Where We Are . . . Where We're Going, E/MJ Mining Guidebook," *Engineering and Mining Journal,* pp. 161–183, June 1972.
11. Niimura, M., Konada, T., and Kojima, R., "Control of Emissions at Onahama Copper Smelter," paper represented at Joint Meeting MMIJ-AIME, Tokyo, May 24–27, 1972.

12. U.S. General Accounting Office. "Report to the Congress," EMD-78-38, June 1978.
13. Young, D., Contreras, L. E., et al. "Expected Present Worths of Cash Flows Under Uncertainty," *The Engineering Economist,* Vol. 20, No. 4, Summer 1975.
14. Grant, E. L. and Ireson, W. G. *Principles of Engineering Economy,* 5th ed. New York: Ronald Press, 1970.
15. White, J., Agee, M., and Case, K. *Principles of Engineering Economic Analysis.* New York: John Wiley & Sons, 1977.
16. Kellog, H. and Henderson, J. "Energy Use in Sulfide Smelting of Copper," in (J. Yannopoulos, ed.) *Extractive Metallurgy of Copper,* AIME, 1976.
17. National Bureau of Standards. "Waste Heat Management Guidebook," NBS Handbook 121, February 1977.
18. International Revenue Service. "Public Law 96–223 Crude Oil Windfall Profit Tax Act of 1980," Internal Revenue Cumulative Bulletin 180–3. Washington, D.C.: U.S. Government Printing Office, 1980.
19. "Tax Credits That Could Save Industry Billions," *Business Week,* April 7, 1980.
20. Hu, S. "Copper Commodity Model and Energy Issues," unpublished Ph.D. thesis, University of Pennsylvania, 1978.
21. A. A. Mathews Inc. "Capital and Operating Cost Estimation System for Mining and Benefication," Vol. 4. Mining Report No 1953–02, Phase II Report for U.S. Bureau of Mines, June 1976.

8
Industrial Energy Conservation Modeling

INTRODUCTION

Chapter 6 provided information on how to identify energy-inefficient processes, whereas costs and benefits of alternative conservation measures were assessed in Chapter 7. This chapter includes details on how to obtain energy data and how to transform these data for the evaluation of investments and regulations in conservation. Some of the energy conservation data discussed in this chapter may be obtained from Chapters 12, 13, and 14 and Appendix H, which enumerate details of conservation technologies commercially available for retrofitting and new installation.

The main purpose of this chapter is to provide industrial energy planners with a systematic method (modeling) for evaluating conservation efforts. Conservation efforts include energy management programs, computerization of energy monitoring, waste recovery and utilization technologies, and new energy-efficient processes that can be used to reduce energy consumption and replace inefficient facilities identified by the methods discussed in Chapter 7.

The main contribution of energy conservation modeling will be to provide a method of evaluating conservation investments that does not require actual surveys of energy consumption and production costs at each facility. In modeling energy conservation, standard engineering data concerning production processes are used to estimate pre-conservation energy consumption and production costs. Methods provided in Chapter 7 for costing alternative conservation measures are used to estimate energy uses and production costs for postconservation production possibilities. A mathematical representation is constructed to simulate given multi-stage production processes so that conservation efforts and policies can be evaluated. This representation will be validated, and its application will also be discussed.

In brief, this chapter covers methodology, definition of production systems, model construction, data preparation, model exercise and interpretation of model results, model validation/verification, and future model improvements and other considerations.

METHODOLOGY

In contrast to the econometric aggregate approach, a micro approach is used here to evaluate costs and benefits of energy conservation policies for a multi-plant, multi-process, single-product, energy-intensive industry. An appropriate model is constructed to simulate production processes in order to assess the market penetration of conservation measures in existing facilities of selected companies.

As shown in Fig. 8–1, an industrial firm uses a series of input factors such as labor, material, energy, capital, and land to make products. The industrial conservation model is constructed to simulate the process of decision-making and the technological relationships between the input and the output of production processes. Since industrial firms consist of multi-stage, multi-plant, and multi-product production lines, simulation models usually are complex. To build an industrial energy conservation model, we need a method for systematic treatment of all factors and roles involved. Figure 8–2 shows the methodology of constructing an industrial energy conservation model. The method basically consists of six parts: (1) identifying the objective

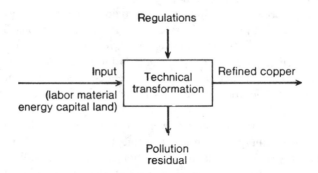

Fig. 8–1. A basic industrial production process.

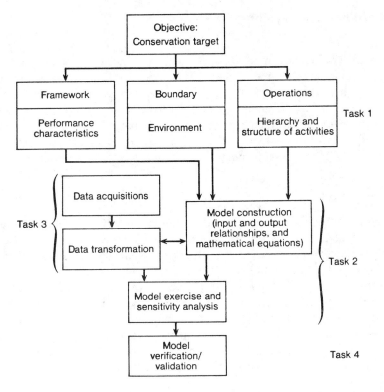

Fig. 8-2. Tasks required for constructing an industrial energy conservation model.

of the model; (2) defining the production system (i.e., framework, boundary, and operations rule); (3) constructing the model; (4) preparing required data; (5) exercising the model; and (6) verifying/validating the model. These elements are summarized below and discussed in detail in the remaining sections.

Objective of the Model

The objective of the model determines the boundary and environment of the production systems to be considered. If the model is going to be used by individual firms to evaluate their energy conservation policies, then the production systems would include the firm and affected surroundings. On the other hand, if the model is to be used by the Department of Energy, then the systems' boundary would be extended to include the United States and possibly the Organization of Petroleum Export Countries (OPEC) and major industrial nations.

Definition of the Production System

A production system is defined through its boundary, framework, and operations rule. The systems framework sets the performance criteria and underlying rules by which the production system is operated. The systems boundary delineates the "territory" of production processes. In an industrial production system, the primary boundary is the occupancy of the entity; however, the secondary boundary is much broader. The operations rule indicates the hierarchy and structure of activities within the production system and how the order "flows" from the top ranks to the bottom ones and vice versa.

Model Construction

After the boundary, framework, and operations rule of a production system have been determined, a set of mathematical equations can be developed to translate them into an operable mathematical model which expresses relationships between the input and the output of the production system.

There are two basic types of mathematical equations, one for describing the structure of the system and the relationships among system components, and the other, for describing the objective and measuring the performance of the system. These mathematical equations can be linear or nonlinear, static or dynamic, deterministic or probabilistic. In formulating these mathematical equations, data availability should be considered, since it is the key to successful model construction.

Data Preparation

Responding to the need of model construction, data must be acquired and transformed for use in construction, testing, and exercising of the model. Data preparation includes data acquisition and model parameter estimation. Standard engineering production data as well as market price data are obtained and transformed into the set of data required for the conservation model.

Model Exercise

Model exercise is the process of combining each of the elements involved in a model and obtaining a quantitative output for a given set of input. The methods used for model exercise include various mathematical techniques and simulation. The exact method chosen depends on the amount and accuracy of the available data, the accuracy of the desired results, and the amount of resources available for performing the analysis. Model exercise also includes sensitivity analysis. The purpose of sensitivity analysis is to test the stability of the model's behavior and the valid range of model applications.

Model Verification and Validation

Model verification and validation are similar; both are used to ensure that the model constructed represents the function of the real production system. Model verification is sometimes defined as a method of ensuring the accuracy of forecasts and computer codes by the model, whereas validation is broadly used as a way to ensure the correctness of the underlying assumptions, the structure of the model, forecasts, and computer codes. The two terms are often used interchangeably.

DEFINITION OF PRODUCTION SYSTEMS

In thermodynamics, a specified region that can be separated from everything else by a well-defined surface is called a system. In production processes, the definition of a system has been broadened. As discussed in the previous section, a production system is defined through three elements: boundary, framework, and operation rule. The framework sets the performance criteria for a system, the boundary delineates the environment, and the operations rule describes the structure and hierarchy of component activities. We will discuss the definition and then use a copper production system as an example.

System Definition

A system is a distribution of the members in a dimensional domain, an entity (conceptual or physical) consisting of interdependent parts. The whole as a system is more than the sum of its parts. In a system, it is significant that the parts are arranged; this is in contrast to the fact that, in aggregates, the parts are added. The term *whole* is designated as the concrete organized object, while the organization itself, the way parts are arranged, is called a *system*.

Open Systems and Closed Systems. A system is closed if no material enters or leaves it; it is open if there are import and export and, therefore, changes of components. Production systems are open systems; they maintain themselves in exchanges of energy and materials with the environment, and in the continuous breaking down and replenishing of their components.

A closed system must, according to the second law of thermodynamics, eventually attain a time-independent equilibrium state, with maximum entropy and minimum free energy, in which the ratio between its phases remains constant. An open system may attain a time-independent state in some sense in which the system remains constant and balanced as a whole and in its phases, although there is a continuous flow of materials into and out of the system; this is called a steady state. For it to perform work, however, the system must not be in equilibrium, but must be tending toward it; to continue this way, the system must maintain a relatively steady state. Therefore, both social organizations and production processes are flagrantly open systems, in that the input of energy and materials and the conversion of output into further refined energy and materials consist of transactions between the organization and its environment.

Purpose of the Open System. The distinctive organization of open systems shows itself in the goal-directedness and directiveness of their activities. For survival and competition, open systems import energy and materials from the external environment, transform them, and export products to the environment. Imports include information about the environment; systems then furnish signals to their structure about their own functioning in response to the environment.

To survive, open systems must move through the entropic process; they require negative entropy. The entropic process is a universal law of nature in which all forms of organization move toward disorganization and retirement.

To compete, open systems must move toward differentiation and elaboration: Diffuse global patterns are replaced by more specialized functions; through differentiation, a system moves closer to a steady state. A system can reach the same final state from different initial conditions and by a variety of paths.

Boundary of an Open System. The boundary condition of an open system depends on the interaction between itself and the environment. In a physical system, the primary boundary is clearly the occupancy of the entity; however, it is difficult to delineate the secondary boundary, in which the impacts of the system are too strong to be ignored.

Framework of an Open System. The framework of an open system sets fundamental criteria of performance, which guide the course of the system's goal pursuing. The framework also entails the objective/motivation of the system.

Operations Rule of an Open System. The operations rule of an open system determines horizontal and vertical relationships among system components and their functions, authority, and responsibility. The operations rule of an open system also entails the means of delivering and taking an order or message.

A Primary Copper Production System

The example of primary copper production has been used throughout this book to illustrate the methods discussed. Here it is used to define a production system.

Boundary of a Primary Copper Production System. The copper production system consists of primary refined copper production from ore, secondary copper recovery from scrap, and imports. About 51% of U.S. copper is supplied by primary producers, 45% from secondary recovery, and 4% from imports. With respect to boundary, there are two types of copper production system:

- U.S. copper industry: the boundary of the copper production system would include world copper production, U.S. primary refined production, and U.S. secondary recovery.
- Individual companies: the boundary of an individual firm depends on its business. There are four types of firms:
  - Primary refined copper producers, who integrate mining, milling, smelting, and refining processes and partially recover copper scrap in the smelting and refining processes.
  - Independent copper producers (small percentage)
  - Secondary copper producers, who primarily recover copper from scrap but do not produce copper from ore.
  - Copper importers, who import copper and copper scrap.

As shown in Fig. 8–3, if we consider the boundary of a primary copper production company, the system is the whole process of mining, beneficiating, smelting, and refining. It may also include marketing, if the producer markets his own products. However, because primary refined copper producers also use scrap in their production (shown in Fig. 8–4), their secondary boundary may cover the domain of secondary copper recovery systems. Depending on the proportion of scrap used in primary copper production, the importance of the secondary boundary varies. For example, if a primary refined copper producer uses more scrap than ore in its production, the domain of the secondary copper recovery system becomes the primary boundary for the primary copper production system.

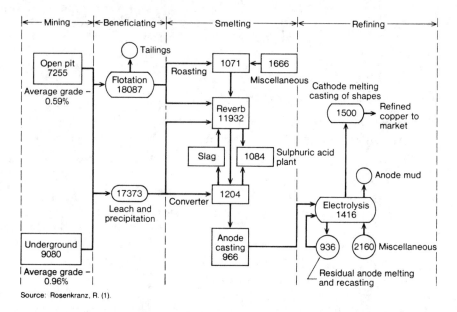

Source: Rosenkranz, R. (1).

Fig. 8-3. Copper production system and energy consumed at each stage of production (Btu/lb). Source: Rosenkranz, R.[13]

In addition, the boundary of a production system may also depend on the length of the time interval considered. For example, if the time interval is more than 20 years, the boundary of a production system may have to cover any predictable process changes due to technological innovations.

As discussed in Chapter 6, the system consisting of mining, beneficiating, smelting, and refining is fairly representative of primary copper production.

Framework of a Primary Copper Production System. The framework of the production system sets the fundamental performance criteria of the system, which guide the course of the system's pursuit of its goal. The framework thus entails the objective/motivation of a production system. The objective/motivation of a copper production system can be one or a combination of the following:

- Minimum cost
- Maximum profit
- Maximum or stable market shares
- Minimum risk
- Maximum employment stability

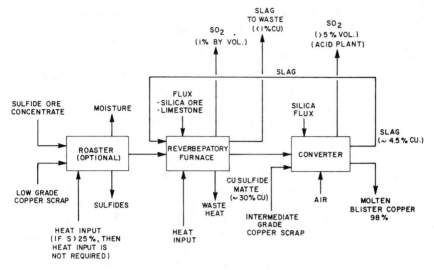

CHART A: SMELTING PROCESS

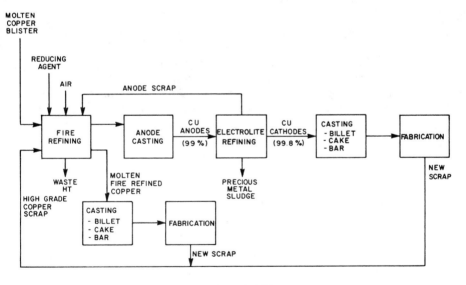

CHART B: REFINING PROCESS

Fig. 8–4. Scrap use by primary producers.

In general, small- or medium-size firms pursue maximum profit, minimum cost, and maximum market shares because they look for growth. For large companies, the goals tend to be maintaining market shares and minimizing risk. Utility companies usually attempt to minimize their investment risk because of a fixed rate of return. Maximum employment stability is one objective of some Japanese firms because of their lifetime-employment policies.

Operations Rule of a Primary Copper Production System. Over 80% of domestic ores are processed by flotation concentration (beneficiating), smelting, and electrolytic refining. Most of the remaining fraction (low-grade ores recovered by leaching followed by cementation or electrowinning, high-grade ores charged directly into converter or reverberatory, etc.) is at least partially processed according to the above basic processing sequence. In addition, significant proportions of scrap material are processed by primary copper smelters or refineries. With current ore content and production technology, one ton of refined copper is produced from three tons of copper ore.

Detailed operations rules of each process in a primary copper production company were discussed in Chapter 6.

MODEL CONSTRUCTION

The main task of model construction is to translate the boundary, framework, and operations rule of a primary refined copper production system into mathematical representations that can be used to simulate refined copper production and, subsequently, to evaluate conservation policies and conservation technologies.

There are two common approaches in constructing an industrial energy conservation model.* One is the mathematical programming method, and the other the market penetration method. The two approaches can be complementary to each other, or they can be used independently to evaluate conservation policies. Because mathematical programming usually leads to the choice of optimal (extreme) points as problem solutions, it is likely that a single conservation technology will take over most markets because of its slightly lower costs. The market penetration approach would keep this from happening. However, the market penetration method requires either historical market shares or judgmental market share data of given technologies. For most new technologies, both kinds of data are difficult to obtain.

Mathematical Programming Method
The mathematical programming method can be described as maximizing a firm's profit subject to production, marketing, transportation and environmen-

---

* See Appendix G for other industrial models.

tal constraints, the availability of energy conservation technologies, and conservation requirements. The problem can be simply formulated as follows:

Maximize: profit $= f$ (unit price, production cost, quantity)   (8–1)

Subject to:

1. Quantity $= g_1$ (capacity utilization, energy conservation technologies, sales commitments, transportation, pollution controls)   (8–2)
2. Energy consumption reduction $= \%$ of a given year's consumption   (8–3)
3. Price $= g_2$ (quantity, competition)   (8–4)

where f, g_1, and g_2 are symbols for functions. In this formulation, production facilities with conservation technologies are considered as new facilities. This mathematical programming problem can be linear, nonlinear, mixed-integer, static, dynamic, or probabilistic, depending on the complexity and time interval required for the conservation model. For example, a probabilistic nature can be incorporated into this programming problem if the uncertainty of model parameters warrants this inclusion.

Market Penetration Method

The market penetration method is generally used in projecting the path of a given conservation technology's adoption by industry over a time interval. Figure 8–5 shows an S-shaped curve for technology penetration over 25 years. From the consumer's point of view, this curve is sometimes called the learning curve. The curve indicates that a conservation technology starts with a very low market share because it is new, then gains markets rapidly, and finally levels off when approaching a saturated market share (e.g., 0.4 as shown in Fig. 8–5).

The above market penetration path can be expressed in mathematical form. The market share limitation of a given technology at the time interval t is M_t, and

$$M_t \leq a \tag{8–5}$$

where a is a fraction of the market for the given technology at a given time interval t. This fraction may be judgmental.

Consequently, Equation 8–5 can be incorporated into the mathematical programming problem shown in Equations 8–1 to 8–4 so that the solution of this programming problem will indicate reasonable market shares for the conservation technologies considered.

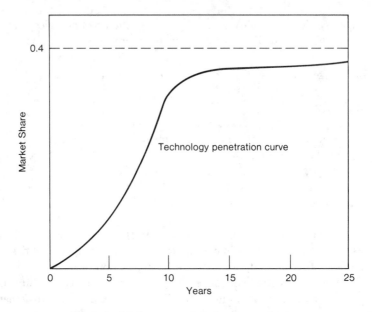

Fig. 8–5. Market penetration of a given technology.

Structure of an Energy Conservation Model

If we do not consider a reduction of primary copper production or an increase of refined copper imports as a viable conservation measure for evaluating alternative conservation strategies, an energy conservation model can be formulated as follows (simplified from the problem of Equations 8–1 through 8–5):

Minimize: cost $= f$ (cost of production factors, production quantity)   (8–6)
Subject to:
1. Production quantity $= g$ (capacity, capacity utilization rate, conservation technologies, sales commit-
ments, transportation, pollution controls)   (8–7)
2. Energy consumption reduction $= \%$ of a given year's
consumption   (8–8)
3. Market penetration limits for conservation technolo-
gies $= \%$ of the market replaced   (8–9)

where f is a cost function, and g is a set of production functions for capacities and other constraints.

The difference between the model consisting of Equations 8–6 through 8–9 and that of Equations 8–1 through 8–5 is that the former do not include copper production price effects in determining conservation technologies. In other words, we assume the producer will produce the same amount of copper with or without energy conservation measures. Therefore, the producer will choose the least-cost production combinations. (The reader not interested in modeling details may skip to Chapter 9.)

Functions in Equations 8–6 through 8–9 need not be linear. For example, unit production cost function in Equation 8–6 for each process looks like a flat u curve as shown in Fig. 8–6; the unit production cost for a given production facility is very high owing to high allocated fixed costs when the capacity of the facility is mostly unutilized (at 10%), decreases gradually when the capacity is being utilized (at 80%), but increases when the capacity is over-utilized (at 100% or more).

In order to use linear programming computer packages commercially available in most computers, the nonlinear equations are sectioned so that they can be approximated by appropriate step functions. Using Fig. 8–6 again to illustrate the approximation, we section the unit cost function by the

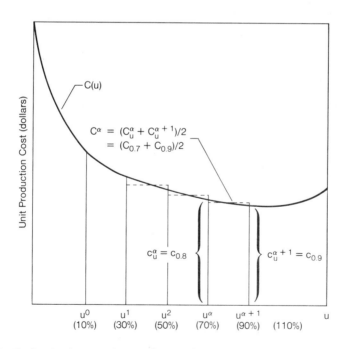

Fig. 8–6. Sectioned unit production cost function for a given mine. (For smelter the α should be changed to β, and for refining it should be γ.)

following: 0–10%, 10–30%, 30–50%, 50–70%, 70–90%, and 90% and above of capacity utilization. In each sectioned region, a constant unit cost is approximated by averaging production costs in this range. For example, c^α is the approximated unit cost for production in the range of 70% to 90% capacity utilization; it is computed by averaging the two unit production costs—one at 70% and another at 90% utilization rate as shown in the following equation:

$$c^\alpha = \frac{1}{2}(c_{.7} + c_{.9}), \text{ where } .7 \leqslant u \leqslant .9 \tag{8-10}$$

where c^α is the approximated unit cost in the αth range of capacity utilization rate (u), which is between the 70% and 90% utilization rates; $c_{.7}$ and $c_{.9}$ are the unit production costs at 70% and 90% utilization rates, respectively.

To transfer this approximation into a linear programming problem that can be solved through a computer package, the following equations are used to generalize Equation 8–10:

$$c(u) = \sum_{1}^{\alpha} c^\alpha \cdot A^\alpha \tag{8-11}$$

subject to:

$$A^\alpha u^\alpha \leqslant u \leqslant A^\alpha u^{\alpha+1}, \text{ for } \alpha = 1, 2 \ldots \alpha \tag{8-12}$$

$$\sum_{1}^{\alpha} A^\alpha = 1; A^\alpha = 0 \text{ or } 1, \text{ for } \alpha = 1, 2 \ldots \alpha \tag{8-13}$$

where A^α is the utilization coefficient. It is equal to one when the utilization rate is in the right range (i.e., $u^\alpha \leqslant u \leqslant u^{\alpha+1}$); otherwise it is zero.

Equation 8–12 can be rewritten in the following form:

$$A^\alpha u^\alpha q < uq < A^\alpha u^{\alpha+1} q, \text{ for } \alpha = 1, 2, \ldots \alpha$$

or

$$A^\alpha q^\alpha \leqslant x \leqslant A^\alpha q^{\alpha+1} \text{ for } \alpha = 1, 2, \ldots \alpha \tag{8-12'}$$

where $q^\alpha = u^\alpha q$, $x = uq$, and $q^{\alpha+1} = u^{\alpha+1} q$.

Equation 8–12' transforms capacity utilization rates into production quantities. Equations 8–11, 8–12', and 8–13 will then be used to represent the step function that approximates the nonlinear unit production cost function in each facility. These equations transform the nonlinearity of each cost function by a step function.

By the same method, functions in the constraint Equations 8–7, 8–8, and 8–9 can be also sectioned within a given range as follows, if they are nonlinear:

$$A^{\alpha}q^{\alpha} \leqslant \phi(x) \leqslant A^{\alpha}q^{\alpha+1} \tag{8-14}$$

where q^{α}, x, and $q^{\alpha+1}$ have explained in Equation 8–12′, and $\phi(x)$ is the production function set governing the relationship between inputs and outputs of the given production system. $\phi(x)$ is linearized within the range.

Copper Energy Conservation Model—An Example

A primary copper firm produces refined copper through four stages: mining, beneficiating (milling), smelting, and refining. In smelting and refining, scrap is also used. Because of the bulk of ores, mines and mills are usually built together.

As discussed in Chapter 6, there are three promising energy conservation measures:

- Recycling: using scrap in smelting and refining
- Changing processes: replacing existing reverberatory smelters with flash smelters
- Retrofitting: installing existing reverberatory smelters with oxygen plants

If a producer has m mines/mills, n smelters, and p refineries, and each of the above three measures is treated as new facilities, he will have $(m+1) \times (n+1) \times (3n + 1)p$ production paths to choose from. These combinations are shown in Fig. 8–7.

Equations 8–7, 8–8, 8–9, 8–10, 8–12′, 8–13 and 8–14 are applied to formulate a mixed-integer programming problem. We call this problem a copper energy conservation model.

Procedure to Formulate the Model. We propose the following four-step procedure to energy planners to formulate an energy conservation model:

- Specify the production systems and subsystems, production flow chart, location of facilities, transportation links, and applied energy conservation options.
- Determine valid constraints of facility capacity, pollution control levels, sales contract commitments, energy conservation requirements, and market limits for conservation options.

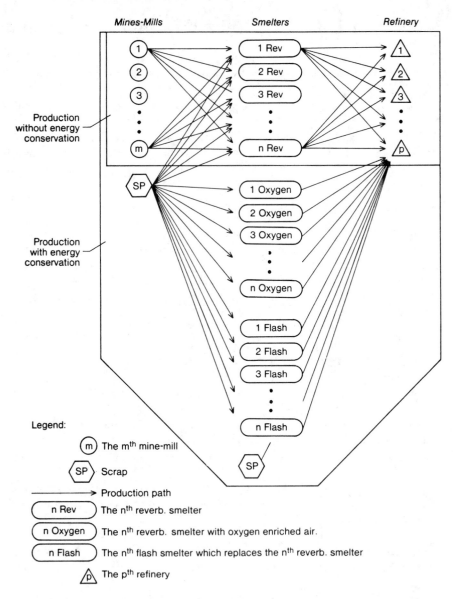

Fig. 8–7. Production pattern possibilities with the three energy conservation measures (recycling, retrofitting, and changing processes).

- Formulate a mathematical program problem by using Equations 8–6 through 8–9, disregarding the nonlinearity in the production and cost functions.
- Linearize the cost and production functions that are nonlinear in the mathematical program problem by using Equations 8–10, 8–12′, 8–13, and 8–14. The mathematical program will become a mixed-integer program after this linearization process.

Company A is used as an example to illustrate the use of the above procedure. Figure 8–8 shows the energy conservation model for Company A.

Minimize $$C_T = \sum_{i=1}^{5} \sum_{j=1}^{13} \sum_{k=1}^{3} c_{ijk} \cdot X_{ijk}$$ (D-1-1)

Subject to $$\sum_{j=1}^{12} \sum_{k=1}^{3} X_{1jk} \leq 283.3$$ (D-1-2)

$$\sum_{j=1}^{12} \sum_{k=1}^{3} X_{2jk} \leq 59.8$$ (D-1-3)

Mining–Milling
Capacity
Utilization $$\sum_{j=1}^{12} \sum_{k=1}^{3} X_{3jk} \leq 36.2$$ (D-1-4)

$$\sum_{j=1}^{12} \sum_{k=1}^{3} X_{4jk} \leq 749.7$$ (D-1-5)

$$\sum_{j=1,5,9} \sum_{i=1}^{5} \sum_{k=1}^{3} X_{ijk} \leq 108$$ (D-1-6)

Smelting
Capacity
Utilization $$\sum_{j=2,6,10} \sum_{i=1}^{5} \sum_{k=1}^{3} X_{ijk} \leq 108$$ (D-1-7)

$$\sum_{j=3,7,11} \sum_{i=1}^{5} \sum_{k=1}^{3} X_{ijk} \leq 113.4$$ (D-1-8)

$$\sum_{j=4,8,12} \sum_{i=1}^{5} \sum_{k=1}^{3} X_{ijk} \leq 270$$ (D-1-9)

$$\sum_{i=1}^{5} \sum_{j=1}^{13} X_{ij1} \leq 186$$ (D-1-10)

Refining
Capacity
Utilization $$\sum_{i=1}^{5} \sum_{j=1}^{13} X_{ij2} \leq 276$$ (D-1-11)

$$\sum_{i=1}^{5} \sum_{j=1}^{13} X_{ij3} \leq 103$$ (D-1-12)

Total
Refined Copper
Production $$\sum_{i=1}^{5} \sum_{j=1}^{13} \sum_{k=1}^{3} X_{ijk} = 481$$ (D-1-13)

Fig. 8–8. A mathematical programming problem for Company A.

Total
Energy $$\sum_{i=1}^{5}\sum_{j=1}^{13}\sum_{k=1}^{3} e_{ijk} \cdot X_{ijk} = K * (27530) \qquad\qquad (D\text{-}1\text{-}14)$$
Use

$$x_{ijk} \geq 0 \quad \text{for} \quad \begin{array}{l} i = 1, 2, 3, 4, 5 \\ j = 1, 2, \ldots 13 \\ k = 1, 2, 3 \end{array}$$

where c_T = total project cost

X_{ijk} = production quantity in thousand short tons mined at i^{th} mine, smelted at j^{th} smelter, and refined at k^{th} refinery

e_{ijk} = unit energy use in billion Btu for a thousand tons of copper mined at i^{th} mine, smelted at j^{th} smelter, and refined at k^{th} refinery

c_{ijk} = unit production cost in thousand dollars

$(1 - K)$ = a fraction, for energy conservation

Fig. 8–8. A mathematical programming problem for Company A *(continued)*.

Specification of the System.   Company A has four mines/mills, four smelters, and three refineries. It is highly integrated and self-dependent.

Figure 8–7 shows all production possibilities for primary copper companies. The alternative conservation options for Company A are: scrap—an alternate to ore; oxygen-retrofitted smelters; flash smelters; and scrap—an alternate to blister (product-in-process of smelting). Transportation links were identified when facilities had been located.

Determination of Valid Constraints.   Table 8–1 shows the capacity of process facilities for Company A. There is no constraint assumed for the capacity of transportation. A sales commitment is incorporated in Equation D-1-13 in Fig. 8–8.* The energy conservation (reduction) requirement is shown in Equation D-1-14 in Fig. 8–8. The energy reduction level is indicated by $(1 - K)$, where K is the fraction of original energy consumption, a desired consumption level.

Because the three conservation options will reduce pollution levels, there is no additional constraint for environmental controls when replacing the existing production possibilities; that is, these conservation options will meet original environmental requirements. In order to see the path of market penetration of alternative conservation options, the model has not contained mar-

---

* The 1974 original production level was used because we intend to evaluate conservation without reducing production.

Table 8–1. The Capacity of Facilities of Company A.

Facility	Production Capacity (Thousand Short Tons)
Mine-Mill	
I	283.3
II	59.8
III	36.2
IV	749.7
Smelter	
I	108.0
II	108.0
III	113.4
IV	270.0
Refinery	
I	186.0
II	276.0
III	103.0

ket share constraints. Market shares of a given conservation option resulting from the mathematical program will be compared with judgmental data. If the resulting market shares are unreasonably high, then an upper limit will be established as a market penetration constraint for that technology.

Formulation of a Mathematical Program Problem. Using Equations 8–6 through 8–9, a mathematical program problem has been formulated for Company A's copper energy conservation model as shown in Fig. 8–8. The model transforms the network problem shown in Fig. 8–7 into mathematical equations. Figure 8–8 can be briefly explained as follows:

- Equation D-1-1 is the objective function to minimize production costs by examining all production possibilities (including conservation options). The costs include those for production and transportation (see Eq. 8–15).
- Equations D-1-2 and D-1-12 are constraints for production of each facility of the process of mining–beneficiation, smelting, and refining in Company A. Undesired transportation links between production facilities are given a very high transportation cost; thus, these links will not be economical and were not chosen for production.
- Equation D-1-13 states the refined copper production requirement for Company A in order to meet the sales commitment.
- Equation D-1-14 requests that the company's total energy consumption

be K percent of the original consumption level. In other words, Company A will have to conserve $(100 - K)$ percent of its present energy consumption.

In Fig 8–8, the cost function (c_{ijk}) is not linear. Linear programming computer packages cannot be used to solve this problem. Thus, we need to linearize them.

Linearization of Nonlinear Cost Functions. Methods exhibited in Equation 8–10, 8–12', 8–13, and 8–14 are applied to linearize nonlinear equations and thus transform the mathematical program problem into a mixed-integer program problem. The process of linearizing cost functions is illustrated below.

The unit cost of production paths via the ith mine–mill, j^{th} smelter, and k^{th} refinery is symbolized as c_{ijk} and expressed in the following equation:

$$c_{ijk} = c_{iM} + C_{iF} + T_{ij} + c_{jS} + T_{jk} + c_{kR} \qquad (8\text{–}15)$$

where c_{iM} = the ith mine's production cost function, c_{iF} = the ith mill's production cost function, c_{jS} = the jth smelter production cost function, c_{kR} = the kth refinery production cost function, and T_{ij}, T_{ik} = transportation costs between the ith mine–mill and the jth smelter, and between the ith mine and the kth refinery.

Using Equations 8–11, 8–12, 8–13, and 8–14 for α (if the facility is a mine/mill), β (if the facility is a smelter), and γ (if the facility is a refinery), sections of the production cost functions can be approximated as:

$$c_{ijk} = \sum_{1}^{\alpha} (c_{iM}{}^{\alpha} + c_{iF}{}^{\alpha}) \cdot A_i{}^{\alpha} + T_{ij}$$

$$+ \sum_{1}^{\beta} (c_{jS}{}^{\beta} \cdot A_j{}^{\beta}) + \sum_{k}^{\gamma} (c_{kR}{}^{\gamma} \cdot A_{kR}{}^{\gamma}) + T_{jk} \qquad (8\text{–}16)$$

and subject to:

$$A_i{}^{\alpha}j_i{}^{\alpha}q_i \leq u_iq_i \leq A_i{}^{\alpha}u_i{}^{\alpha+1}q_i, \text{ for } \alpha = 1, 2 \ldots \alpha, \qquad (8\text{–}17)$$

$$A_j{}^{\beta}u_j{}^{\beta}q_j \leq u_jq_j \leq A_j{}^{\beta}u_j{}^{\beta+1}q_j, \text{ for } \beta = 1, 2 \ldots \beta, \qquad (8\text{–}18)$$

$$A_k{}^{\gamma}u_k{}^{\gamma}q_k \leq u_kq_k \leq A_k{}^{\gamma}u_k{}^{\gamma+1}q_k, \text{ for } \gamma = 1, 2, \ldots \gamma \qquad (8\text{–}19)$$

$$A_\eta{}^{\xi} = \begin{cases} 1 \text{ where } \mu_\eta{}^{\xi} \leq \mu_\eta \leq \mu_\eta{}^{\xi+1} \\ 0 \text{ elsewhere} \end{cases}$$

and

$$\sum_{1}^{\xi} A_\eta{}^\xi = 1$$

for $\eta = i, j, k;$ $\xi = \alpha, \beta, \gamma;$ and
$i = 1, 2, 3, \ldots m;$ $j = 1, 2, 3, \ldots n;$ $k = 1, 2, 3, \ldots p.$

Symbols in Equations 8–16 through 8–19 are self-explanatory, when referring to Equations 8–11, 8–12, and 8–13. Symbols q_i, q_j, and q_k refer to the value of q when it is related to mine, smelter, and refinery, respectively.

The above sets of cost equations and the restrictions are used to transform the mathematical program in Fig. 8–8 into the mixed-integer program in Fig. 8–9. There are two major differences between Figures 8–8 and 8–9:

- Production of each facility in Fig. 8–9 is confined to a particular region of capacity utilization rate. Each capacity utilization rate is associated with a unit cost of production, which is in turn to be used in the objective function.
- There are additional constraints (Equation set D-1-15) to make sure that only one capacity utilization rate is used at one facility at one time.

DATA PREPARATION

Data preparation includes data acquisition and transformation. Data acquisition is defined as the obtaining of necessary data from various data sources, and data transformation is the estimating of parameters for the equations in an energy conservation model by using the data acquired.

A discussion of the coefficients that need to be estimated in an energy conservation model is followed by details on how to obtain and transform the data required.

Coefficients Needed in an Energy Conservation Model

Basically there are two types of coefficients needed in an energy conservation model, unit production cost and unit energy requirement of any given possible production paths, shown in Fig. 8–9.

From Fig. 8–9, the unit cost of the production path via the ith mine-mill, j^{th} smelter, and k^{th} refinery is c_{ijk} as defined in Equation 8–15, repeated here:

$$c_{ijk} = c_{iM} + c_{iF} + T_{ij} + c_{jS} + T_{jk} + c_{kR} \qquad (8\text{–}15)$$

Minimize

$$C_T = \sum_{i=1}^{5} \sum_{j=1}^{13} \sum_{k=1}^{3} c_{ijk} \cdot X_{ijk} \tag{D-1-1}$$

Subject to

$$A_1{}^{\alpha} \cdot \phi_1{}^{\alpha} \cdot 283.3 \leq \sum_{j=1}^{12} \sum_{k=1}^{3} X_{1jk} \leq 283.3 \cdot \phi_1{}^{\alpha+1} \cdot A_1{}^{\alpha} \tag{D-1-2}$$

$$A_2{}^{\alpha} \cdot \phi_2{}^{\alpha} \cdot 59.8 \leq \sum_{j=1}^{12} \sum_{k=1}^{3} X_{2jk} \leq 59.8 \cdot \phi_2{}^{\alpha+1} \cdot A_2{}^{\alpha} \tag{D-1-3}$$

Mining–Milling
Capacity
Utilization

$$A_3{}^{\alpha} \cdot \phi_3{}^{\alpha} \cdot 36.2 \leq \sum_{j=1}^{12} \sum_{k=1}^{3} X_{3jk} \leq 36.2 \cdot \phi_3{}^{\alpha+1} \cdot A_3{}^{\alpha} \tag{D-1-4}$$

$$A_4{}^{\alpha} \cdot \phi_4{}^{\alpha} \cdot 749.7 \leq \sum_{j=1}^{12} \sum_{k=1}^{3} X_{4jk} \leq 749.7 \cdot \phi_4{}^{\alpha} \cdot A_4{}^{\alpha} \tag{D-1-5}$$

for $\alpha = 1, 2, 3, 4, 5, 6, 7, 8$

$$A_1{}^{\beta} \cdot \phi_1{}^{\beta} \cdot 108 \leq \sum_{j=1,5,9} \sum_{i=1}^{5} \sum_{k=1}^{3} X_{ijk} \leq 108 \cdot \phi_1{}^{\beta+1} \cdot A_1{}^{\beta} \tag{D-1-6}$$

$$A_2{}^{\beta} \cdot \phi_2{}^{\beta} \cdot 108 \leq \sum_{j=2,6,10} \sum_{i=1}^{5} \sum_{k=1}^{3} X_{ijk} \leq 108 \cdot \phi_2{}^{\beta+1} \cdot A_2{}^{\beta} \tag{D-1-7}$$

Smelting
Capacity
Utilization

$$A_3{}^{\beta} \cdot \phi_3{}^{\beta} \cdot 113.4 \leq \sum_{j=3,7,11} \sum_{i=1}^{5} \sum_{k=1}^{3} X_{ijk} \leq 113.4 \cdot \phi_3{}^{\beta+1} \cdot A_3{}^{\beta} \tag{D-1-8}$$

$$A_4{}^{\beta} \cdot \phi_4{}^{\beta} \cdot 270 \leq \sum_{j=4,8,12} \sum_{i=1}^{5} \sum_{k=1}^{3} X_{ijk} \leq 270 \cdot \phi_4{}^{\beta+1} \cdot A_4{}^{\beta} \tag{D-1-9}$$

for $\beta = 1, 2, 3, 4, 5, 6, 7, 8$

$$A_1{}^{\gamma} \cdot \phi_1{}^{\gamma} \cdot 186 \leq \sum_{i=1}^{5} \sum_{j=1}^{13} X_{ij1} \leq 186 \cdot \phi_1{}^{\gamma+1} \cdot A_1{}^{\gamma} \tag{D-1-10}$$

Refining
Capacity
Utilization

$$A_2{}^{\gamma} \cdot \phi_2{}^{\gamma} \cdot 276 \leq \sum_{i=1}^{5} \sum_{j=1}^{13} X_{ij2} \leq 276 \cdot \phi_2{}^{\gamma+2} \cdot A_2{}^{\gamma} \tag{D-1-11}$$

$$A_3{}^{\gamma} \cdot \phi_3{}^{\gamma} \cdot 103 \leq \sum_{i=1}^{5} \sum_{j=1}^{13} X_{ij3} \leq 103 \cdot \phi_3{}^{\gamma+3} \cdot A_3{}^{\gamma} \tag{D-1-12}$$

for $\gamma = 1, 2, 3, 4, 5, 6, 7, 8$

Total
Refined Copper
Production

$$\sum_{i=1}^{5} \sum_{j=1}^{13} \sum_{k=1}^{3} X_{ijk} = 481 \tag{D-1-13}$$

Total
Energy
Use

$$\sum_{i=1}^{5} \sum_{j=1}^{13} \sum_{k=1}^{3} e_{ijk} \cdot X_{ijk} = K * (27530) \tag{D-1-14}$$

$$
\begin{aligned}
\phi_{\eta}{}^{1} &= 25\% & \phi_{\eta}{}^{5} &= 70\% \\
\phi_{\eta}{}^{2} &= 40\% & \phi_{\eta}{}^{6} &= 80\% \\
\phi_{\eta}{}^{3} &= 50\% & \phi_{\eta}{}^{7} &= 90\% \\
\phi_{\eta}{}^{4} &= 60\% & \phi_{\eta}{}^{8} &= 100\%
\end{aligned}
\quad \text{for } \eta = i, j, k
$$

Fig. 8–9. A mixed-integer programming problem for Company A.

$$A_\eta{}^\xi = \begin{matrix} 1 \\ 0 \end{matrix} \quad \begin{matrix} \text{when} \quad \phi_\eta{}^\xi \leqslant \phi_\eta \leqslant \phi_\eta{}^{\xi+1} \\ \text{otherwise} \end{matrix} \qquad \begin{matrix} \eta = \text{i, j, k} \\ \text{for} \quad \xi = \alpha, \beta, \gamma \end{matrix}$$

$$\text{(D-1-15)}$$

$$\sum_{\alpha=1}^{8} A_i{}^\alpha = \sum_{\beta=1}^{8} A_j{}^\beta = \sum_{\gamma=1}^{8} A_k{}^\gamma = 1 \quad \text{for} \quad \begin{matrix} i = 1, 2, 3, 4, 5 \\ j = 1, 2, \ldots 13 \\ k = 1, 2, 3 \end{matrix}$$

$$c_{ijk} = \sum_{\alpha=1}^{8} (c_{iM}{}^\alpha + c_{iF}{}^\alpha) \cdot A_i{}^\alpha + \sum_{\beta=1}^{8} c_j{}^\beta \cdot A_j{}^\beta + \sum_{\gamma=1}^{8} c_k{}^\gamma \cdot A_k{}^\gamma \quad \text{for} \quad \begin{matrix} i = 1, 2, 3, 4, 5 \\ j = 1, 2, \ldots 13 \\ k = 1, 2, 3 \end{matrix} \qquad \text{(D-1-16)}$$

$$A_i{}^\alpha, A_j{}^\beta, A_k{}^\gamma = 1, \text{ or } 0$$

$$X_{ijk} \geqslant 0 \quad \text{for} \quad \begin{matrix} i = 1, 2, 3, 4, 5 \\ j = 1, 2, \ldots 13 \\ k = 1, 2, 3 \\ \alpha, \beta, \gamma = 1, 2, \ldots \end{matrix}$$

where C_T = total cost

$L_\eta{}^\xi$, $U_\eta{}^\xi$ = lower and upper bounds at i^{th} facility and ξ^{th} capacity utilization rate:

$L_\eta{}^\xi = A_\eta{}^\xi \cdot u_\eta{}^\xi \cdot q_\eta$ and $U_\eta{}^\xi = A_\eta{}^\xi \cdot u_\eta{}^\xi \cdot q_\eta$

$\phi_\eta{}^\xi$ = capacity utilization rate

X_{ijk} = production quantity in thousand short tons mined at i^{th} mine, smelted at j^{th} smelter, and refined at k^{th} refinery

e_{ijk} = unit energy use in billion Btu for a thousand tons of copper mined at i^{th} mine, smelted at j^{th} smelter, and refined at k^{th} refinery

$c_k{}^\gamma$, $c_j{}^\beta$, $c_{iF}{}^\alpha$, $C_{iM}{}^\alpha$, and $c_{ijk}{}^{\alpha\beta\gamma}$ = in unit of thousand dollars; its meaning was defined in Equation 8–15

$(1 - K)$ = a fraction, for energy conservation

Fig. 8–9.   A mixed-integer programming problem for Company A *(continued)*.

where c_{iM} = the ith mine's production cost function, c_{iF} = the ith mill's production cost function, c_{jS} = the jth smelter production cost function, c_{kR} = the kth refinery production cost function, and T_{ij}, T_{ik} = unit transportation cost between facilities (to be defined in Equation 8–27).

The unit energy requirement of the production path via the ith mine–mill, j^{th} smelter, and k^{th} refinery is symbolized as e_{ijk} and expressed as follows:

$$e_{ijk} = e_{iM} + e_{iF} + TE_{ij} + e_{jS} + TE_{jk} + e_{kR} \qquad (8\text{–}20)$$

where $e_{im} =$ unit energy requirement in Btu per ton of refined copper produced at the ith mine, $e_{iF} =$ unit energy requirement in Btu per ton of refined copper produced at the ith mill, $e_{jS} =$ unit energy requirement in Btu per ton of refined copper produced at the jth smelter, $e_{kR} =$ unit energy requirement in Btu per ton of refined copper produced at the kth refinery, and TE_{ij}, $TE_{jk} =$ unit transportation energy requirements between facilities (to be defined in Equation 8–28).

In order to estimate the above two types of coefficients, the following data are required for each production facility:

- Unit production cost and unit energy requirement for existing facilities
- Unit production cost and unit energy requirement for conservation options
- Unit transportation cost and transportation energy requirement

Unit Production Costs and Unit Energy Requirements for Existing Facilities

Unit production costs and energy requirements for existing facilities can be calculated through two approaches. The first approach is an actual survey of facilities to determine their unit production costs and unit energy requirement. The second is an estimation method using published standard engineering capacity cost data, which are adjusted with capacity utilization rates and ages.

The actual survey approach is the more accurate of the two, but it is far more expensive and difficult in practice. The estimation approach is accurate enough for an energy conservation model because the model compares only relative merits of existing production paths and paths with alternative conservation options. The more important advantage for the estimation method is that it is much easier and less expensive.

Unit Production Costs for Existing Facilities. For convenience, a unit production cost function is used to generate the unit production cost for a given facility at a given capacity utilization rate. A *unit production cost function* is defined as a relationship between a facility's output (or input), expressed in tons of copper, and incurred unit cost, expressed in dollars per ton of copper produced for various levels of production. To accomplish this task we will use the *capacity production cost matrices* that were collected by Silverman[1] for mining, flotation, smelting, and refining. These matrices show the relationships between production factor costs such as electricity, fuel, natural gas, labor, material, and capital cost and the output (or input in flotation) capacity. (See Tables 8–2 through 8–5.)

Table 8–2.   Annual Surface Mining Production Factor Cost Matrix (in 1970 Dollars).

Output Capacity (In 1000 Tons of Ore/Yr), q^0	Electricity E_M^v	Fuel F_M^v	Production Factor Costs (in 1000's/Yr)				
			Material		Labor		Capital* K_M^f
			Variable M_M^v	Fixed M_M^f	Variable L_M^v	Fixed L_M^f	
Col. 1	Col. 2	Col. 3	Col. 4	Col. 5	Col. 6	Col. 7	Col. 8
10,710	34,785	161,759	1,391,421	35,677	1,203,807	300,951	1,355,004
14,280	65,973	207,295	1,691,624	43,374	1,512,079	378,019	1,333,962
16,065	100,759	235,887	1,595,391	40,907	1,256,239	314,059	1,382,366
28,560	95,961	385,199	3,177,718	81,479	2,280,399	570,099	4,013,650
35,700	179,627	539,799	3,422,345	87,752	3,166,318	791,579	3,089,268
53,550	149,939	966,494	6,531,421	167,472	3,486,798	871,699	4,839,492

* Annual allocated capital cost is assumed 14% of the total capital cost, in which 7% is for depreciation allowance and 7% for interest payments. (Assumed life span of the equipment in mining is 15 years.)

Source: Silverman.[1]

Table 8–3. Annual Flotation Concentrating Production Factor Cost Matrix (in 1970 Dollars).

| Input Capacity (In 1000 Tons of Ore/Yr), q^0 | Electricity E_F^v | Gas G_F^v | Material | | Labor | | Capital* K_F^f |
| | | | Variable M_F^v | Fixed M_V^f | Variable L_F^v | Fixed L_F^f | |
Col. 1	Col. 2	Col. 3	Col. 4	Col. 5	Col. 6	Col. 7	Col. 8
1,785	239,303	17,992	852,620	63,978	1,048,134	184,964	1,609,440
2,140	317,272	29,987	1,214,561	91,137	897,004	158,294	2,285,112
5,355	761,693	149,939	1,664,033	124,864	1,811,859	319,739	3,486,744
8,210	1,001,597	179,927	3,029,752	227,345	2,456,243	433,454	6,132,456
14,280	2,081,163	119,951	4,512,488	338,605	3,649,981	644,114	9,569,052
25,700	2,746,895	419,831	8,323,792	624,596	5,415,683	955,709	14,481,876

Production Factor Costs (in 1000's/Yr)

*Annual allocated capital cost is assumed 12% of the total cost, in which 5% is for depreciation allowance and 7% for interest payments. (Assumed life span of the equipment is 20 years.)

Source: Silverman.[1]

Table 8-4. Annual Smelting Production Factor Cost Matrix (in 1970 Dollars).

Output Capacity (In 1000 Tons of Copper/Yr), q	Electricity E_S^v	Gas G_S^v	Production Factor Costs (in 1000's/Yr)				Capital* K_S^f
			Material		Labor		
			Variable M_S^v	Fixed M_S^f	Variable L_S^v	Fixed L_S^f	
Col. 1	Col. 2	Col. 3	Col. 4	Col. 5	Col. 6	Col. 7	Col. 8
86	252,244	1,601,249	1,099,679	164,319	2,508,716	927,881	3,912,720
100	296,589	1,712,066	1,156,229	172,769	2,628,509	972,188	4,485,816
152	254,477	1,582,055	1,631,336	243,762	4,351,960	1,609,630	5,532,900
268	415,883	3,772,788	1,725,209	257,789	7,536,515	2,787,478	6,981,680
442	590,161	6,874,484	2,571,718	384,279	13,732,671	5,079,203	9,334,560
1,822	1,564,812	33,759,585	4,242,808	633,983	58,543,738	21,653,160	19,379,640

* See note in Table 8-3.

Source: Silverman.[1]

Table 8–5. Annual Electrolytic Refining Production Factor Cost Matrix (in 1970 Dollars).

| Output Capacity (In 1000 Tons of Copper/Yr), q | Electricity E_R^v | Gas G_R^v | Material M_R^v | Labor | | Capital* K_P^f |
| | | | | Variable L_R^v | Fixed L_R^f | |
Col. 1	Col. 2	Col. 3	Col. 4	Col. 5	Col. 6	Col. 7
100	229,809	182,979	609,299	1,973,329	886,568	2,763,516
163	459,619	397,599	806,099	3,718,890	1,670,805	4,324,200
300	1,103,758	835,177	1,201,899	7,288,187	3,274,403	6,842,916
442	1,604,229	1,320,199	1,329,082	15,080,632	6,775,355	8,707,860
1,203	5,459,289	4,169,195	2,084,399	64,291,324	28,884,507	19,337,964
1,876	9,587,593	4,286,796	2,421,699	100,828,875	45,299,953	23,670,360

Production Factor Costs (in 1000's/Yr)

* See note in Table 8–3.

Source: Silverman.[1]

Procedures. The following steps were taken to develop the unit production cost function for each facility:

1. A correlation analysis was made between each production factor cost column and the output (or input) capacity (Col. 1) of Tables 8–2 through 8–5. These relationships are called *capacity production factor cost functions*, and are shown in Tables 8–6 through 8–9 for mining, flotation, smelting, and refining, respectively. These relationships also include a price index to take into account the change in value of the dollar: I_E is the price index for electricity, I_F the price index of fuel, I_G the price index for natural gas, I_L the price index for labor, I_M the price index for material, and I_K the price index for capital. Price indices (1970 = 1) for prices of electricity, fuel, gas, labor, material, and capital may be obtained from the *Minerals Yearbook*,[2] or other data sources.[3,4] For use in this example, these prices are raised to 1974 dollars by 1974 price indices.

2. The unit production cost function for each facility can be obtained through the following general equation:

$$c_u = C_c{}^v/q + (C_c{}^f/q)/u \qquad (8\text{–}21)$$

where c_u is the unit cost of production factors in dollars per ton of refined copper produced at the utilization rate of u (expressed in a fraction of the

Table 8–6. Capacity Production Factor Cost Functions in Surface Mining Process (Costs in Thousands of 1970 Dollars).

Factor	Equations Annual Output Expressed in 1000 Tons of Ore, q^0	Annual Output Expressed in 1000 Tons of Copper,* q	Coefficient of Determination, R^2
Electricity (variable), $E_M{}^v$	$E_M{}^v = .0315(q^0)^{.794} \cdot I_E$†	$E_M{}^v = .0515(q/\alpha)^{.794} \cdot I_E$	.80
Fuel (variable), $F_M{}^v$	$F_M{}^v = 0.2(q^0)^{.969} \cdot I_F$	$F_M{}^v = .02(q/\alpha)^{.969} \cdot I_F$	.96
Labor			
Variable, $L_M{}^v$	$L_M{}^v = 2.5306(q^0)^{.662} \cdot I_L$	$L_M{}^v = 2.5306(q/\alpha)^{.662} \cdot I_L$	.93
Fixed, $L_M{}^f$	$L_M{}^f = .06326(q^0)^{.662} \cdot I_L$	$L_M{}^f = .06326(q/\alpha)^{.662} \cdot I_L$	.93
Material			
Variable, $M_M{}^v$	$M_M{}^v = .1848(q^0)^{.95} \cdot I_M$	$M_M{}^v = .1848(q/\alpha)^{.95} \cdot I_M$	.97
Fixed, $M_M{}^f$	$M_M{}^f = .0047(q^0)^{.95} \cdot I_M$	$M_M{}^f = .0047(q/\alpha)^{.95} \cdot I_M$	.97
Capital (fixed), $K_M{}^f$	$K_M{}^f = .1572(q^0)^{.957} \cdot I_K$	$K_M{}^f = .1572(q/\alpha)^{.97} \cdot I_K$	.92

* α in the equations is ore grade of individual mines expressed in fractions.
† I_E is a price index (I) for electricity (E); other price indices are similarly expressed.

Table 8–7. Capacity Production Factor Cost Functions in Flotation Process (Costs in Thousands of 1970 Dollars).

	Equations		Coefficient of Determination, R^2
Factor	Annual Input Expressed in 1000 Tons of Ore, q^0	Annual Input Expressed in 1000 Tons of Copper,* q	
Electricity (variable), $E_F{}^v$	$E_F{}^v = .2471(q^0)^{.929} \cdot I_E$	$E_F v = .2471(q/\alpha)^{.929} \cdot I_E$	.99
Gas (variable), $G_F{}^v$	$G_F{}^v = .0065(q^0)^{1.04} \cdot I_G$	$G_F{}^v = .0065(q/\alpha)^{1.04} \cdot I_G$	.85
Labor			
Variable, $L_F{}^v$	$L_F{}^v = 6.6379(q^0)^{.663} \cdot I_L$	$L_F{}^v = 6.6379(q/\alpha)^{.663} \cdot I_L$	.98
Fixed, $L_F{}^f$	$L_F{}^f = 1.1184(q^0)^{.663} \cdot I_L$	$L_F{}^f = 1.1184(q/\alpha)^{.663} \cdot I_L$	.98
Material			
Variable, $M_F{}^v$	$M_F{}^v = 2.1070(q^0)^{.805} \cdot I_M$	$M_F{}^v = 2.1070(q/\alpha)^{.805} \cdot I_M$	.97
Fixed, $M_F{}^f$	$M_F{}^f = .1581(q^0)^{.805} \cdot I_M$	$M_F{}^f = .1581(q/\alpha)^{.805} \cdot I_M$	.97
Capital (Fixed), $K_F{}^f$	$K_F{}^f = 4.2087(q^0)^{.803} \cdot I_K$	$K_F{}^f = 4.208(q/\alpha)^{.803} \cdot I_K$	.98

* See notes in Table 8–6.

Table 8–8. Capacity Production Factor Cost Functions in Smelting Process (Costs in Thousands of 1970 Dollars).

	Equation	Coefficient of Determination, R^2
Factor	Annual Output Expressed in 1000 Tons of Copper,* q	
Electricity (variable), $E_S{}^v$	$E_S{}^v = (15.5491 q^{.603}) \cdot I_E$	.96
Gas (variable), $G_S{}^v$	$G_S{}^v = (7.4992 q^{1.042}) \cdot I_G$	.97
Labor		
Variable, $L_S{}^v$	$L_S{}^v = (2.18084 q^{1.052}) \cdot I_L$	.999
Fixed, $L_S{}^f$	$L_S{}^f = (8.0662 q^{1.052}) \cdot I_L$	.999
Material,		
Variable, $M_S{}^v$	$M_S{}^v = (156.99198 q^{.444}) \cdot I_M$	.975
Fixed, $M_S{}^f$	$M_S{}^f = (23.4584 q^{.444}) \cdot I_M$	.975
Capital (fixed), $K_S{}^f$	$K_S{}^f = (3384.38 q^{.521}) \cdot I_K$	.996

* See notes in Table 8–6.

Table 8-9. Capacity Production Factor Cost Functions in Electrolytic Refining Process (Costs in Thousands of 1970 Dollars).

Factor	Equation Annual Output Expressed in 1000 Tons of Copper,* q	Coefficient of Determination, R^2
Electricity (variable), $E_R{}^v$	$E_R{}^v = (.7658q^{1.225}) \cdot I_E$	.998
Gas (variable), $G_R{}^v$	$G_R{}^v = (.8553q^{1.102}) \cdot I_G$	.997
Labor		
Variable, $L_R{}^v$	$L_R{}^v = (3.2848q^{1.38}) \cdot I_L$	.997
Fixed, $L_R{}^f$	$L_R{}^f = (1.4758q^{1.38}) \cdot I_L$	
Material, (variable), $M_R{}^v$	$M_R{}^v = (75.9579q^{.466}) \cdot I_M$	.987
Capital (fixed), $K_R{}^f$	$K_R{}^f = (98.001q^{.736}) \cdot I_K$	.996

* See notes in Table 8–6.

capacity of the facility). $C_c{}^V$ is the cost of variable production factors (in terms of dollars) used for production at full capacity, estimated by substituting the value of a given plant capacity in the set of equations in Table 8–6 through 8–9. $C_c{}^f$ is the cost of fixed production factors (in terms of dollars) used for production at full capacity estimated in the same manner as $C_c{}^V$. Quantity q is the annual capacity of plants expressed in tons of refined copper.

An Example. An example is given below to demonstrate the computation of a unit production cost function for a given mine.

We assume that the ore grade of a mine is .861% and its capacity is 59.8 thousand short tons of ore. Price indices of the year 1975 for electricity, fuel, labor, material, and capital (major equipment) were obtained from the *Minerals Yearbook*[2] and other sources.[3,4] Substituting the above information into the appropriate set of equations in Table 8–6 yields:

$$c_u = 4477.23/59.8 + (1136.2/59.8)/u$$

The above tedious computation was done by computer. This method was applied to obtain facility production cost functions for all facilities for mining, flotation, smelting, and refining. Any given utilization rate will result in a unit production cost at that rate.

Unit Energy Requirements for Existing Facilities. The unit energy requirement for a facility is expressed in Btu per ton of refined copper produced. It is determined by (1) the nature of the process involved—mining, flotation, smelting, or refining process (production stage); (2) the copper content of the input to the process; and (3) the capacity utilization rate (i.e., the percent of nominal capacity used).

Examination of standard engineering data for copper processes as shown in Tables 8–10 through 8–13 indicates that energy is considered to be a variable cost in copper production. Consequently, we can assume that the energy requirement in production processes (not including space heating and cooling, and lighting) is independent of capacity utilization rates. In order to compute unit energy requirements, the following information is needed:

- Energy requirement at various production stages (e.g., mining)
- Adjustment for the copper content at each production stage

Table 8–10. Surface Mining Energy Requirement for Various Capacities.

Silverman's Data			Converted Data Used for Regression	
Output (Tons/Day of Ore)	Electricity* (kWh/Yr)	Fuel* (Gal/Yr)	Output Capacity (1000 Tons/Yr of Ore)	Total Energy Required (Billion Btu/Yr)
30,000	4,969,400	1,010,988	10,710	191.10
40,000	9,424,800	1,295,600	14,280	276.19
45,000	14,394,240	1,474,300	16,065	351.85
80,000	13,708,800	2,407,500	28,560	474.26
100,000	25,661,160	3,367,500	35,700	729.78
150,000	21,420,000	6,040,600	53,550	1,057.09
180,000	27,417,600	4,621,800	64,260	921.73

* These are variable costs.

Capacity Energy Requirement Function. The energy requirement at various production stages (processes) is indicated by a capacity energy requirement function, which is defined as a relationship between the energy required when the plant is operated at the full capacity level and the capacity expressed in tons of ore, products-in-process, or refined copper, as the case may be. Point data for energy requirements at various capacity levels are collected and shown in Table 8–10 through 8–13. It should be noted that there are some small amounts of coal, coke, heavy fuel, or other forms of energy also used in industry that are not shown in these tables.

These data points are then used to estimate a capacity energy requirement

Table 8–11. Flotation Concentrating Energy Requirement for Various Plant Capacities.

Silverman's Data			Converted Data Used for Regression	
Input (Tons/Day of Ore)	Electricity* (kWh/Yr)	Fuel* (Mcf/Yr)	Input Capacity (1000 Tons/Yr of Ore)	Total Energy Required (Billion Btu/Yr)
5,000	34,186,320	25,704	1,785	376.53
6,000	45,324,720	42,840	2,140	508.25
15,000	108,813,600	214,200	5,355	1,334.98
23,000	143,085,600	257,040	8,210	1,730.06
40,000	297,309,600	171,360	14,280	3,220.83
72,000	392,414,400	599,760	25,700	4,636.28

* These are variable costs.

Table 8–12. Smelting Energy Requirement* for Various Plant Capacities.

Silverman's Data			Converted Data Used for Regression	
Output (Tons/Yr of Copper)	Electricity† (kWh/Yr)	Fuel† (Mcf/Yr)	Output Capacity (1000 Tons/Yr of Copper)	Total Energy Required (Billion Btu/Yr)
86,345	36,035,000	2,287,500	86.00	2,727.37
100,400	42,370,000	2,445,811	100.00	2,955.46
152,125	36,354,000	2,260,080	152.00	2,702.37
268,000	59,412,000	5,389,700	268.00	6,165.10
442,400	84,309,000	9,820,700	442.00	10,988.38
1,821,778	223,545,000	49,228,000	1,822.00	52,011.95

* Includes costs of operating an acid plant (SA or DA) without by-product credits.
† These are variable costs.

Table 8–13. Electrolytic Refining Energy Requirement for Various Plant Capacities.

Silverman's Data			Converted Data Used for Regression	
Output (Tons/Yr of Copper)	Electricity* (kWh/Yr)	Fuel* (Mcf/Yr)	Output Capacity (1000 Tons/Yr of Copper)	Total Energy Required (Billion Btu/Yr)
100,000	32,830,000	261,400	100	605.65
163,000	65,660,000	568,000	163	1,257.90
300,000	157,680,000	1,193,112	300	2,844.58
442,000	229,176,000	1,886,000	442	4,291.00
1,203,000	779,900,000	5,956,000	1,203	14,126.03
1,875,992	1,369,659,000	6,124,000	1,876	20,337.78

* These are variable costs.

function for each of the four production processes. Ordinary least-square schemes are used to approximate the following general functional form:

$$E_j = \alpha q_j^{-\beta} \qquad (8\text{–}22)$$

where E_j is the capacity energy requirement of the jth production process at the capacity level q, and α and β are estimated coefficients. The results are shown in Table 8–14.

Adjustment for the Copper Content. The two equations (Equations 1 and 2) for mining and flotation in Table 8–14 are expressed in billion Btu/1000 tons of ore. For consistency with the rest of the equations, these two equations should be converted into the unit of billion Btu/1000 tons of refined copper.

This is accomplished by using the following equation, which incorporates the ore concentration γ_j at the jth mine into Equations 1 and 2 of Table 8–14:

$$E_j = \alpha \gamma_j^{-\beta} (\gamma_j q^0)^\beta \qquad (8\text{–}22')$$

E_j is the total energy required in billion Btu to produce an annual q^0 where q^0 is expressed in thousand short tons of ore. Since $\gamma_j q^0$ is equal to q_j expressed

Table 8–14. Capacity Energy Requirement Functions for the Four Stages of Operation (General Form of Equation: $E_j = \alpha(q_j)^\beta$).

Process		
Mining*	$E_M = \ .0436(q_M^0)^{.916}$	(1)
Flotation	$E_F = \ .362(q_F^0)^{.942}$	(2)
Smelting†	$E_S = 24.903 q_S^{1.004}$	(3)
Refining‡	$E_R = \ 2.775 q_R^{1.196}$	(4)

* Open-pit.
† Reverberatory furnace.
‡ Electrolytic refining.
Legend: E_M = energy requirements in billion Btu/1000 tons
   of ore
E_F = energy requirements in billion Btu/1000 tons
   of ore
E_S = energy requirements in billion Btu/1000 tons
   of refined copper
E_R = energy requirements in billion Btu/1000 tons
   of refined copper
q_M = quantity of ore in 1000 tons
q_F = quantity of ore in 1000 tons
q_S = quantity of refined copper in 1000 tons
q_R = quantity of refined copper in 1000 tons

in thousand short tons of refined copper, Equations 1 and 2 of Table 8–14 can be rewritten in the following functional form:

$$E_j = \alpha \gamma_j^{-\beta} q_j^{\beta} \qquad (8\text{–}23)$$

Various ore grades in various mines and mills of Company A are substituted into Equation 8–23 to obtain adjusted capacity energy requirement functions for these mines and mills, as shown in Table 8–15.

Unit Energy Requirements for the Four Production Processes. Because energy required in copper production in the four processes as shown in Tables 8–10 through 8–13 is considered to be a variable cost, the unit energy requirement (e_j) of a given production stage (j) is the dividend of capacity energy requirement (E_j) of the given stage divided by the capacity as shown in Equation 8–24:

$$
\begin{aligned}
e_j &= E_j/q_j \\
&= (\alpha q_j^{-\beta})/q_j \quad \text{(substitution of Eq. 8–22)} \\
&= \alpha q_j^{-(\beta+1)} \qquad\qquad\qquad\qquad\qquad\qquad (8\text{–}24)
\end{aligned}
$$

By using Equation 8–24, capacity data of facilities can be substituted into appropriate equations in Tables 8–14 and 8–15 to obtain unit energy required for each facility of various production processes.

Table 8–15. Capacity Energy Requirement Functions in Mining and Flotation for Company A (Billion Btu/yr).

I. Mining (Open Pit; M)

Mine

I	$E_{1M} = 2.849 q_{1M}^{.916}$
II	$E_{2M} = 3.396 q_{2M}^{.916}$
III	$E_{3M} = 3.74 q_{3M}^{.916}$
IV	$E_{4M} = 4.382 q_{4M}^{.916}$

II. Flotation (F)

Mine

I	$E_{1F} = 26.685 q_{1F}^{.942}$
II	$E_{2F} = 31.911 q_{2F}^{.942}$
III	$E_{3F} = 35.336 q_{3F}^{.942}$
IV	$E_{4F} = 41.446 q_{4F}^{.942}$

Unit Production Costs and Energy Requirements for Energy Conservation Options

Energy savings and cost changes for the three alternative energy conservation options for copper production were discussed in Chapter 7. They will be used in this section to determine unit production cost and energy requirements for production paths with energy conservation options.

In general, the unit production cost (c^T) and unit energy requirement (e^T) for a process with a given energy conservation option are expressed as in Equations 8–25 and 26:

$$c^T = c + \Delta c^T \tag{8-25}$$

$$e^T = e - \nabla e^T \tag{8-26}$$

where c and e are the unit production cost and energy requirement of the original process before conservation, and Δc^T and ∇e^T are incremental production cost and unit energy savings of a given conservation option. The c and e in Equations 8–25 and 8–26 were calculated in the previous subsection, and Δc^T and ∇e^T have been discussed in Chapter 7 and are summarized in Table 8–16.

Table 8–16. Summary of Energy Savings and Cost Increase of Implementing Various Conservation Technologies in Primary Refined Copper Production.

Conservation Technologies*	Energy Savings	Cost Increases**
Recycling (using scrap)		
Low-grade scrap	$e_M + e_F$	$Y_{2S} - (C_M - C_F)$
High-grade scrap	$e_M + e_F + e_S$	$Y_2 - (C_M + C_F + C_S)$
Retrofitting (oxygen plants in smelters)	.07 e_S	\$1.42/ton of copper
New process (flash smelting)	.6 e_S	.07 C_S

where: e_M, e_F, and e_S are unit energy requirements for mining, flotation, and smelting, respectively; C_M, C_F, and C_S are unit production costs for mining, flotation, and smelting, respectively; and Y_2 and Y_{2S} are prices for No. 2 heavy (high-grade) and No. 1 composition (low-grade) scraps, respectively.

* Detailed definition of these technologies are discussed in Chapter 7.

** In constant 1974 price structure.

Transportation Costs and Energy Requirements between Processing Facilities

The transportation cost per ton of refined copper between the ith and j^{th} facilities is expressed as:

$$T_{ij} = ad_{ij} \tag{8-27}$$

where a is the unit cost per ton-mile of transporting copper products, and d_{ij} is the distance between the ith and j^{th} facilities. In this study, a is assumed to be \$0.03/ton-mile.

The energy requirement per ton of copper transported between the ith and j^{th} facilities is expressed as:

$$TE_{ij} = fd_{ij} \tag{8–28}$$

where f is the unit energy requirement per ton-mile of transporting copper products, and d_{ij} is the distance between the ith and j^{th} facilities. However, in this study, the energy requirement for transporting products between mills and smelters or between smelters and refineries is excluded partially because these energy requirements in transporting products are insignificant in quantity when compared with unit energy consumption in copper production, and partially because most of this energy is consumed by public transporters (e.g., truckers).

MODEL EXERCISE AND INTERPRETATION OF MODEL RESULTS

Model exercise is defined as the process of inputting numerical exogenous data to a model such as an energy conservation model in order to obtain numerical results regarding policies. Interpretation of model results is defined as the process of relating and comparing to the real world the results from model exercises regarding a given policy.

Model Exercise

Depending on the characteristics of the input data, two primary methods for model exercise are used by most modelers. The more common of the two is deterministic simulation, which uses point input data to generate model results. The second method is probabilistic simulation, in which a set of random numbers is used to generate input data, and the results of models are expected values. The deterministic simulation is easier to run and to understand, but its results may not be predictive because of its use of point estimates of input data. On the other hand, the probabilistic simulation, mainly a Monte Carlo simulation, is more difficult in practice, and it is hard to explain the method's results, but it provides an expected outcome of future events.

Procedure Used to Run a Deterministic Simulation. Exercising an energy conservation model deterministically involves two activities: prepa-

ration of data for use in a computer linear programming (LP) package, and use of a computer LP package to solve the LP problem in the conservation model. The procedure used to run a deterministic simulation of an energy conservation model consists of eight steps, as shown in Fig. 8–10. The preparation of data consists of steps 1 through 5, and the use of a computer package includes steps 6 through 8. These steps are as follows:

- Step 1: Forecast a set of price indices for all production factors, which were shown in Tables 8–6 through 8–9 (1970 price index = 1).
- Step 2: Update unit production costs for all facilities and alternative conservation options.

The original unit production costs of all facilities and alternative conservation options discussed in the previous section are adjusted into a given year's dollars by that year's price indices (1970 prices = 1). Price indices predicted in step 1 are used to update these costs to the desired year's dollar values.

- Step 3: Estimate an optimal capacity utilization rate for production quantity in each facility. One may start from the neighborhood of the facility's present capacity utilization rate, if this information is available.

Because an energy conservation model usually contains a mixed-integer program (MIP) problem when avoiding nonlinearity in unit cost functions (see Fig. 8–9), a convenient way to solve this MIP problem is to employ a trial-and-error routine. In this routine, an optimal utilization rate is estimated for every facility, and then the MIP problem is transformed into an LP problem. Benders and others also suggested similar approaches.[5]

Note that when each facility's optimal utilization rate is estimated, summation of production quantities from all plants under this set of conjectured capacity utilization rates should be greater than or equal to the required total refined copper production.

- Step 4: Calculate unit production cost (c_{ijk}) and energy requirement (e_{ijk}) for each production path and corresponding valid upper and lower boundaries for each facility's production at the conjectured capacity utilization. Equations to calculate these data were discussed in the previous section.
- Step 5: The MIP becomes an LP problem. Prepare a coefficient matrix to be used in the IBM MPS/360 package.

This step prepares a coefficient matrix that can be accepted by the IBM MPS/360 package. The package is used to solve LP problems. Appendix B

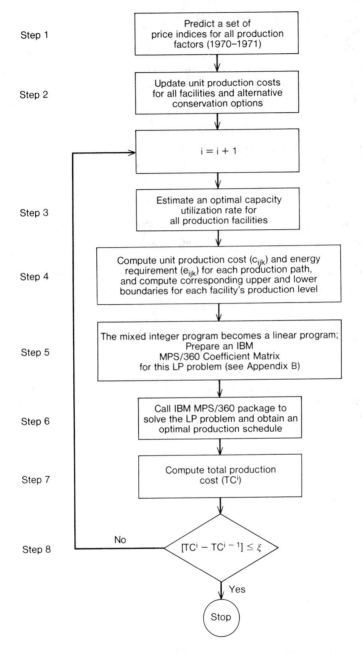

Step 1 — Predict a set of price indices for all production factors (1970–1971)

Step 2 — Update unit production costs for all facilities and alternative conservation options

$i = i + 1$

Step 3 — Estimate an optimal capacity utilization rate for all production facilities

Step 4 — Compute unit production cost (c_{ijk}) and energy requirement (e_{ijk}) for each production path, and compute corresponding upper and lower boundaries for each facility's production level

Step 5 — The mixed integer program becomes a linear program; Prepare an IBM MPS/360 Coefficient Matrix for this LP problem (see Appendix B)

Step 6 — Call IBM MPS/360 package to solve the LP problem and obtain an optimal production schedule

Step 7 — Compute total production cost (TC^i)

Step 8 — No — $[TC^i - TC^{i-1}] \leq \xi$

Yes

Stop

Fig. 8–10. Flow chart for exercising an energy conservation model.

discusses a computer routine for the preparation of this coefficient matrix. In using this routine to prepare a matrix, all constraint equations in the LP problem must be in the form of $ax \leq b$; that is, the left-hand side of an equation must be smaller than or equal to the right-hand side, the constant. Any equations not in this form must be converted into this form (For a detailed method, see the discussion in Appendix B.)

- Step 6: Call the IBM MPS/360 package to solve the above LP problem, and obtain the company's optimal production scheme and total cost (TC^i). The LP subroutine to call this package is shown in Appendix B.
- Step 7: Repeat steps 3 to 6 to generate other sets of production patterns and total cost (TC^{i+1}).
- Step 8: Stop looping steps 3 to 6, if $TC^i - TC^{i-1} \leq \epsilon$, where ϵ is a small number considered to be acceptable.

In the algorithm of solving the mixed-integer programs in Fig. 8–9, two or three iterations usually are needed to generate a set of reasonable optimal production patterns.

Procedure Used to Run a Monte Carlo Simulation. As noted earlier, a Monte Carlo procedure is generally used to run a probabilistic simulation. A Monte Carlo simulation is defined as an inductive technique that employs random processes to infer certain properties from observed behavior in a controlled setting.[6]

Using a random number generator and a controlled setting with reasonable ranges of production factor price, the Monte Carlo procedure is used to project price indices for production factors involved in energy conservation models for industries. Clearly, the major difference between this procedure and the previous one (a deterministic simulation) is that the price indices in this procedure are not fixed point estimates; here they are determined by controlled ranges and random processes.

The procedure consists of the following 11 steps:

- Step 1: Estimate the range of price indices for all production factors, and assign each data point in this range a set of random numbers. For example, if the range of the natural gas price index is between 100% and 200%, and the random number is between 1 and 1000, then the assigned price index of natural gas is 100% if the outcome of the random number generation is between 1 and 10, 101% if the outcome is between 11 and 20, etc.

- Step 2: Draw random numbers from a random number generator, and determine price indices for all production factors.
- Steps 3–9: These are the same as steps 2–8 in the deterministic simulation (shown in Fig. 8–10).
- Step 10: Repeat steps 2–9 as many times as desired.
- Step 11: Average results of outcomes of the model from step 10.

We obtain the expected results of the model from a Monte Carlo simulation. The disadvantage of this simulation is that it is laborious, and it can be so complicated that it is difficult to interpret the outcome of a simulation.

Interpretation of Model Results

The interpretation of energy conservation model results is defined as the way the results of models are arranged to indicate implications of a given policy. For example, in Chapter 7 we indicated that under the 1974 price structure for energy and nonenergy production factors, the alternative conservation measures examined were not economically favorable. If the Department of Energy or the Congress were to impose a certain percentage reduction of current energy consumption as a mandatory energy conservation target, what would be the trade-offs between cost changes and conservation targets (percentage reduction)? Furthermore, what kind of energy conservation options would be implemented?

Detailed answers to the above questions for both individual firms and the nation will be discussed in Chapters 10 and 11. This subsection presents a brief illustration of interpretation of results from the copper energy conservation model for Company A. Table 8–17 summarizes the model results under three (5%, 10%, 15% energy reduction) conservation targets. The implication for a mandatory conservation target program is that, under the 1974 price structure for energy and nonenergy production factors in primary copper production, conservation will slightly increase production costs for Company A, estimated at less than 0.1% per 1% of energy saved. It can be further interpreted that conservation will slightly increase unit refined copper production costs, estimated at less than .4¢/lb in the 1974 price structure.

The interpretation of model results for other policies such as energy tax credits is discussed in detail in Chapter 11.

MODEL VALIDATION AND VERIFICATION

Model validation and model verification are interchangeable terms in most cases. However, in a strict sense, model validation includes model verification.

Table 8–17. Economics of Energy Conservation for Primary Copper Production for Company A.

10^9 Btu	% of Original Consumption	Conservation Methods Implemented† (Production in 10^3 Short Tons)			Cost Increase, ΔC (10^6)	% of Original Cost	Average Cost Increase ($\Delta C/\nabla E$: $/10^6$ Btu)
1.376	5	(A) 7.2	(B) 189.4	(C) 292.5	2.544	.51	1.85
2.753	10	(A) 7.2	(B) 293.0	(C) 188.9	4.033	.81	1.46
4.129	15	(A) 7.2	(B) 396.6	(C) 85.3	5.523	1.12	1.34

† (A)—scrap; (B)—flash smelting; and (C)—tonnage oxygen.

Model validation is defined as any activities taken to make sure that the model represents the functions of the entity from which it is abstracted. Therefore, the validation consists of examining model formulation (model structure, underlying assumptions, data bases, and computer coding) and testing model predictive creditability (accuracy and sensitivity). Model verification may be limited to examination of computer coding, data bases, and model structure. Here, the broad definition of model validation is used. Theoretically, the purpose of model validation is to prove that the model under study is valid in any aspect.

In practice there is no need and almost no way to prove that a model is absolutely valid; however, one should ensure that the model is statistically sound and valid for the entity it represents. Thus, to make validation a meaningful and useful task, its scope must be limited to the specific needs and requirements of the user and, ultimately, to the performance requirement of the end use of the model output. The overall assessment of the validity of a model should therefore be made with respect to the user's need and priorities, just as the model should be constructed to fill the user's requirements.

Methodology for Model Validation and Verification

The methodology for validating a model consists of a series of tests, each evaluating a different attribute of the model reflecting its confidence. These tests include criticisms of underlying economic assumptions or mathematical approximation, evaluation of the seriousness of limitations to model applicability, and, most important, determination of the statistical confidence of the projected results of the model. The tests can be categorized into two kinds of criteria:

- Model formulation: these tests seek to determine how well the design of the model approximates the course of the entity's behavior that had been expected. Such tests question the underlying economic and engineering theories that constitute the backbone of the model's structure.
- Predictive creditability: these tests determine the confidence and applicability of the output regardless of the model structure. The model is considered as an input–output "black box" that is capable of predicting cause–effect behavior or output when it receives a given set of input data.

Under certain circumstances, the two types of criteria may be contradictory. Owing to some structural change in the energy and economy parameters, models that were determined to be "well-formulated" and sophisticated may turn out to be relatively poor predictors of events of interest to decision-makers; such changes must be incorporated in the improvement of models.

Validation of Model Formulation

We need to check underlying theories, data bases, and computer coding to validate the formation of a given model.

Underlying Theories. Major assumptions and statistical structures are examined to see if they are contradictory to a priori economic, mathematical, or engineering theories. Better formulas that represent the behavior of the entity should also be sought.

Two questions can be asked in examination of the validity of model formulation:

- How reasonable is the mathematical approximation used in expressing the final form of the model?
- Will the structure of the model become incorrect after several transformations?

Data Bases Prepared. Data constitute one of the most important factors in model performance. The following questions should be answered:

- Are data available?
- Are data sources reliable?
- Are data consistent?
- Is the quality of the data sufficient for use in the model under study?
- Can the accuracy of the data be verified, and the caveats of the data be observed?

The importance of data accuracy depends on the requirement of the model.

Computer Coding. Because almost every large-scale model must use computers, computer coding of the model becomes critical. Computer coding translates the model into languages that computers can understand. The following questions should be asked to verify computer coding:

- Are the computer codes faithful to the formulation of the model?
- Is the programming language the most suitable and economic choice for the model?
- Are there better ways to implement and operate the computer version of the model? Can it be truncated or partitioned?
- Is the computer capacity sufficient to operate the model?

Predictive Creditability

In examining the predictive creditability of a model, two aspects are considered—accuracy and sensitivity. Accuracy of a prediction from a model is defined as the closeness of its projection of the future event to the event itself. Sensitivity of forecasts is defined as the ability of the projection to pick up some turning points of events, such as recession and booming periods.

Accuracy of Forecasts. The criteria to validate accuracy of a model's forecasts are closeness of forecasts to actual events and identifiability of forecast errors. There is no forecast that can perfectly match actual events; however, if the sources of forecast error can be identified, the forecast is still creditable. The following questions can be asked to examine the model's predictive accuracy:

- How well does the model forecast?
- Which forecast variables are most/least reliable?
- Where are the sources of forecast errors? What is the range?
- What are the criteria to be used for evaluating accuracy of model forecasts?

Sensitivity of Forecasts. In examining sensitivity of the forecasts of a given model, the ability of the forecasts to pick up the turning points of events such as recession and booming is highlighted. This may imply also the capability of the model to capture structural changes of the entity it represents. The following questions should be asked:

- How sensitive is the model structure to changes in its basic assumptions or the data it is based on?
- How well does the model forecast the turning points of events such as recession or boom?
- What are the confidence limits of model forecasts under various scenarios of imperfect or distorted data?

Convenient Model Validity Tests

The components of model validation have been detailed above, but, for convenience, a simple index may sometimes be needed to indicate the overall effectiveness of a given model, instead of evaluation of the detailed components.

There are a few tests that serve this purpose, but it may be difficult to use only a single test for validation because results of various tests may be contradictory. Some of the more frequently used tests suitable for evaluating energy conservation models utilize comparison of forecasts with actual events, turning point errors, and comparative errors.

Comparison of Forecasts with Actual Events. This is the most commonly used method. The estimated or simulated observations of a given model are compared with actual events. In comparison, "backcasting" is used. Backcasting, an analog to forecasting, is defined as a method in which historical input data are fed into the given model to reproduce historical events. The reproduced outcomes are then compared with actual observations. If large observations are available, the following two single-point indices are used to indicate this comparison, mean absolute percentage error (%E.) and mean squared error (M.S.E.):

$$\%\,E. = \frac{1}{n} \sum_{t=1}^{n} \frac{|y_t - \hat{y}_t|}{y_t} \times 100\%$$

$$M.S.E. = \frac{1}{n} \sum_{t=1}^{n} (y_t - \hat{y}_t)^2$$

where y_t = actual observation values, and $\hat{y}_t$ = forecast or estimated values.

Turning Point Errors. As discussed earlier, the ability to pick up the turning points of fluctuations in values of model outputs is one of a model's major functions. Some statistical indices to describe turning point errors are the number of turning points missed, turning points falsely predicted, and under- and overpredictions.

Comparative Errors. A given conservation model can be validated by comparing the model's forecast errors with the forecast errors achieved from a naive or statistical model or from judgmental or other econometric forecasts. A simple autoregressive moving average model can be used to forecast; its results are then compared with the results from the given energy conservation models. The indices measuring the previous two tests can be used to indicate this comparison.

Validity of the Copper Energy Conservation Model

The method of comparison of forecasts with actual events, previously discussed, is used to test the validity of the copper energy conservation model.

The energy conservation model is used to backcast 1974 production for Companies A and B. As shown in Table 8–18, the average costs resulting from the simulations of the copper energy conservation models are almost

Table 8–18. Comparison between Average Costs Simulated by MIP and those Calculated from Annual Reports ($¢$/lb refined copper).

Company	Simulated by MIP	Calculated from 1974 Annual Reports	% Error
A	56.50	58.23	−3.0
B	68.33	67.12	1.8

identical with those calculated from 1974 annual reports from Companies A and B.

FUTURE MODEL IMPROVEMENT AND OTHER CONSIDERATIONS

The energy conservation model discussed above is a static model that does not take into account uncertainty and feedback from the previous conservation experiences. However, the model may become too complicated to be understood if both uncertainty and feedback are integrated into it; there is a trade-off between model complexity and ease of understanding. These problems are discussed below.

Uncertainty

Uncertainty regarding future prices of production factors and sales affects the effectiveness of an energy conservation model in inferring implications of energy policies. To discuss uncertainty, we need to first examine the structure of the model. An energy conservation model can be simplified as a linear programming problem in the following cost-minimization form:

$$\text{Minimize:} \quad c'x$$
$$\text{Subject to:} \quad Ax \leq b, \; x \geq 0$$

where c' is the unit cost vector; x, production quantity vector; A, technological transformation coefficient matrix; b, vector of capacity, sales commitment, and other constraints.

There are three types of uncertainty to be considered in the above formation:

- Uncertainty about input prices: this translates into uncertainty about the vector of unit cost coefficients c. This problem can be resolved by using the Monte Carlo simulation discussed in the previous section.
- Uncertainty about demand: this is uncertainty about the vector of coefficients b (for sales commitment and capacity). This problem can also

be resolved by using a Monte Carlo simulation to generate the range and probability for uncertain sales or capacity (e.g., facility failure).

- Uncertainty about the technical coefficients: this is really uncertainty about the matrix A. However, since the vector x refers to activity levels, and the vector c refers to unit costs for the different activities, even if there is no uncertainty regarding the input prices as such, uncertainty about the technical coefficients A translates itself into uncertainty about the vector of cost coefficients c as well. This problem can be solved indirectly through a Monte Carlo simulation.

Although a Monte Carlo simulation can resolve the uncertainty problem in the energy conservation model, as discussed, it is difficult to control and explain the results of such a simulation. However, if one of the above three types of uncertainty can be defined with a known probability distribution, two principal approaches—chance-constrained programming (CCP) and two-stage methods—can be used to solve this "known" uncertainty.

Chance-Constrained Programming Method. This method, which was first introduced by Charnes, et al.,[7] is illustrated below.

Assume that, for producing copper, we have two ores with different concentrations, a_1 and a_2. The cost per unit ore are c_1 and c_2; capacity restriction is d.

Here the problem is:

Minimize: $2X + 4y$

Subject to: $.5X + .3y \geq b$ (demand constraint)

$X + y \leq 4$ (capacity constraint)

Assume b is random with uniform distribution over $(1.2, 1.6)$, the data are monthly, and the ore has to be ordered 1 month in advance.

In the CCP or reliability approach, we are required to satisfy the demand with a high probability, say 0.9; that is:

$$\text{Prob} (.5X + .3y \geq b) \geq 0.9$$

This transforms to the constraint $.5X + .3y \geq 1.56$, and the problem becomes a deterministic linear programming problem.

Two-Stage Method. This method is also called the penalty method or the stochastic problem with recourse.[8] It was first introduced by Dantzig and Beale.[9,10] Using the same example, we are assumed to have a 2-year contract to exactly meet the customer's demand. We could produce more than the demand and dispose of the surplus at a lower price of \$2.00; or, we could produce less than the demand and buy the difference in the open market at \$4.00

Hence, after determining the demand b, the second-stage program (also called the recourse program) can be formulated:

$$\text{Minimize:} \quad (-2Z_1 + 4Z_2)$$
$$\text{Subject to:} \quad Z_2 - Z_1 = b - 0.5X - 0.3y$$
$$Z_1 \geqslant 0, \; Z_2 \geqslant 0$$

If the solution to this second-stage program is denoted by $Q(X,y,b)$, then minimizing the total expected costs (per month) yields the two-stage program:

$$\text{Minimize:} \quad [2X + y + E_b Q(X,y,b)]$$
$$\text{Subject to:} \quad X + y \leqslant 4$$
$$X \geqslant 0, \; y \geqslant 0$$

where $E_b Q(X,y,b)$ is the expected cost at a given b.

There are other methods not discussed here, such as Tintner's passive and active approaches.[11]

Feedback from Previous Conservation Experience

Early conservation experiences, after suitable time periods, should help decision-makers determine how to invest in conservation technologies and programs. It may be important for an energy conservation model to incorporate this learning process, that is, a pattern of pursuing a prudent and optimal policy through sequential chains of previous decisions. Investment plans in energy conservation must be constantly revised in light of new information regarding regulations, energy prices and availability, and so on.

This learning process is illustrated in Fig. 8–11. Depending on the prices of refined copper and other macroeconomic variables such as the industrial production index, a goal for profit and cost of a company is set. In facing anticipated energy price increases, supply uncertainty, and government energy policies, the company decision-maker then sets investment plans to ensure meeting long-range production requirements with a safety margin to allow for uncertainty. However, the actual energy prices, supply, and policies in each subsequent year will affect the production schedule. The resultant deviation, if any, will then be fed back to the goal to be corrected, and the investment plan is thus revised. A model with this learning process will definitely predict better in the time path of investments in energy conservation. However, this formation of the learning process requires a substantial and expensive revision

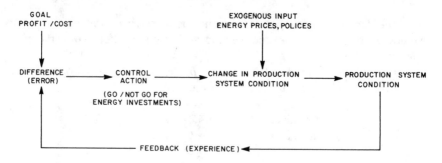

Fig. 8–11. Learning process in an energy conservation model.

of the original energy conservation model. (The interested reader can refer to an EPRI report for methods of incorporating a learning process into the energy conservation model.[12])

Other Considerations

The above two improvements, incorporating uncertainty and feedback controls in an energy conservation model, will make the model more complex. The tradeoff between model complexity and ease of understanding should be carefully evaluated.

In addition, even if it is determined that a more complex model that incorporates uncertainty and feedback controls is warranted, the quality of the data should be considered. The unit production cost and unit energy requirement data which were estimated from standard engineering capacity data can be improved if an actual survey for these unit data is conducted to validate their accuracy. The reason for this survey is that data for individual companies' facilities may not be identical to the standard facility in the open literature, owing to their age and locality.

CONCLUSIONS

Many industrial energy planners are hurt by having inadequate actual energy and production data in planning their energy conservation efforts. Surveying or monitoring systems to obtain these data are usually extremely costly. This chapter provides a practical method (i.e., energy conservation modeling) that does not need surveys and can be used with sufficient accuracy for determining alternative energy conservation investments and their impacts. The method employs computer linear program packages that are available anywhere in the United States.

In energy conservation modeling, data required for energy consumption and production costs are estimated for existing facilities and conservation options from standard engineering capacity data publicly available through the U.S. Bureau of Mines or in other publications. Methods used to prepare the data, construct the model, exercise it, interpret its results, validate it, and improve it were discussed. The main emphasis of the energy conservation model is to ensure that the model is useful and easy to understand.

REFERENCES

1. Silverman, B. "The Copper Industry: Available Data," unpublished Master's thesis, University of Pennsylvania, 1977.
2. U.S. Department of the Interior, Bureau of Mines. *Minerals Yearbook,* Vol. I: *Metals and Minerals (except fuels).* Washington, D.C.: U.S. Government Printing Office. Annual.
3. A. A. Mathews, Inc. "Capital and Operating Cost Estimation System for Mining and Benefication," Mining Report No. 1953–02, Phase II Report, for U.S. Bureau of Mines, Vol. 4, June 1976.
4. U.S. Department of Labor, Bureau of Labor Statistics. *Employment and Earnings Statistics for the United States.* Washington, D.C.: U.S. Government Printing Office. Annual.
5. Driebeek, N. "An Algorithm for the Solution of Mixed Integer Programming Problems," *Management Science,* Vol. 12, No. 7, 576–587, 1966.
6. Smith, V. K. *Monte Carlo Methods.* Lexington, Mass.: Lexington Books, 1973.
7. Charnes, A. W., Cooper, W., and Symonds, G. H. "Cost Horizons and Certainty Equivalents: An Approach to Stochastic Programming of Heating Oil," *Management Science,* Vol. 4, 235–263, 1958.
8. Walkup, D. W. and Wets, R. "Stochastic Programs with Recourse," *SIAM Journal of Applied Mathematics,* Vol. 15, 1299–1314, 1967.
9. Dantzig, G. B. "Linear Programming Under Uncertainty," *Management Science,* Vol. 1, pp. 197–206, 1955.
10. Beale, E. M. L. "On Minimizing a Convex Function Subject to Linear Inequalities," *Journal of the Royal Statistical Society,* B Series, Vol. 17, 173–184, 1955.
11. Tintner, G. "Stochastic Linear Programming with Applications to Agricultural Economics," in H. A. Antosiewicz (ed.), *Proceedings of Second Symposium in Linear Programming,* Washington, D.C., pp. 197–228, 1955.
12. Electric Power Research Institute (EPRI). *Supply Model with Feedback Features,* Vol. 2: "A Review of Adaptive Economic Modeling," prepared by Economic Dynamics, EPRI EA-1357, March 1980.
13. Rosendranz, R. "Energy Consumption in Domestic Primary Copper Production," U.S. Bureau of Mines Information Circular 8698. Washington, D.C.: U.S. Government Printing Office, 1976.

9

Legislative, Institutional, and Environmental Impacts on Industrial Energy Conservation

INTRODUCTION

In the preceding chapters, energy conservation was dealt with primarily in terms of economic considerations. There are other factors that can drastically change a decision made for economic reasons. For example, such energy and environmental legislation as the National Energy Act and the Clean Air Act have forced many industries to change their production processes; and price and supply regulations from the Economic Regulatory Administration and the Federal Energy Regulatory Commission may have prevented large industrial boilers from using oil and natural gas.

Although industry is the "taker" of these factors, its decision-making process in balancing these factors and economic considerations can be very different from that of the government, which has the broader responsibility of overseeing the nation's interests. We will discuss legislative, institutional, and environmental impacts on industrial energy conservation in this chapter, analyze industry's decision-making with respect to economic considerations in Chapter 10, and examine government's policy-making process with respect to economic, socio-political, and environmental considerations in Chapter 11.

This chapter is divided into four sections: energy legislation, institutions, environmental regulations, and industry impacts versus government impacts.

ENERGY LEGISLATION

The National Energy Act of 1978 is the center piece of energy legislation affecting industrial energy conservation. It consists of the following five acts:

- Energy Tax Act of 1978
- Natural Gas Policy Act (NGPA) of 1978
- Powerplant and Industrial Fuel Use Act (FUA) of 1978
- National Energy Conservation Policy Act (NECPA) of 1978
- Public Utility Regulatory Policies Act (PURPA) of 1978

The National Energy Act provides tax credits, fuel price and use regulations, coal-substitution for oil and gas, and evaluation of appliance efficiencies. The act either supplements or replaces the earlier Energy Policy and Conservation Act and the Energy Conservation and Production Act.

Other legislation affecting industrial energy conservation includes the following:

- Energy Policy and Conservation Act (EPCA) of 1975
- Energy Conservation and Production Act (ECPA) of 1976
- Crude Oil Windfall Profits Tax Act of 1980
- Energy Security Act of 1980

ECPA is the basic legislation for state energy conservation programs. It provides supplemental industrial energy conservation plans for states. Under this act, increased funds were provided to the states to support conservation programs (e.g., energy audits and public education). The Crude Oil Windfall Profits Tax Act provides tax credits for cogeneration and other unconventional fuel-producing technologies. The Energy Security Act allows conservation loan subsidies.

All of the above nine acts are discussed in detail in the following sections.

The Energy Tax Act of 1978

A variety of tax credits for investment by business is provided for in this act. An additional 10% investment tax credit is provided for investment in the following industrial and end-use energy technologies.

Alternative Energy Property. This applies to boilers and other combustors that use coal or an alternative fuel, equipment to produce alternative fuels, pollution control equipment, equipment for handling and storage of alternative fuels, and geothermal equipment. The credit is *not* available to utilities.

Specially Defined Energy Property. This applies to equipment to improve heat efficiency of existing industrial processes, including heat exchangers and recuperators.

Recycling Equipment. This applies to equipment for the recycling of waste materials (e.g., biomass fuels).

Depreciation Allowance for Early Retirement of Oil- or Gas-Fired Boilers. If the taxpayer demonstrates that an oil- or gas-fired boiler will be retired earlier than originally anticipated, the useful life may be decreased, and the remaining basis subjected to straight-line depreciation over that shortened remaining life.

The act specifies the regular or normal investment tax credit to be 10% through December 31, 1980, and 7% thereafter. An additional 10% investment tax credit, applicable to the three types of industrial energy equipment listed above, would be available from October 1, 1978 through December 31, 1982. The tax credit for geothermal equipment has been raised to 15% through 1985 by the Crude Oil Windfall Profits Tax Act of 1980.

In addition, states are increasingly allowing some types of financial incentive or tax credit for the purchase of solar and other renewable energy devices. A recent study by Common Cause showed that about 40 states had such incentives.

The act thus provides a substantial financial incentive for industries to invest in retrofitting and recycling conservation technologies. These technologies may not save a significant portion of energy used in industry.

The Natural Gas Policy Act of 1978

There are two provisions in the NGPA that significantly affect industrial energy conservation: (1) incremental pricing of natural gas for industrial use, and (2) natural gas curtailment for industrial use during emergency. Because large numbers of industrial boilers use natural gas, as discussed in Chapter 2, this act is detailed.

Incremental Pricing. The NGPA sets a series of maximum prices for various categories of natural gas, including gas sold in both the interstate and intrastate markets. Thus, it places intrastate and interstate gas markets on the same footing. Also, price controls on new gas and certain intrastate gas will be lifted as of January 1, 1985. Certain high-cost gas will be deregulated approximately one year after NGPA's enactment.

Residential consumers will be protected by the passing on of a greater portion of the gas price increase to industrial users. However, this incremental pricing to industrial users cannot result in industrial gas prices higher than the regional cost of substitute fuels as determined by the Federal Energy Regulatory Commission (FERC).

Affected Users. The first to notice the effects of incremental pricing will be industrial boiler fuel users. FERC proposed the initial incremental pricing rule in 1979 to cover only industrial boiler fuel uses. The commission failed to give any indication of whether to eventually include industrial process and feedstock uses.[2]

Incremental Pricing Mechanism. The incremental pricing mechanism is contained in Section 204 of NGPA. It requires interstate pipelines to establish a special incremental pricing account for all costs of gas that are to be incrementally priced. The pipeline operators then establish a surcharge to its rate, which must be passed on directly to all incrementally priced ultimate customers of the natural gas. Distributors are prohibited from offsetting the incremental surcharge to their industrial customers by reducing other elements of their rates. Therefore, gas purchased either from pipelines or from distributors will share an identical surcharge.

Price Ceilings. Because Congress was aware that too high a price increase might result in forcing too many industrial customers out of gas uses too suddenly, a ceiling on the extent that gas supplies will be incrementally priced is included in the act. The ceiling is the Btu-equivalent price of No. 2 or No. 6 fuel oil in the region, depending on the commission's decisions. Maximum gas price ceilings are shown in Table 9–1.

However, the NGPA does not discuss what would happen to incremental pricing if the above ceiling were reached in a given region. Two consequences follow from this case. On the one hand, FERC or state agencies could allocate all costs beyond the ceiling to nonindustrial customers in order to attract industrial customers. On the other hand, they could allocate all additional costs to industrial customers and continue incremental pricing beyond the ceiling.

Natural Gas Curtailment Provisions. Under the act, the president may declare an emergency if a gas shortage that endangers supplies for "high-priority" users exists or is imminent. (Such high-priority use is defined in the following paragraphs.) During an emergency, the president may authorize certain emergency sales of gas. If these emergency sales are not sufficient to protect high-priority users, he may allocate certain supplies of gas, as necessary.

Table 9-1. Maximum Gas Price Ceilings Set by the Natural Gas Policy Act.

Section of the Act	Price per Million BTU's[1]	Category of Gas	Date of Deregulation
102	$1.75 + inflation[2] and escalation[3] ($2.07)[4]	NEW NATURAL GAS —new Outer Continental Shelf (offshore) leases (on or after 4/20/77) —new onshore wells 1) 2.5 miles from the nearest marker well[5] 2) if closer than 2.5 miles to a marker well, 1000 feet deeper than the deepest completion location of each marker well within 2.5 miles —new onshore reservoirs	1/1/85
		Gas from reservoirs discovered after 7/27/76 on old (pre-4/20/77) Offshore Continental Shelf	not de-regulated
103	$1.75 + inflation ($1.97)[4]	NEW ONSHORE PRODUCTION WELLS (wells, the surface drilling of which began after 2/19/77, that are within 2.5 miles of a marker well and not 1000 feet deeper than the deepest completion location in each marker well within 2.5 miles) —gas produced above 5000-foot depth —gas produced from below 5000-foot depth	7/1/87 1/1/85
104	$1.45 + inflation ($1.63)[4] $ .94 + inflation ($1.06)[4] $ .295 + inflation ($ .33)[4]	GAS DEDICATED TO INTERSTATE COMMERCE BEFORE THE DATE OF ENACTMENT (rates previously set by FPC) —from wells commenced from 1/1/75 to 2/18/77 —from wells commenced from 1/1/73 to 12/31/74 —from wells commenced prior to 1/1/73	not de-regulated

105	GAS SOLD UNDER EXISTING INTRA-STATE CONTRACTS —if contract price is less than Section 102 price it may escalate, as called for by contract, up to Section 102 price —if contract price exceeds Section 102 price, then contract price plus annual inflation factor or Section 102 price plus escalation applies, whichever is higher	contract price[6]	1/1/85—if unescalated contract price exceeds $1.00 by 12/31/84 if lower, not deregulated
106	SALES OF GAS MADE UNDER "ROLLOVER" CONTRACTS (an expired contract that has been renegotiated) —interstate	$ .54 or other applicable FERC price + inflation ($ .61)[4]	not deregulated
	—intrastate	the higher of expired contract price or $1.00 + inflation ($1.13)[4]	1/1/85 if more than $1.00
107	HIGH COST NATURAL GAS —production from below 15,000 feet from wells drilled after 2/19/77	$1.75 + inflation + escalation[3] ($2.07)[4]	deregulated on effective date of FERC incremental pricing rule called for by the Act

Table 9-1 (continued)

Section of the Act	Price per Million BTU's[1]	Category of Gas	Date of Deregulation
	applicable rate under the Act or higher incentive rate as set by FERC	—gas produced from geopressurized brine, coal seams, Devonian shale —gas produced under other conditions the FERC determines to present "extraordinary risks or costs"	(approximately one year after enactment) not deregulated
108	$2.09 + inflation (after 5/78) + escalation[3] ($2.21)[4]	STRIPPER WELL NATURAL GAS (natural gas not produced in association with crude oil, which is produced at an average rate less than or equal to 60,000 cubic feet per day over a 90-day period)	not deregulated
109	$1.45, or other "just and reasonable" rate set by FERC, + inflation ($1.63)[4]	OTHER CATEGORIES OF NATURAL GAS —any natural gas not covered under any other section of the bill —natural gas produced from the Prudhoe Bay area of Alaska	not deregulated

[1] Under the NGPA, if natural gas qualifies under more than one price category, the seller may be permitted to collect the higher price. The ceiling prices set by NGPA do not include state severance taxes.

[2] These prices include an "annual inflation adjustment factor," in order to adjust prices for inflation. The price for a given month is arrived at by multiplying the price for the previous monthly equivalent of the annual inflation factor. Since most of the prices set by NGPA are as of April 20, 1977, the adjustment for inflation begins in May 1977.

[3] These prices to escalate monthly, in addition to the inflation adjustment factor, by an annual rate of 3.5% until April 1981, after which they will escalate by 4%.

[4] The estimated maximum ceiling price as of October 1978, due to operation of inflation and escalation adjusters.

[5] A marker well is any well from which natural gas was produced in commercial quantities after January 1, 1970, and before April 20, 1977, with the exception of wells the surface drilling of which began after February 19, 1977.

[6] The average price reported to FERC for intrastate gas sales contracted for during the second quarter of 1978 was approximately $1.90.

Source: DOE.[21]

Curtailment Priorities. Interstate gas supplies needed for certain agricultural and industrial uses generally will not be curtailed unless the gas is needed to serve high-priority users.

The NGPA defines the high-priority users as (1) residences, including apartment buildings; (2) commercial establishments using less than 50 Mcf/day; (3) schools, hospitals, and similar "human needs" institutions; and (4) any other use, the curtailment of which would result in endangerment of life, health, or physical property (includes plant protection). However, general industrial uses, including feedstock and process uses, are omitted from the above definition.

Declaration of an Emergency. Under Title III of the Act, the president must first declare a "natural gas supply emergency" if he finds that a severe shortage of supply to "high priority uses" exists or is imminent in any region of the United States. The emergency period is limited to 120 days' duration, but this period may be extended by the president.

Impact on Industrial Conservation. The ultimate impacts of the Natural Gas Policy Act appear to be adverse for industrial nonagricultural users of natural gas. Natural gas for industrial uses will be incrementally priced. It will be curtailed in supply sooner than it would have been before passage of the act. This means natural gas conservation will be more prevalent in the industries whose major fuel is natural gas.

The Powerplant and Industrial Fuel Use Act (FUA) of 1978

The purpose of the Powerplant and Industrial Fuel Use Act of 1978 is to encourage use of coal and certain other fuel, and to force certain power plants and industrial plants to switch back to coal from oil and natural gas. It attempts to reverse the trend set by these plants in the 1960s as they switched from burning coal to using oil or natural gas in order to meet new and stringent air pollution requirements. The FUA revises some of the original provisions of the Energy Supply and Environmental Coordination Act* (ESECA) of 1974, which expired on December 31, 1978. ESECA originally banned the use of oil and gas in industrial boilers larger than 100 MMBtu/hour. DOE, however, is required under FUA to grant an exemption when the user can demonstrate that the use of a mixture of oil and coal or other alternate fuel is not economically or technically feasible.

The provisions of the FUA that are important to industrial energy conservation are summarized as follows:

---

* 15 U.S.C. §§791, *et seq.*

- Prohibition against use of oil or natural gas in new electric utility generation facilities or in new industrial boilers with a fuel heat input rate of 100 million Btu/hour or greater, unless exemptions are granted by DOE.
- DOE authority to require existing coal-capable facilities, individually or by categories, to use coal and to require non-coal–capable units to use coal–oil mixtures.
- Limitation of natural gas use by existing utility power plants to the proportion of total fuel used during 1974–1976, and a requirement that there be no switches from oil to gas. There is also a requirement that natural gas use in such facilities cease by 1990 (with certain exceptions).
- Supplemental authority to prohibit use of natural gas in small boilers for space heating and in decorative outdoor lighting and to allocate coal in emergencies.

One important point deserves attention: Permanent exemptions are provided for industrial cogeneration facilities, if a firm can demonstrate that cogeneration benefits are not obtainable unless petroleum or natural gas is used in such a facility. For details, see Chapter 13.

The National Energy Conservation Policy Act (NECPA) of 1978

Most of the provisions of this act relate to residential, commercial, and government buildings. The following provisions of NECPA affect industrial energy conservation:

- Energy-efficiency labeling of industrial equipment
- Industrial recycling targets and reporting requirements
- Mandatory minimum energy-efficiency standards for major home appliances, such as refrigerators and air conditioning units
- Authority for the secretary of transportation to increase the civil penalties on automobile manufacturers from $5 to $10 per car for each one-tenth of a mile a manufacturer's average fleet mileage fails to meet the EPCA automobile fleet average fuel economy standards

In particular, the act gives DOE authority over industrial equipment such as compressors, fans, blowers, refrigeration equipment, electric lighting, steam blowers, oven furnaces, kilns, evaporators, and dryers.

The secretary of energy is directed to examine the feasibility of establishing standards and to establish test procedures and labeling procedures. Every equipment manufacturer would have to label the efficiency of industrial equipment in a common manner.

The program for industrial recycling targets and reporting requirements is patterned like existing conservation reporting programs. The secretary of energy is required to set targets for the increased use of energy-saving recovered material in each of the following industries: the metals and metal products industry, the paper and allied products industry, the rubber industry, and the textile mill product industry. These industries will have to report their progress in meeting established goals on a regular basis similar to that of the efficiency reporting program.

NECPA also amends Title III of the Energy Policy and Conservation Act of 1975 (discussed later in this section), expanding the information reporting requirements, so all companies that consume at least one trillion Btu/year in each of the ten most energy-consuming industries must report their energy consumption figures to DOE each year and show the actions they are taking to conserve energy.

The reporting requirement of NEPA forces the ten greatest energy-consuming industries to voluntarily invest in energy-conserving equipment and measures.

The energy-efficiency standards for major home appliances and penalties on fleet mileage will enforce the need for producing energy-efficient equipment. If industries do not produce energy-efficient products, they may be forced out of markets such as the automobile industry.

The Public Utility Regulatory Policies Act (PURPA) of 1978

The Public Utility Regulatory Policies Act of 1978 has three broad purposes:

- Conservation of energy supply by electric utilities
- Optimization of efficient use of facilities and resources by electric utilities
- Equitable rates to electric customers

Of the provisions designed to achieve these purposes, the one for cogeneration is the most important to industrial energy conservation. PURPA requires FERC to set rules "favoring industrial cogeneration facilities, and requiring utilities to buy or sell power from qualified cogenerators at just and reasonable rates." Details of the implication of PURPA will be discussed in Chapter 13.

The Energy Policy and Conservation Act (EPCA) of 1975

EPCA was the most significant energy conservation legislation until the passage of the National Energy Act of 1978. Key programs initiated by the act established a voluntary industrial energy conservation program and set efficiency standards for appliances and automobiles.

Voluntary Business Energy Conservation Program. The act directed the federal energy administrator to identify and establish a priority ranking of each major energy-consuming industry, and set voluntary energy-efficiency improvement targets for the ten most energy-consuming industries in the United States. Each target "shall be based upon the best available information" and "be established at the level which represents the maximum feasible improvement in energy efficiency which such industry can achieve by January 1, 1980."

EPCA also required that the top energy-consuming companies in the United States report to the Federal Energy Administration (FEA) or the Commerce Department on the efforts they were making to improve their energy efficiency. FEA identified each corporation that consumed at least one trillion Btu of energy per year, and was among the 50 most energy-consumptive corporations in each industry selected.

These companies were required by law to report their energy efficiencies to FEA or industry trade associations. Based on reports by industry trade associations, DOE (the successor to FEA) reported the actual improvements in industrial energy efficiency, relative to 1972, through 1978, compared with the 1980 targets (shown in Table 9–2). Table 9–2 shows that, in a few industries such as transportation equipment, machinery other than electrical, and chemi-

Table 9–2.   Energy-Efficiency Improvements in the Ten Most Energy-Intensive Industries in the United States.

SIC Code	Industry	Energy Consumption, 10^{15} Btu/Year (Quads)	% Reduction in Energy Use Relative to 1972 Base Year	
			Actual 1978	Goal 1980
33	Primary metals industry	4.7	9	9
28	Chemicals and allied products	4.1	17	14
29	Petroleum and coal products	3.5	16	12
32	Stone, clay, and glass	1.5	13	16
26	Paper and allied products	1.2	14	20
20	Food and kindred products	1.0	17	12
37	Transportation equipment	0.6	21	16
35	Machinery except electrical	0.6	29	15
34	Fabricated metal products	0.45	22	24
22	Textile mill products	0.45	19	22
	Composite	18.1	16	13
	Approximate overall level of savings, 1978 achievement			
	Quads	2.9		
	Millions BOE/D	1.4		

Source: Department of Energy.

cal and allied products, actual efficiency improvements in 1978 were better than the 1980 targets. Overall, the actual 1978 improvement (a 16% reduction) exceeded the 1980 target (a 13% reduction). The 1980 targets might have been set too low because of underestimation of energy price effects.

Appliance and Fuel Efficiency Standards. EPCA also directed FEA to establish energy-efficiency targets for consumer products other than automobiles, including 13 kinds of appliances; and to set fuel efficiency standards for all passenger cars. The act required manufacturers to label their products according to energy efficiency and report to FEA on their progress toward designated targets. The appliance efficiency standards were further enhanced by the National Energy Conservation Policy Act of 1978. The fuel efficiency standards were reenforced in the Energy Tax Act of 1978, which specified a gas-guzzler tax to penalize automobile fleets that do not meet the yearly standards established in the Energy Tax Act.

State Energy Conservation. EPCA also funded state energy conservation programs that were to meet a goal for achieving a certain level of energy conservation by the year 1980. States were required to achieve a minimum energy conservation goal of 5% of the state's projected 1980 energy consumption. Each year, states were to revise their state energy conservation plans and submit them to DOE for approval. As long as the mandatory provisions of the state conservation plan were addressed, the states were free to add measures that they felt would reduce energy consumption by a minimum of 5% in 1980. As a result, many states developed clearinghouses and audit programs, and even legislated certain measures to improve industrial energy efficiency.

The Energy Conservation and Production Act (ECPA) of 1976

ECPA provided for supplemental energy conservation plans for the states. More funds were provided to the states to support additional energy conservation programs in such areas as intergovernmental relations, public education, and providing for Class A and Class C audits. Class A audits are on-site and review the potential of energy conservation in a building or plant. Class C audits consist of guidebooks or workbooks that allow plant or building managers to conduct their own audits. The provisions for audit are an important adjunct to industrial energy conservation.

The Crude Oil Windfall Profits Tax Act of 1980

The primary purpose of the Crude Oil Windfall Profits Tax Act is to tax oil companies $227 billion from their profits defined as windfall in the act.

However, the act also provides additional tax credits for investments in alternate energy sources and conservation. The following tax credits are important to industrial energy conservation:

- A new 10% credit through 1982 for cogeneration equipment
- A more liberal 10% credit extended three years through 1985 for biomass equipment to convert waste into fuel

The Energy Security Act of 1980

The main purpose of the bill is to authorize $20 billion to be used by the Synthetic Fuels Corporation during the next five years to produce 500,000 b/d of oil equivalent by 1987. However, the act also provides $1.45 billion for gasohol and urban waste programs and $1 billion for solar and conservation loan subsidies. The effects of this act on industrial energy conservation remain to be seen.

Concluding Comments on Energy Legistion

Because of relatively inexpensive energy prices, the U.S. government has been relying on energy legislation to achieve energy conservation. Examples are voluntary conservation target, energy-efficiency standards and tax credits. Particularly, the United States has been, compared with other industrial societies, extremely generous in awarding tax credits for saving energy. For example, Britain and Japan provide specific incentives for industry, but not for individuals. In contrast, France and West Germany offer tax credits to individuals but little to business. Canada does not give federal tax incentives, and the Dutch still must wait for their first energy credits.[3]

INSTITUTIONS

A number of federal and state agencies are particularly concerned with industrial energy conservation. Some of these agencies and their major duties related to industrial energy conservation are listed below:

- Department of Energy—overall responsibility for industrial energy conservation, technologies, and regulations
- Department of Commerce (DOC)—development of energy savings measurement methods and standards
- Department of Agriculture (USDA)—specified interest in agricultural industries (e.g., poultry producers)
- Department of Transportation (DOT)—development of R&D programs on engine efficiency and fuel economy

- Department of Housing and Urban Development (HUD)—development of community dual energy use (cogeneration) systems
- Environmental Protection Agency (EPA)—test and publication of automobile fuel efficiency and regulation of environmental protection
- State and local energy agencies—enforcement of conservation laws and implementation of state and local energy programs

Because DOE is the most important federal agency for overall industrial energy conservation policies, it is discussed first. Other concerned federal agencies are then briefly described. Finally, state and local agencies, which have a major role in the implementation of industrial energy conservation policies, are considered in some detail.

Department of Energy

The Department of Energy was created under the Department of Energy Organization Act (Public Law 95–91) and organized on October 1, 1977. DOE is an amalgamation of ERDA, FEA, the Federal Power Commission (FPC), and some parts from the Departments of Interior, Defense, Commerce, and Housing and Urban Development and the Interstate Commerce Commission (ICC).

DOE has been reorganized continuously. Figure 9–1 shows the recent organization chart. The department had approximately 20,000 employees and a $9 billion annual budget in FY 1980. The major DOE offices that affect industrial energy conservation are discussed below. (Because of the Reagan Administration's intention to abolish DOE, the existence of DOE, at the time of this writing, is to be determined by Congress.)

Office of the Assistant Secretary for Conservation and Renewable Energy. As shown in Fig. 9–2, the Office of the Assistant Secretary for Conservation and Renewable Energy includes four major offices: Conservation, State and Local Assistance Programs, Renewable Energy, and Alcohol Fuels. The Conservation and State and Local Assistance Programs substantially affect industrial energy conservation. Most industrial energy conservation projects are included in the Office of Industrial R&D Programs (OIR&D) under the deputy assistant secretary as for conservation.

Office of Industrial R&D Programs (OIR&D). The OIR&D had a budget of $60 million in 1980 and $60.5 million in 1981. The functions of the original OIR&D are summarized as follows:*

---

* At time of this writing, the existence of this office and programs is to be determined by Congress.

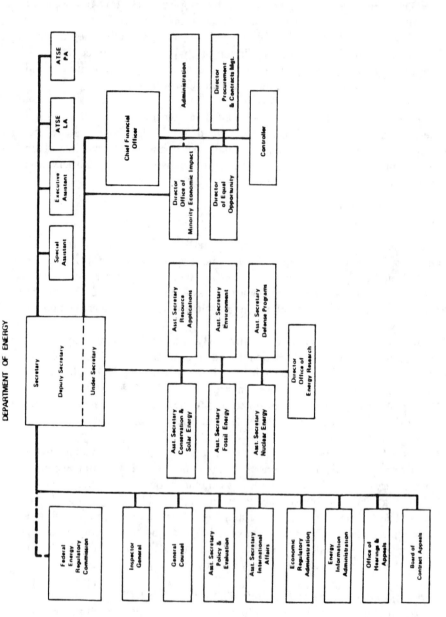

Fig. 9–1. Organization chart of the Department of Energy. Source: Department of Energy. Released on February 25, 1981. It is still in the process of reorganization.

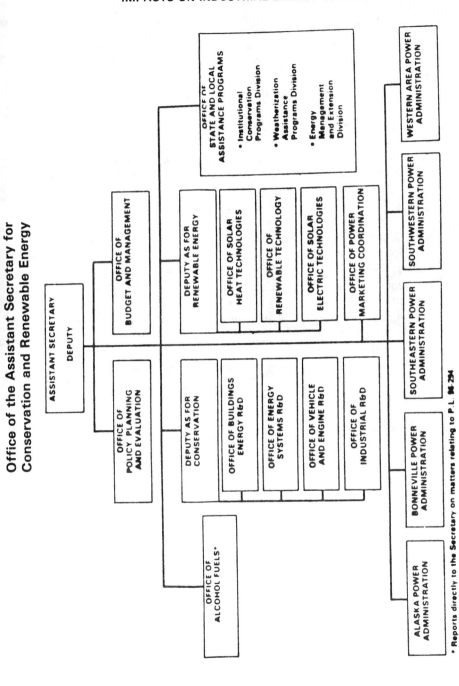

Fig. 9–2. Organization chart of the Conservation and Renewable Energy Office in DOE. Source: Department of Energy.

- Performs applied research, development, and demonstrations of high-risk, high-payoff technology cost-shared with the private sector and designed to conserve scarce fuels in industry and agriculture.
- Identifies and analyzes existing but inadequately utilized industrial energy conservation technologies and implements appropriate federal action.
- Conducts a variety of activities directed at stimulating expeditious implementation of new and existing technologies by private industry. These include market analyses, product-specific commercialization plans and implementation, and industrial information dissemination and assistance.
- Operates an Industrial Reporting Program mandated by the Energy Policy and Conservation Act to monitor the progress of energy conservation within the most energy-intensive industries.
- Conducts the activities mandated by the National Energy Act including establishing targets for utilization of recovered materials in four industries, the second law (of thermodynamics) study, the study of feasibility of equipment standards, and definitions of equipment performance for additional tax credits.
- Coordinates closely with the Office of Advance Conservation Technologies* regarding research and demonstration activities of a more basic nature that have the potential for contributing significantly to energy efficiency problems in several industrial processes or systems, but are of a more high-risk, long-term-payoff nature than projects normally undertaken by the Office of Industrial R&D Programs.

There were four program elements in the original Office of Industrial R&D Programs: waste energy reduction, industrial process efficiency, industrial cogeneration, and implementation and deployment. Detailed distribution of management responsibilities and budgets to various locations by major contracts and project activities is shown in Table 9–3. The FY 1980 budget authorizations for waste energy reduction, industrial process efficiency, and industrial cogeneration were $16.5 million, $20.7 million, and 11.3 million, respectively. The FY 1981 budget was largely reduced because of the Reagan Administration's energy policy.

Economic Regulatory Administration (ERA).**   ERA is the successor agency to FEA. ERA organizes and manages an intervention program on behalf of the secretary of energy before FERC and other federal and state regulatory agencies.

---

* This office, as well as DOE, is subject to change.
** The existence of ERA is in question.

Table 9-3. Industrial Energy Conservation Projects in DOE's Office of Industrial Programs. ($000)

Projects and Distribution of FY 1980 $ (BA)*

Program Element	Management Location	Major Contracts	Waste heat recovery	Combustion effic. impr.	Program support	Waste to fuel	Waste to feedstock	Material recycling	Totals
Waste energy reduction	Headquarters	Tasks Under $3000	$2,462	$791	$4,860	$1,188	$899	$153	$10,353
	Idaho	Tasks Under $3000	2,502	—	1,200	400	—	—	4,102
	Bartlesville	Tasks Under $3000	—	—	—	—	850	—	850
	San Francisco	Tasks Under $3000	—	—	—	950	—	—	950
	Chicago	Tasks Under $3000	195	—	—	—	—	—	195

Program Element	Management Location	Major Contracts	Steel	Aluminum	Glass	End product	Textiles	Paper	Chem. & petroleum	Agri. & food	Totals
Industrial process efficiency	Headquarters	Hydropyrolysis	—	—	—	—	—	$3,000	—	—	3,000
	Headquarters	Tasks under $3000	2,998	850	800	256	246	520	331	2,580	8,581
	Oak Ridge	Formcoke	5,000	—	—	—	—	—	—	—	5,000
	Oak Ridge	Tasks under $3000	2,125	—	—	—	—	—	—	—	2,125
	Idaho	Tasks under $3000	1,118	—	—	—	—	—	—	851	1,969

Program Element	Management Location	Major Contracts	Studies & analyses	Bottoming demo's	Topping demo's	Adv. sys. dev.	Totals
Industrial cogeneration	Headquarters	Low speed diesel	—	—	3,440	—	3,440
	Headquarters	Tasks under $3000	2,750	—	180	1737	4,667
	Idaho	Organic rankine	—	3,000	—	—	3,000
	Idaho	Tasks under $3000	—	143	—	—	143

Program Element	Management Location	Major Contracts	Legislated progms.	Tech. dissem.	Indus. report.	Envir. plans & analy.	Totals
Implementation and deployment	Headquarters	Tasks under $3000	1,924	7,536	240	100	9,800

Source: Department of Energy.

* Most projects have been cut substantially.

ERA regulates oil pricing, oil allocation, conversion of oil- and gas-fired utility and industrial facilities to coal (mandated by the Powerplant and Industrial Fuel Use Act), natural gas curtailment priorities (mandated by the Natural Gas Policy Act), natural gas allocations, and natural gas imports/exports; and is responsible for regional coordination of electric power system planning and reliability and energy planning programs. The above regulations strongly affect the direction of industrial energy conservation efforts due to the availability and prices of oil and natural gas. Also, ERA standards and guidelines regarding PURPA affect electricity prices as well as energy conservation.

Federal Energy Regulatory Commission (FERC). FERC has taken over the role of the Federal Power Commission and part of the responsibility of the Interstate Commerce Commission. FERC is composed of five members appointed by the president and approved by the Senate. No more than three members of the commission may come from the same political party. FERC is intended to be independent of DOE. When submitting its budget to Congress, DOE is required by law to indicate the difference between the amount requested by FERC and that asked by DOE. Also, the law requires that whenever FERC submits any legislative recommendation, testimony, or comments to the secretary of energy, the president, or the Office of Management and Budget, it shall concurrently send copies to Congress.

FERC has sole responsibility to consider and take final action without further review by the secretary of energy and/or other executive branch official in certain areas; for example, it is to:

- Establish and enforce rates and charges for the transportation or sale of natural gas
- Issue certificates of public convenience and necessity to connect or disconnect natural gas
- Establish, review, and enforce natural gas curtailments
- Regulate mergers and securities acquisitions
- Establish rates or charges for the transportation of oil by pipeline
- Issue and renew hydroelectric licenses
- Establish, review, and enforce rates and charges for the transmission and wholesale of electric energy, including cogeneration
- Consider major oil pricing and allocation proposals submitted by the secretary of energy to Congress under the Emergency Petroleum Act of 1973. If FERC proposes changes, the secretary may issue the rule only if it conforms to FERC's conclusions

FERC will thus affect industrial energy conservation in cogeneration and in prices and availability of natural gas and oil.

Other Concerned Federal Agencies

In the beginning of this section, five important nonenergy federal agencies were identified: the Departments of Commerce, Agriculture, Transportation, and Housing and Urban Development, and the Environmental Protection Agency. Their effects on energy conservation are described below.

Department of Commerce. DOC is a liaison with business and industry to promote energy conservation through energy management. Its major division responsible for energy conservation is the National Bureau of Standards. The Bureau develops measurement methods and standards related primarily to appliance efficiency and energy conservation in industry, building, and communities. For example, in 1975, the Bureau prepared waste heat management handbooks for FEA.[11]

Department of Agriculture. USDA is primarily interested in agriculture-related industries (e.g., the poultry industry). It provides educational materials on energy conservation for these industries. Demonstration and research projects for efficient energy use in these industries used to belong to the Office of Industrial R&D of the Department of Energy before 1981.

Department of Transportation. DOT promotes use of rail and water-freight carriers, and conducts research and development programs on engine efficiency, fuel economy, and high-performance advanced batteries for electric vehicles. The Office of Transportation Program of DOE had parallel responsibilities and coordinated with DOT in energy projects before 1981.

Department of Housing and Urban Development. Major HUD activities in industrial energy conservation are in community dual energy use (cogeneration systems) and solar applications. Dual energy systems are also called integrated utility systems. Details of these systems will be discussed in Chapter 14.

Environmental Protection Agency (EPA). The purpose of EPA is to reduce and control pollution systematically by implementing a series of monitoring regulations, standard-setting, and research and development programs. Industrial energy conservation activities are mostly managed by the Offices of Air and Waste Management, and Research and Development. Their enforcement of the Clean Air Act and its amendments substantially affects the implementation of certain industrial energy conservation technologies. In addition, as part of its air pollution program, EPA tests and publicizes fuel efficiency for automobiles. This publication of fuel economies has provided a means for the public to put strong pressure on the U.S. automobile industry to produce energy-saving motor vehicles.

State and Local Energy Agencies

A major responsibility for enforcement of industrial energy conservation resides in state and local energy policy agencies. Their effects on industrial energy conservation are as significant as those of federal agencies. Some other state and local authorities are also involved with industrial energy conservation, but they play a less significant role. For instance, building code enforcement agencies are concerned with plant insulation requirements; and police or highway patrols enforce the 55-mph speed limit. Only some significant examples of state and local energy agencies are discussed below. This is not a comprehensive discussion of all state energy agencies. For more information, the reader is referred to the U.S. Department of Energy report *Source Book of Energy Conservation Programs.*[4]

Ohio Department of Energy. Ohio has one of the best-organized nonmandatory industrial energy conservation programs.[5] The Ohio Revised Code provides for tax exemptions for energy conservation, solid waste energy conversion, and thermal efficiency improvement facilities. The Ohio Department of Energy promotes this exemption.

The department supports demonstration projects for combustion efficiency and electronic energy management systems, and provides technical assistance on waste heat recovery, waste oil recycling, boiler condensate recycling, and the use of wood waste for industry. Its planned expenditure was about $.9 million in 1979.

California Energy Resources Conservation and Development Commission. California has emphasized the legislative/regulatory approach, in particular, for utilities. In May 1979, the commission adopted regulations requiring electric utilities voluntarily to implement load management programs for industrial customers. Electric utilities are required to submit annual progress reports on surveys of facilities and detailed program evaluation measures.

In addition, through another legislative/regulatory provision,[6] the commission requires the State Air Resource Board to:

- Inventory potential cogeneration projects to be constructed by 1987 in cooperation with the Public Utility Commission (PUC)
- Inventory resource recovery projects planned or proposed to be constructed by 1987, in cooperation with the State Solid Waste Management Board, and solid waste district and regional planning agencies
- Prepare revisions in the State Implementation Plan required by the Clean Air Act Amendments of 1977 to provide for the mitigation of air quality impact for these projects

Finally, the bill requires air pollution control districts to issue permits for the construction of cogeneration and resource recovery facilities if certain conditions are met. It also requires the PUC to make cogeneration projects the highest priority for the purchase of natural gas. California has been pioneering in conservation and has adopted a much more comprehensive energy plan than many other state plans.

Pennsylvania Energy Department Authority. New legislation was introduced to create a state cabinet-level department with a budget of $2 million in 1979.

Pennsylvania is an industrial state with high potential for coal development. Most industries are in primary metals and chemical and allied products. Past state industrial energy conservation efforts were mostly limited to information transfer, except a recycling program for paper, glass, aluminum, and motor oil.[7] These efforts would not significantly change state industrial energy-use patterns.

New Jersey Department of Energy/Board of Public Utilities. In August 1978, New Jersey DOE established the state's boiler efficiency standards for all fossil-fuel-fired large boilers, except those operated by electric and gas public utilities subject to the jurisdiction of the New Jersey Board of Public Utilities. All large boilers are required to operate at a combustion efficiency such that neither shall the percent oxygen be higher than 1.25 times the optimum percent oxygen value, nor shall the temperature of the flue gases be higher than 1.15 times the optimum temperature value.

The optimum temperature is obtained from performance characteristic curves for a load condition. A large boiler is defined as any steam boiler, steam generator, hot water boiler, or hot oil unit whose rated capacity exceeds either 499 ft^2 of heating service of 100 boiler horsepower or 4 million Btu/hr input regardless of temperature or pressure conditions.

Initial performance characteristics curves are determined for each larger boiler under various operating conditions, and are redetermined every five years, when the fuel type or any component of the boiler that could change its combustion efficiency has changed, or at the request of the state. Large boilers are required to be tested for efficiency each week by the state, and records are required to be maintained by the plant where the boiler is located for a period of at least five years. Such reports are to be made available to officials of the New Jersey Department of Energy and/or the New Jersey Department of Labor and Industry.[8]

The New Jersey Board of Public Utilities is allowing all utilities to charge higher summer and demand rates and allowing some utilities to charge interruptible rates for industrial customers. It has also moved toward pricing natural gas in parity with fuel oil, pioneered in load control, and so on.

New York State Energy Office. The New York State Energy Office focused on information transfer, audits, data surveys, and technical seminars.[9] In particular, the state provided industrial boiler seminars and waste heat recovery workshops for operators and managers. More than a thousand people attended these seminars. No long-term state plan that would significantly change the state's industrial energy consumption patterns were identified.

Summary of State and Local Energy Agencies. Although energy conservation has been an important issue for states, only a few states have cabinet-level energy departments. In addition, even for the industrial states, most common state programs are limited to audits, surveys, and information transfer. The following statistics were summarized from various reports:[4,10]

- Thirty-two states offer industrial on-site audits.
- Thirty-seven states established energy information clearinghouses.
- More than 25 states set up energy advisory committees.
- Twelve states established efficiency standards for boilers or other industrial equipment.
- Very few states have initiated legislative and regulatory energy programs.

Clearly, many of the federal legislative/regulatory programs regarding industrial energy conservation have not been implemented in state and local programs owing to lack of coordination between federal and state agencies. Industry should expect more legislative and regulatory programs from the state and local agencies when energy shortages become more imminent, and as state and local agencies gain more authority. At this time, impacts of state and local energy programs upon industry have been largely focused on information transfer, data collection, and energy audit.

ENVIRONMENTAL REGULATIONS

Environmental protection is a necessity for a clean atmosphere and healthy society. However, some environmental regulations, such as those for waste heat recovery devices, cogeneration equipment, and power plant reject heat utilization systems, prevent implementation of industrial energy conservation technologies. In addition, many pieces of equipment for environmental protection require substantially larger amounts of energy to operate (e.g., scrubbers for air pollution).

In this section, we will discuss effects of pollution and general environmental regulations for protecting the atmosphere, in particular, the Clean Air Act. The impact of environmental regulations on implementation of individual industrial energy conservation technologies will be discussed in Chapters 12, 13, and 14.

Health and Climatic Effects of Pollution

Currently, there are insufficient data for establishing optimum standards for pollution because of lack of biological information concerning the relation between public health and pollution. Effects of pollution on health are often measured in terms of the cost of ill health, and such factors as the days lost from work or productive life lost, although this method of measurement tends to be objected to on the basis that human suffering and misery should not be assessed in financial terms. In addition to the above biological aspect, the psychological aspect of pollution impact on the public health is also an important factor to consider.

There are two major types of pollution: air pollution and thermal pollution.* Because most industrial energy conservation technologies are involved with waste heat recovery, reject heat utilization, and process efficiency improvements, it is likely that the level of one or more of the two types of pollution will increase. The health aspects of the two pollution types are detailed below.

Air Pollution. In many regions, most air pollution results from burning fossil fuels. Combustion of fossil fuels produces air pollutants such as carbon monoxide, particulates, hydrocarbons, and oxides of sulfur and nitrogen. The annual total of these pollutants is estimated to be more than 200 million tons.[12] The distribution of total emissions among pollution sources is estimated as follows: automobiles, 40%; stationary plants, 20%; industrial processes, 14%; and agricultural and solid wastes, 26%. Among the stationary plants, coal-fired plants contribute over two-thirds of the total of 20%. Examination of the above distribution indicates that air pollution produced from the sources of candidate technologies for industrial conservation may reach a level between 34% and 60%. It will be 60% if pollutants from agricultural and solid waste are all included. Industrial conservation efforts that increase air pollution levels will be subject to environmental considerations. The effects of air pollutants are presented in the following paragraphs.

Particulate Matter. Particulates may present a vehicle on which absorbed gases can be carried into the human body. The current federal standard is 75 micrograms per cubic meter. Studies indicate that if the particulates are

---

* Other types of pollution are radiative pollution, nonthermal water, and noise, which are not discussed here because of their insignificance in relation to industrial energy conservation.

sufficiently close to this level and compounded with gaseous pollutants, significant health hazards can be expected.[13]

Sulfur Oxides. Laboratory experiments indicate that oxides of sulfur at prevailing ambient air levels, if present alone in the atmosphere, represent no significant health hazard. When particulates are also present, increased morbidity and mortality may be demonstrated when SO_2 levels rise over .25 ppm (but these data are controversial). The present standard is .14 ppm (365 micrograms per cubic meter), for a maximum 24-hour concentration.

Nitrogen Oxides. A variety of toxic effects may be anticipated from oxides of nitrogen, particularly NO_2. An epidemiological study by C. M. Shy and others[18] indicated that increased respiratory infections among school children were associated with increased levels of NO_2. The primary national ambient air quality standard is 100 micrograms per cubic meter (.05 ppm) on an annual average basis (also a controversial figure).

Carbon Monoxide. Carbon monoxide has both health and climatic effects. It may decrease sensory perception and discrimination, impair the oxygen-carrying capacity of the blood, and thus poison the brain. The present industrial threshold value limit is 50 ppm.

Carbon dioxide also acts like a climatic modifier by changing the absorption property of the atmosphere so as to cause a general warming of the troposphere (the so-called greenhouse effect). This warming effect can result in other climatic changes through interaction of clouds and water vapor content in the atmosphere. The ambient background CO_2 concentration has increased. As a result of this increase, it is estimated that there will be a mean global increase in surface temperature of about .5°C by the year 2000. The current ambient background CO_2 concentration is about 323 ppm.[14]

Hydrocarbons. The hydrocarbon component of pollution in ambient urban air is related to oxidants through complex photochemical processes. The combination of the two may produce eye irritation and deterioration in athletic performance. However, there is no evidence of permanent health effects on normal populations by the mixture. The present EPA standard for oxidant–hydrocarbon is 160 micrograms per cubic meter (.08 ppm) as a maximum one-hour concentration.[19]

Thermal Pollution. Thermal discharges in water from power plants and industrial facilities create thermal pollution problems. Many of these discharges contain poisonous chemicals. There is limited information available on specific interactions to determine where a body of water (e.g., lake or river) can accept given amounts of physical and chemical inputs. In particular, such information is seldom available where water is used in industrial processes or energy production. The full capacity of water to accept heat or chemical waste has not been quantified. More research in this area is needed. Thermal pollution is an area of concern because most industrial conservation technologies will be involved with heat and steam, and will significantly affect reject heat and waste contents in discharged water. Pollution control in this area will be discussed further in Chapters 13 and 14.

Key Environmental Legislation

There are two key pieces of environmental legislation affecting industrial energy conservation: the Clean Air Act of 1970 for air pollution control and the Federal Water Pollution Control Act (FWPCA) for water pollution control. FWPCA was further amended by Congress with the Clean Water Act of 1977.

Clean Air Act of 1970. The Clean Air Act of 1970, as amended in 1977 (42 U.S.C. 7401, *et seq.*), requires any new fossil-fired plant to meet the emission control and the prevention of significant deterioration (PSD) provision of the act. Emission control requirements include the Best Available Control Technology (BACT) for clean air in attainment areas, and the Lowest Achievable Emission Rate (LAER) for nonattainment areas.* A strict interpretation of the Clean Air Act would prevent the siting of all new air-polluting facilities in nonattainment areas. Once existing nonattainment areas came into compliance, new facilities could be sited as long as the new pollutants did not interfere with maintenance of the standards or prevention of significant deterioration of air quality requirements.

However, in December 1976, EPA announced an offset policy setting forth conditions under which new facilities could be sited in nonattainment areas while conforming to the requirements of the Clean Air Act. The policy allows new sources to be located in nonattainment areas as long as, among other things, the new pollutants are more than offset by a reduction in emissions of the same pollutants from existing facilities in the same area. In addition, individual states which have the responsibility to implement Clean Air Act requirements can set stricter new source regulations than those of the federal government.

---

* EPA established primary and secondary standards for six classes of pollutants—sulfur dioxide, particulate matter, carbon monoxide, hydrocarbons, nitrogen oxides, and photochemical oxidants. In many areas of the country, neither the primary nor the secondary standards have been attained. These areas are called nonattainment areas.

Furthermore, the Clean Air Act Amendments of 1977 required each state to establish a State Implementation Plan (SIP). All existing plants, including converting units, are subject to the SIPs.

The effect of the environment-protection requirement is substantial. In particular, revised New Source Performance Standards (NSPS) for new coal-fired steam generators require a 90% reduction in uncontrolled SO_2 emissions and a 99.5% reduction in uncontrolled particulate emissions. Therefore, because the two available types of SO_2 removal equipment, limestone wet scrubbers and flue gas desulfurization (FGD) units, are usually 90% effective, the NSPS virtually requires necessary redundancy for 100% availability (reliability) of SO_2 removal efficiency and mandatory baghouses for particulate removal. Baghouses are required to be 99.5% effective with 100% reliability. For some industrial energy conservation equipment, such as coal burning cogeneration facilities, it would be difficult to overcome this environmental constraint if the efficiency standards were not relaxed for these types of equipment. The deadline for meeting EPA limits on concentrations of major air pollutants were extended by the 95th Congress to 1982 for most areas and 1987 for those with severe automobile pollution. The limits on concentrations were established by EPA under the requirement of the original 1970 Clean Air Act for the seven most common air pollutants: hydrocarbons, carbon monoxide, nitrogen dioxide, lead, ozone, particulates and sulfur dioxide.

The Clean Air Act has been waiting for an overhaul since it was expired on September 30, 1981. Like the 1974 Clean Air Act Amendments which ended up as the 1977 Clean Act Amendments, the overhaul of the act is likely to be a prolonged process, because of the subsequent factors: significant improvements in air quality, cost/benefit debates of pollutant standards, the new regulatory reform movement and need for secure domestic energy supply.

Federal Water Pollution Control Act and Clean Water Act of 1977. Congress enacted P.L. 92–5000 in 1972 to provide far-reaching amendments for FWPCA. In 1977, Congress further amended FWPCA with the Clean Water Act of 1977 (P.L. 95–217). The legislation also initiated the National Pollutant Discharge Elimination System (NPDES), which is used for the issuance of discharge permits to various categories of point source dischargers. Discharge limitations for these permits were based on phases and levels of technologies, and then attainment dates were specified.[15] (For details see Chapter 14.)

EPA was thus required by law to publish effluent limitations for the classes of treatment technologies and to publish performance standards for "new sources." A "new source" was defined as any source for which construction was commenced after the publication of NSPS.

The effluent limitations affect mostly issuance of thermal discharge permits and the use of cooling lakes. They may limit the applications of reject heat utilization from power plants or industrial plants.

INDUSTRY IMPACTS VERSUS GOVERNMENT IMPACTS

We have repeatedly emphasized that industry is a profit-maximizer. The major incentive for industrial conservation efforts has been the dramatic increase in energy prices and concern over availability of fuel, particularly oil and natural gas. The reason is that fuel price and availability affect industry's production costs and profits.

Despite recent increases, American natural gas and domestic oil prices have been low because of past oil and present gas price controls. On the other hand, government has subsidized energy production through generous depletion allowances and other means at a total of approximately $130 billion over the past three decades, according to a recent Battelle Pacific Northwest Laboratories study. These two factors have not made conservation investments as economically attractive as they would otherwise have been, because energy market prices do not reflect the full replacement costs. Conservation simply can not compete with subsidized energy sources.

Costly Energy Legislation

Logically, the government should have completely removed price controls on natural gas, so that prices could reflect their market values. Because price decontrols are politically unpopular, the government in the past,* instead, had amassed volumes of energy legislation and energy tax credits and hoped industry will conserve energy not because of energy costs, but because of noneconomic regulations and tax incentives. These pieces of energy legislation are costly both to industry and the government.

It is costly for industry to report conservation progress, label product energy efficiencies, convert oil- and natural-gas-fired boilers and power plants back to coal, and so forth. These actions are required by various pieces of energy legislation. For the government, tax credits and subsidies for conservation sources are expensive. For example, Table 9–4 shows the federal government's estimated tax loss due to the tax credit provisions in the Crude Oil Windfall Profits Tax Act of 1980. The total loss directly related to industry can be as high as $4.4 billion between 1980 and 1990. Figure 9–3 exemplifies another would-be costly operation: gasoline rationing (considered by the Carter Administration, but not instituted).

---

* The Reagan administration's energy policy/plan is tied to the free market. The effect of this plan will be positive.

Table 9–4. Tax Loss Due to the Crude Oil Windfall Profits Tax Act.

	Estimated Revenue Loss 1980–90 (Millions of Dollars)
For industry $4,429	
Biomass equipment to convert waste into fuel	$ 648
Cogeneration equipment	356
Coal gasification equipment	277
Geothermal and ocean thermal equipment	34
Small-scale hydroelectric facilities	1,797
Solar or wind equipment	1,058
Production of shale oil and similar products	160
Production of engine-fuel alcohol	99
For consumers .... $ 600	
Home insulation, storm windows, thermostats	67
Solar, wind, and geothermal energy equipment	533
Total	$5,029

Source: Data collected from *Business Week*.[20]

Conflicts Between Regulations

It would be easier for industrial energy conservation efforts if regulations among various agencies were more consistent. There are often clashes between energy and environmental goals. For example, pollution control equipment increased total energy consumption in copper production by 6% from 1972's 166 × 10^{12} Btu to 1976's 176 × 10^{12} Btu, even though refined copper production was reduced by 7.7%.[16] The newly proposed EPA emissions standards for new fossil-fueled power plants could increase utility oil consumption by .4 to .7 million barrels/day in 1990, according to DOE's estimation.[17] In order to encourage cogeneration, many states (e.g., California) have to revise emissions standards. Also, public utility regulations and laws banning the sale of electricity need to be modified.

Offsetting Efforts of Various Energy Legislative Programs. Table 9–5 shows an example of offsetting efforts of various energy legislative programs. The data were obtained from an Energy Information Administration study.[1] The table shows forecasts for 1990 of the effects of individual programs on U.S. oil imports. The entry shown in each case is the difference between the forecast without the program in question and the current programs case.

Fig. 9–3. Sample of Gasoline Rationing Coupons. Source: DOE photo.

Table 9–5. Effects of Energy Legislative Programs on 1990 Forecasts of Oil Imports.

Forecast Changes Due to:	Oil Imports (Quadrillion Btu)
Price controls and related programs	
NGPA	−0.1
Windfall profits tax	0.5
Subtotal	0.4
Conservation and related programs	
Demand management	−0.6
FUA	1.2 to 0.2
Subtotal	0.6 to −0.4
Supply side programs	
Subsidies for new technologies	−0.2
Leasing restrictions	0.5
Subtotal	0.3
Taxes and tax subsidies	
Oil and gas	−1.2
Coal	—
Electric utilities	0.1
Subtotal	−1.1
Sum of changes	0.2 to −0.8

For example, the entry under "Oil Imports" for the NGPA (−0.1 quadrillion Btu) means that the NGPA is forecast to lower U.S. oil imports by 0.1 quadrillion Btu in 1990 as compared to the current programs forecast. Note that these forecasts do not include the interaction of various programs.* Examination of the three legislative programs previously discussed (NGPA, Windfall Profits Tax, and FUA) indicates the following:

- NGPA will reduce oil imports in 1990 by 0.1 quad Btu
- Windfall Profits Tax will increase oil impacts in 1990 by 0.5 quad Btu
- FUA will increase oil imports in 1990 by 0.2 to 1.2 quad Btu

The reduction of oil imports resulting from NGPA is largely offset by the substantial increase in imports due to the Windfall Profits Tax and FUA. If reduction of oil imports in 1990 is the goal of the National Energy Policy, then the Windfall Profits Tax and FUA will not be helpful.

Effectiveness of Legislation and Institutions

The effectiveness of these noneconomic factors in industrial energy conservation can be examined by studying the past performance of some energy conservation efforts.

Housekeeping Conservation Measures. The U.S. General Accounting Office (GAO) reported that most industries surveyed only used housekeeping measures to conserve energy. These measures require little capital investment. Examples are engine tune-ups, light turn-offs, insulation installation, lowered thermostats, steam leak repairs, and so on. Such housekeeping efforts generally can reduce original energy use by 5 to 10%.

The GAO further concluded that "the federal voluntary industrial energy conservation programs had not had a significant impact on the conservation activities of the major companies and industries which we visited."[22] Former Assistant Energy Secretary John F. O'Leary also told a House subcommittee that the voluntary reporting program was in danger of becoming a "public relation–oriented effort on the part of some industry reporting systems" rather than a comprehensive industrial energy audit.

Insignificant Impacts of Policies. The study by the Energy Information Administration[1] reported that government policies will have little effect on the nation's prospects for the next decade. Because many of the policies

---

* The programs examined there prove to have larger effects, taken together, than the sum of the effects of the separate programs.[1]

encouraging conservation and production are often offset by other programs that produce opposite impacts (e.g., Table 9–5), the U.S. energy outlook will be mostly determined by world oil prices and supply, if no consistent U.S. energy policies are formulated soon. Clearly, the price effect on energy conservation will be more significant than energy legislation and programs from energy institutions.

Overgenerous Tax Incentives. As discussed before, because of relatively inexpensive oil and natural gas prices, the federal government must rely heavily on energy credits to encourage conservation and alternate energy production. Examination of other industrial societies' tax incentives indicated that the United States has been too generous in providing tax credits.[3] This is shown in the fact that other industrial societies provide only fast depreciation allowance for business and little tax credits for individuals.

CONCLUSIONS

In this chapter, we have examined impacts of energy legislation, institutions, and environmental regulations on industrial energy conservation. The analysis clearly indicates that these impacts, in many cases, become entangled and cancel each other out. However, sometimes any of them can cause substantial problems for application of particular industrial conservation techniques. Studies of these impacts are needed for applications of any particular conservation measures.

Because of energy legislation, institutions, and environmental regulations are only responses to energy shortages, they often do not catch the essence of energy problems. Overall, they are ineffective because energy problems constantly change. The U.S. energy problems will come from abroad, particularly high oil imports and prices. Oil availability and prices will be the key factors directing future U.S. energy policy. Economics will become the dominant consideration in industrial energy conservation because the legislation, institutions, and environmental regulations will be modified in response to future energy problems caused by high oil imports and oil prices.

REFERENCES

1. Energy Information Administration. "Energy Programs/Energy Markets," Department of Energy, DOE/EIA-0201/16, July 1980.
2. Federal Energy Regulatory Commission (FERC). "FERC Regulations Implementing the Incremental Pricing Provisions of the Natural Gas Policy Act of 1978," Docket No. RM 79–14, January 12, 1979.
3. "Saving Energy through Tax Rewards gets more Play in the U.S. than Abroad," *Wall Street Journal,* May 7, 1978.

4. U.S. Department of Energy, Office of State and Local Programs. *Source Book of Energy Conservation Programs,* Vol. 8. Washington, D.C.: U.S. DOE, May 1980.

5. Ohio Department of Energy. *Ohio Energy Conservation Program Evaluation for 1979.* Columbus, Ohio: Ohio Department of Energy, 1979.

6. California Air Resources Board. *Air Pollution Control in California.* Sacramento, Calif.: California Air Resources Board, 1979.

7. Governor's Energy Council. *1979 Pennsylvania Energy Conservation Plan.* Harrisburg, Pa.: Governor's Energy Council, 1979.

8. N.J. Department of Energy. *New Jersey 1980 Revised Energy Conservation Plan.* Newark, N.J.: New Jersey Department of Energy, 1980.

9. N.Y. State Energy Office. *1980 New York State Energy Conservation Plan and Supplemental Energy Conservation Plan.* Albany, N.Y.: New York State Energy Office, 1980.

10. Common Cause. "The Path Not Taken: A Common Cause Study of State Energy Conservation Programs," Washington, D.C., 1979.

11. National Bureau of Standards. "Waste Heat Management Guidebook," NBS Handbook 121, February 1977.

12. U.S. Environmental Protection Agency, Air Pollution Control Administration, Division of Air Quality and Emissions Data, Durham, North Carolina.

13. Ferris, F., Douglas, J., Waller, R., et al. "Air Pollution and Respiratory Infection in Children," *British Journal of Preventive Social Medicine,* Vol. 20, 1–8, 1966.

14. S.M.I.C. *Inadvertent Climate Modification,* Report of the Study of Man's Impact on Climate. Cambridge, Mass.: MIT Press, 1971.

15. U.S. Environmental Protection Agency. "Development Document for Effluent Limitations, Guidelines and New Source Standards for the Steam Electric Power Generating Point Source Category," EPA 440/1–74 029a, 1974.

16. U.S. Department of Commerce and U.S. Department of Energy. "Voluntary Industrial Energy Conservation," Progress Report 5, DOE/CS-0035, reprinted March 1978.

17. Investor Responsibility Research Center, Inc. "Energy Conservation by Industry," prepared by Scott Fenn, Washington, D.C., January 1979, p. 38.

18. Shy, C. M. et al. "The Chattanooga School Children Study: Effects of Community Exposure to Nitrogen Dioxide. II. Incidence of Acute Respiratory Illness," *Journal of the Air Pollution Control Association,* Vol. 20, 582–588, 1970.

19. "Summary Report of the Cornell Workshop on the Environment (February 22–24, 1972)," supported by the National Science Foundation Research Applied to National Needs (RANN) Program, Ithaca, N.Y., Cornell University, April 1972.

20. "Tax Credits that Could Save Industry Billions," *Business Week,* April 7, 1980, p. 7.

21. U.S. Department of Energy. "The National Energy Act," DOE Information Kit, DOE/OPA-0003, November 1978.

22. U.S. General Accounting Office, "Report to the Congress," EMD-78–38, June 1978.

10

Economics of Energy Conservation Policies—from Industry's Point of View*

INTRODUCTION

United States oil and gas price regulation has made energy-saving investments less attractive than they otherwise might have been. The questions thus arise: At what energy prices will conservation be attractive? What will be the economics of energy conservation policies? This chapter answers these questions from industry's point of view and analyzes U.S. primary copper production to illustrate a methodology and to derive some inferences. In Chapter 11, we will discuss these problems from the point of view of the nation's energy policy.

Background

The energy required to produce copper increased from 81.254 million Btu (MMBtu)/ton in 1963 to 98.864 MMBtu/ton in 1973. During that decade, copper smelting and refining technologies did not change significantly; the increase in energy consumption can be attributed largely to the use of lower grades of ore. Of the energy used in copper production, oil and gas account for about half, and electricity accounts for the other half. The average energy price paid by the copper industry in 1973 was $.97/MMBtu.[1]

---

* A version of this chapter appeared in *Energy Economy.*[17]

Recent rising oil prices and the federal voluntary industrial energy conservation program were expected to halt increases in industrial energy consumption by active energy conservation methods. However, in the *Report to the Congress,* the U.S. General Accounting Office (GAO) found that most industrial energy conservation activities involved only operational changes instead of extensive energy conservation investments.[3] Total industrial energy consumption was reduced by 4.2% from 1972 to 1976. Most of those energy savings came from reductions in production. In particular, total energy consumed for the primary copper companies reporting increased from 166 trillion Btu in 1972 to 176 trillion Btu in 1976 (a 6% increase), despite a drop in refined copper production from 1.95 to 1.80 million short tons.*

In its report the GAO indicated that the large-energy-consumptive industries generally have the knowledge and technical expertise to conserve energy. However, past and current federal regulations covering oil and gas prices have caused energy prices to be lower than they would otherwise have been. Consequently, investments with the primary purpose of saving energy could not compete economically with other investments.

Clearly, the primary question arising from the above discussion is: What are the costs and benefits of energy conservation in copper production? A secondary question logically follows: What would be the energy price when copper companies invest in energy conservation technologies? In this chapter, we will discuss the methodology used to answer these questions, the costs and benefits of alternative conservation measures, the industrial conservation economics and required energy prices.

METHODOLOGY

In contrast to the econometric aggregate approach, a micro approach is used to measure costs and benefits of energy conservation measures in the primary copper industry, which serves as an example of a multi-plant, multi-process, single-product, energy-intensive industry. Three leading companies, which share about 50% of the market, are analyzed in detail. The methodology consists of the following steps:

- Evaluation of costs and benefits of three major conservation measures (recycling, retrofit, and new process)
- Assessment of the market penetration of the three measures on existing facilities of the three leading companies through production simulations

The costs and benefits of implementing each of the three major conservation measures at a standard facility are analyzed. As previously mentioned, most

---

* Part of this increase was due to pollution controls.[4]

operational changes to reduce energy consumption with little or no costs have already been made. Consequently, the alternative measures will increase production costs.

Each company uses four stages (mining, milling, smelting, and refining) to produce refined copper, and owns many facilities for each process; these facilities are located in different areas, and their ages are different. Assuming that each company attempts to maximize profits, a mathematical program is formulated to simulate the company's copper production. The program models the company's facilities, transportation paths, and production patterns. It maximizes the company's profit under constraints of facility capacities, sales commitments, pollution controls, and other institutional regulations. Finally, it simulates various degrees of use of the three energy conservation measures in the company's facilities.

Significance of the Methodology

Since the 1973 Arab oil embargo, most economists as well as national policy-makers have focused on the national aggregate elasticity of GNP with respect to energy. Econometric models such as PILOT, Kennedy-Niemeyer, Wharton, Hudson-Jorgenson, Hnyilicza, and DRI-Brookhaven were completed to estimate the elasticity of GNP based on respective aggregate substitution between energy and other production factors. Both the Kennedy–Niemeyer and the PILOT models, which assumed limited substitution, display very small aggregate elasticities (between 0 and .1). The remaining models, which include higher substitution possibilities, trend toward large, long-run elasticities (between .3 and .5).[16]

Evidently, the substitution between energy and other production factors is the key in determining the elasticity of GNP, but few studies of this substitution have been made. Because each industry uses different production processes and conservation technologies, aggregate substitution for the industrial sector may be inappropriate. This chapter is thus designed to evaluate energy substitution in individual companies within the copper industry and to make some inferences about industrial energy conservation.

ECONOMICS OF ALTERNATIVE CONSERVATION MEASURES*

Alternative Conservation Options

There are several ways to conserve energy in the primary copper industry. One way is to import copper from abroad instead of producing it in this

---

* This topic of this section was discussed in detail in Chapter 7. For the completeness here, however, we summarize some of the results given in Chapter 7.

country. This is certainly a viable choice from the standpoint of energy conservation, but is unlikely, owing to the high deficit in balance of trade. As a result, analysis in this chapter is based on the expectation that the copper industry will reduce its energy consumption without reducing its production. Energy consumed at each stage of copper production is shown in Fig. 10–1.

Battelle Columbus Laboratories, in various reports concerning metallurgical mineral processing, has indicated that there is great potential for energy conservation in the copper industry.[5,6] Three promising options are compared in this section:* (1) using scrap, (2) replacing reverberatory smelters with flash smelting, and (3) installing tonnage oxygen plants in reverberatory smelting processes. These measures are expected to conserve energy but not to reduce copper production. Each alternative produces various costs and benefits to the economy and the environment. Our emphasis here is on the economic impacts.

Using Scrap Copper Instead of Ores

The U.S. Bureau of Mines estimated that more than 11 million short tons of scrap copper were unrecovered in 1970.[15] In 1972, the primary producers consumed 30% of the total yearly unalloyed copper and copper base scrap in the United States. Forty-five percent of the scrap used was low-grade, and 55% was high-grade (No. 1 and 2). The low-grade scrap was fed into the smelting processes to replace part of the ore concentrate, and high-grade scrap was used in refining processes as a substitute for blister copper.

Cost and Benefit. Energy savings due to the use of copper scrap depend on the grade of scrap used. Since the scrap is being introduced at different points in the processes depending on its grade, calculating the energy savings associated with various substitutions must be done on a case-by-case basis. Substitution of high-grade scrap for blister copper conserves almost as much energy as would have been required to produce the blister. Using low-grade scrap in place of ores saves the amount of energy that would have been used in mining and milling an equivalent amount of ore. According to Battelle, these energy savings from low- and high-grade scrap can range from 66 to 92.5% of unit energy requirement to produce equivalent amounts of refined copper from ores and copper-in-process, respectively.[5]

The unit gain (or loss) incurred from recycling high-grade scrap is the difference between the price of the copper content of No. 2 heavy scrap (54.87¢/lb in 1974) and the sum of the unit production cost of mining, milling,

---

* In mining, leaching and hydrometallurgy, which may save large portions of energy used in milling, are in experimental stages as well.[7,10] Therefore, these measures are not discussed.

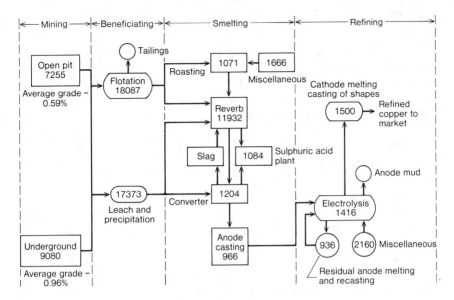

Fig. 10–1. Energy consumption at each stage of copper production (Btu/lb). The average total for the four processes was 49,432 Btu/lb in 1973. Source: Rosenkranz, R.[1]

and smelting an equivalent amount of blister. The unit gain (or loss) incurred from recycling low-grade scrap is the difference between the price of low-grade scrap (No. 1 composition, 43.75¢/lb in 1974) and the cost of producing concentrates from ores (mining and milling costs). Prices for No. 2 heavy scrap and No. 1 composition (red brass) copper scrap are published daily in the *American Metal Market.* Since the scrap market is significantly competitive, we may assume any increases in scrap demand will result in increases in scrap prices.

For the purposes of this chapter, it is impractical to detail the energy savings and costs (or gains) accounted for by using scrap for each process in each facility. (See Chapter 7 for details.) There are too many facilities in each company. Rather, they can be computed individually as needed.

Replacing Existing Furnaces with Flash Smelting

Kellog and Henderson examined energy requirements for alternative sulfide smelting processes and concluded that a high potential exists for energy conservation using flash smelting to replace reverberatory furnaces that do not use oxygen-enriched air.[9] Flash smelting is a process by which copper sulfide ore concentrates are smelted by burning a portion of the iron and sulfur contained in the concentrates while they are suspended in an oxidizing environment. It typically produces high-grade mattes containing 45 to 65% copper.[2]

Two companies have developed flash smelting technology, Outokumpu Oy in Finland and International Nickel Company (INCO) in Canada, and by mid-1973, thirteen copper-smelting installations in the world operated flash smelting furnaces ranging in capacity from 300 tons/day to 1500 tons/day of copper concentrates. One installation is of INCO design; the remaining 12 are of Outokumpu design. For this study we compared a model plant using flash smelting with a model plant using a reverberatory furnace to estimate the costs and benefits of replacing existing furnaces with flash smelting.

Cost and Benefit. Flash smelting requires only about 40% of the energy needed by the existing reverberatory smelters. (See Table 7–7.) Since it produces high-grade mattes, it produces far less air pollution per unit of copper.[2]

Based on 1974 prices for energy, labor, materials, and capital, annualized capital costs and operation costs are analyzed for both the flash and reverberatory smelters. Capital costs are then annualized through straight-line depreciation over a 20-year smelter life cycle. Operation costs of a new flash smelter are 18% less than those of a new reverberatory smelter. However, if an existing reverberatory furnace is replaced with a flash smelter, the annualized costs of demolishing the existing facility and compensating for its remaining value make the flash smelter 7% more costly than the reverberatory smelter. (See Table 7–10.) This fact helps explain the General Accounting Office's finding that most energy conservation activities involved only operational changes. However, if energy prices increase faster than the prices of material, labor, and capital, then the flash smelter may become attractive.

Installing Tonnage Oxygen Plants in Reverberatory Smelting Processes

Oxygen efficiency in a reverberatory furnace is nearly 100%, since essentially all the oxygen in the combustion air is used in burning up the fuel.[8] A principal reason to use tonnage oxygen rather than ambient air in reverberatory smelters is that the capacity of a reverberatory furnace is limited by the amount of nitrogen in the combustion air supply (ambient air generally is 80% nitrogen). Therefore, replacing part (or all) of this air with oxygen will reduce the total volume of nonoxygen gases introduced into the reverberatory furnace and will permit combustion of a larger quantity of fuel, thereby increasing the smelting throughput rate.[13]

Cost and Benefit. In an evaluation study, INCO Ltd., in Canada, equipped one of its coal-burner reverberatory furnaces with four water-cooled oxygen "lances," one below each of the coal-burners, and directed away from the furnace side-wells. The improved fuel efficiency shows that one ton of oxygen is almost equivalent to 0.95 ton of powdered coal, and that the reverberatory smelter equipped with tonnage oxygen plants uses about 7% less energy than a reverberatory smelter without tonnage oxygen plants.[11] (See Table 7–11.)

To estimate economic impacts of this conservation measure, 1974 prices for energy, labor, materials, and capital are used. Installing the reverberatory smelting processes with tonnage oxygen plants under 1974 price conditions will increase the cost of copper produced by $1.42/ton. (See Table 7–12.) This statement again explains the GAO findings previously discussed.

ENERGY CONSERVATION AND REQUIRED ENERGY PRICES

Three leading companies (A, B, and C) were chosen as representative of the industry and were analyzed in detail. The first firm primarily mines, smelts, refines, and sells products by itself. The second one is a partially custom-smelting and refining company that does not produce sufficient ores for itself and uses large portions of its input from scrap or other firms. The third company has large deposits of rich ores and produces more ores than it uses. In 1974, these three companies composed 50% of the estimated recoverable copper reserves, contributed 40% of the total mine production, held 51% of the total smelting works, owned 51% of the total copper refinery capacity, and shared 54% of the total refined copper sales in the United States.[7,16,2]

Primary Copper Production with Energy Conservation Measures

A simple way to evaluate a company's conservation choices is to simulate primary copper production with the objective of pursuing maximum profits under the constraints of energy conservation, facility capacities, current production commitments, and other environmental requirements. The problem can be formulated as follows:

Maximize: Profit $= f$ (unit price, production cost, quantity)
Subject to: (1) Quantity $= g_1$ (capacity, utilization, energy conservation requirements, sales commitments, transportation, pollution controls)
(2) Energy reduction level
(3) Price $= g_2$ (quantity, competition)

where f, g_1, and g_2 are symbols for functions.

Given the three conservation options previously discussed, each primary copper producer will have $(m + 1) \times (3n + 1) \times p$ production paths to choose if a company has m mines–mills, n smelters, and P refineries. These combinations are shown in Fig. 10–2.

Constant 1974 revenues are assumed because the company is assumed not to reduce its production in order to save energy. Therefore, the choice of conservation measures can be approached by a mathematical program that attempts to minimize the company's production costs under production constraints. Because the cost functions for each facility (or process) are not linear with respect to various production levels, they are approximated by step-functions. Consequently, a mixed-integer program is used to generate production schedules. Cost and capacity data for each facility in the industry are generated from open publications,[12,14] with the aid of some confidential data sources.[13] The reader interested in details of mixed-integer programs is referred to Chapter 8.

The cost and benefit relationships for the three conservation measures are applied to appropriate facilities in a company. The mixed-integer program is run through a computer four times to generate data of cost changes with respect to 5, 10, 15, and 20% of energy reduction. These data are then used to estimate a function governing cost changes and energy reductions.

Conservation Economics

Cost changes responding to three levels of energy reduction are shown in Table 10–1. Clearly, each company adopts different combinations of the three conservation measures. The increase in production costs accounted for by the conservation measures is estimated at less than .1% per 1% of energy saved.* Our computation indicates that it is economically unfeasible to maintain the 1974 production level while implementing the three conservation measures to reduce energy consumed by 20%.[11] The reason for the cost increase is that the companies have already made the necessary operational changes to optimize their energy use in the light of expected prices. Any further energy reduction will thus increase production costs. The cost increase rate in the first 5% of energy reduction is greater than in the next 5 or 10% (see curve A, Fig. 10–3.) In the early stage of conservation, companies may be reluctant to change the production process and are likely to adopt only less costly short-run measures, such as using scrap and installing tonnage oxygen plants. But these measures are inefficient in the long run. When more energy reduction is required, companies will have to change their production processes in order to accomplish the conservation goal. The economics of

---

* Small firms in the industry may not be able to enjoy the same tax credit opportunity; therefore, they will be at a slight disadvantage.

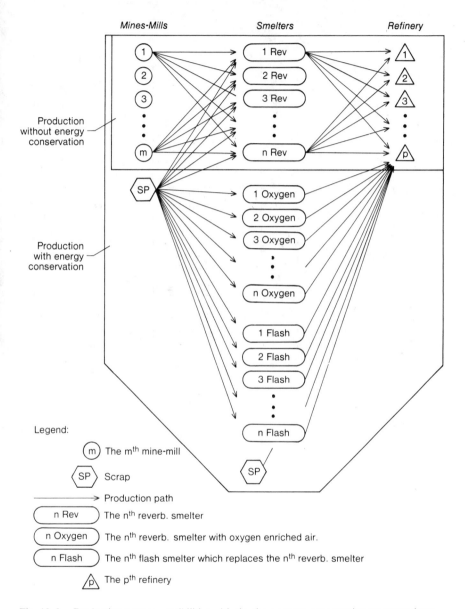

Fig. 10–2.   Production pattern possibilities with the three energy conservation measures (recycling, retrofitting, and changing processes).

Table 10–1.   Economics of Energy Conservation for the Primary Copper Production.*

Company/Industry	Energy Saved (∇E)		Conservation Methods Implemented† (Production in 10³ Short Tons)		Economic Impacts	
	10⁹ Btu	% of Original Consumption			Cost Increased, ΔC ($10⁶)	% of Original Cost
Company A	1.376	5	(A) 7.2 (B) 189.4	(C) 292.5	2.544	.51
	2.753	10	(A) 7.2 (B) 293.0	(C) 188.9	4.033	.81
	4.129	15	(A) 7.2 (B) 396.6	(C) 85.3	5.523	1.12
Company B	.93	5	(A) 7.4 (B) —	(C) 472.0	1.9785	.34
	1.863	10	(A) 7.4 (B) 56.7	(C) 415.3	3.1305	.53
	2.794	15	(A) 7.4 (B) 125.8	(C) 346.2	4.3163	.73
Company C	.763	5	(A) — (B) 89.6	(C) 112.9	1.3727	.40
	1.527	10	(A) — (B) 146.7	(C) 55.8	2.2379	.66
	2.290	15	(A) — (B) 202.5	(C) —	3.1031	.91

* The industry is assumed to be represented by the three leading companies which share 50% of the total market.

† (A)—using scrap; (B)—flash smelting; and (C)—retrofitting, tonnage oxygen

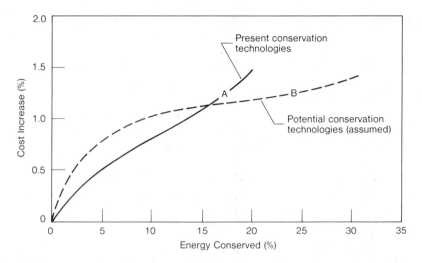

Fig. 10–3. Economics of energy conservation in copper production with present and future conservation technologies under 1974 price conditions.

scale apply, and the rate of cost increase becomes smaller as more efficient long-term conservation measures are introduced. However, conservation limitations exist at near 20% energy reduction for the present conservation technologies. Unless new technologies are invented, additional conservation becomes very costly. (See curve A.) Curve B shows the conservation possibility resulting from the theoretical implementation of potential conservation technologies.

One interesting point deserves our attention. Figure 10–3 indicates that conservation technologies are the key to energy substitution. Since energy conservation potentials vary from industry to industry, each industry's energy substitution possibility is different. Therefore, an aggregate approach to computing substitution for the industrial sector is inappropriate, although many economic models have been based on aggregate substituion elasticities.

Energy Price Increase Required for Conservation

Data concerning energy saved (∇E) and cost increase (ΔC) in Table 10–1 can be used to estimate equations governing the relationship between these two variables for the three companies as follows:

$$\Delta C_A = 1.41 \ \nabla E_A \qquad (10\text{–}1)$$

$$\Delta C_B = 1.63 \ \nabla E_B \qquad (10\text{–}2)$$

$$\Delta C_C = 1.42 \ \nabla E_C \qquad (10\text{–}3)$$

where ΔC_A, ΔC_B, and ΔC_C are conservation costs in thousands (1974 dollars) incurred from installing energy-saving equipment, and ∇E_A, ∇E_B, and ∇E_C are energy savings in billion Btu for companies A, B, and C, respectively.

We will briefly explain the implication of the above equations. Companies A and C have relatively large deposits of rich ores, and therefore have the opportunity to obtain low-cost ores while conserving energy. Their cost increments per billion Btu conserved are almost identical. For company B, because it has low-grade ores, which are energy-consuming, energy conservation mostly comes from improving smelters and using scrap, which is more expensive.

Owing to technological limitations in energy savings, and the resulting high cost of further penetration for scrap copper, the cost increases per billion Btu energy savings would be much higher than those shown in Equations 10–1, 10–2, and 10–3 if more than 20% of the present energy consumption were to be conserved.

If prices for capital, labor, and materials are maintained at the 1974 level, the required average price increase (ΔP) to offset conservation costs increases in Equations 10–1 through 10–3 can be expressed as follows:

$$\Delta P = \frac{\Delta C}{\nabla E} \qquad (10\text{–}4)$$

Equations 10–1, 10–2, and 10–3 are then substituted for Equation 10–4, and the average price increases required for reducing energy currently consumed by up to 20% are computed and shown in Table 10–2 for the three companies. The average 1974 energy price paid by the copper industry is estimated at $1.504/MMBtu.* Clearly, if average energy prices paid by the copper industry increase by nearly 110% of the 1974 price level, then the three leading companies will invest in energy conservation technologies. Since the three companies are leaders in the primary copper industry and share about 50% of the refined copper market, we may conclude that the copper industry is likely to invest in energy conservation technologies at this level of energy price.

OTHER FACTORS TO BE CONSIDERED

The previous inferences were based solely on the economic consideration. However, there are other factors to be considered besides economics, such

---

* The 1973 average price, $0.97MMBtu, times the 1974 energy price index, 1.5503.

Table 10-2. Average Price Increases Required for Reducing within 20% of Energy Currently Consumed in Copper Production. (in 1974 dollars)

Company	Required Price Increase ($/MMBtu)	Percent Increase Compared to 1974 Average Energy Price Level*
A	1.41	93
B	1.63	108
C	1.42	94

* 1974 average energy price was estimated at $1.504/MMBtu.

as conservation targets, efficiency standards, a company's public image, and the uncertainty of future fuel supply and price. In many cases, any of these factors may override the economic consideration. For example, if a copper producer has to comply with a mandatory industrial energy conservation target of 15% reduction in energy use per unit of production, then the producer may have to implement some conservation measures that are economically unfavorable.

It is, therefore, necessary for an industrial firm to consider all factors concerning energy conservation. However, we must emphasize that economic considerations should play an important role if we want the market to determine the allocation of resources. In the next chapter, we will discuss energy management and policy from the point of view of the nation's energy policy.

CONCLUSION

Under 1974 price structures for energy, capital, labor, and materials, it is estimated that energy conservation will increase the production costs of the three leading major primary copper companies by less than .1% of 1974 costs for every percent of energy reduced from 1974 consumption level. This estimation of conservation costs is valid only for saving rates of up to 20% of 1974 energy consumed, the limit at which the current technologies can be employed to reduce energy uses economically.

If prices for all nonenergy production factors are maintained at the 1974 level, a nearly 110% increase in average energy prices paid by the copper industry will represent the threshold level at which investments to save energy become economically attractive to the copper companies. This influence can be altered if other factors (e.g., fuel unavailability, Congress's mandatory conservation program) are emphasized in the consideration.

REFERENCES

1. Rosenkranz, R. "Energy Consumption in Domestic Primary Copper Production," Bureau of Mines Information Circular, 8698.
2. Office of Air and Waste Management. *Background Information for New Source Performance Standards: Primary Copper, Zinc, and Lead Smelters,* Vol. 1: *Proposed Standards.* Research Triangle Park, N.C.: U.S. EPA, 1974.
3. U.S. General Accounting Office. "Report to the Congress," EMD-78-38, June 1978.
4. U.S. Department of Commerce and U.S. Department of Energy, "Voluntary Industrial Energy Conservation," Progress Report 5, DOE/CS-0035, reprinted March 1978.
5. Battelle Columbus Laboratories. "Evaluation of the Theoretical Potential for Energy Conservation in Seven Basic Industries," final report to Federal Energy Administration, NTIS, July 1975.
6. Battelle Columbus Laboratories, "Energy Use Patterns In Metallurgical and Nonmetallic Mineral Processing (Phase 4—Energy Data and Flowsheets, High Priority Commodities)," interior report. USBM Open File Report 80–75. Washington, D.C.: GOP, 1975.
7. Silverman, B. "Copper Resoumetric Model," unpublished Ph.D. dissertation, University of Pennsylvania, 1977.
8. Saddington, R., Curlook, W., and Queneau, P. "Use of Tonnage Oxygen by the International Nickel Co.," pp. 261–269 in (J. Anderson and P. Queneau, eds.) *Pyrometallurgical Processes in Nonferrous Metallurgy.* New York: Gordon and Breach, 1967.
9. Kellog, H. and Henderson, J. "Energy Use in Sulfide Smelting of Copper," in (J. Yannopoulos, ed.) *Extractive Metallurgy of Copper,* AIME, International Symposium of the Met Society, 1976.
10. Phillips, T. "Economic Evaluation of Process for Ferric Chloride Leaching of Chalcopyrite Concentrate," U.S. Bureau of Mines Information Circular 8699. Washington, D.C.: GOP, 1976.
11. Hu, S. D. "The Copper Commodity Model and Energy Issues," unpublished Ph.D. thesis, University of Pennsylvania, 1978.
12. American Bureau of Metal Statistics. *Yearbook of the American Bureau of Metal Statistics.* New York, 1975.
13. A. A. Mathews, Inc. "Capital and Operating Cost Estimation System for Mining and Benefication," Mining Report No. 1953–02, Phase II Report, for U.S. Bureau of Mines, Vol. 4, June 1976.
14. Bennett, H. et al. "An Economic Appraisal of the Supply of Copper for Primary Domestic Sources," Bureau of Mines IC 8598, 1973.
15. Carrillo, F. et al. "Recovery of Secondary Copper and Zinc in the United States," Bureau of Mines Information Circular C 8622. Washington, D.C.: GPO, 1974.
16. Energy Modeling Forum. *Energy and the Economy,* Vol. 1, September 1977.
17. Hu, S. D. and Zandi, I. "The Economics of Energy Conservation Policies: A Study of U.S. Primary Copper Production," *Energy Economy,* Vol. 1, No. 3, pp. 173–179, July 1979.

11

Energy Management and Policy from the Nation's Point of View

INTRODUCTION

In the previous chapter, we based our analysis of energy conservation policies solely on economic considerations and noted that legislative and environmental considerations that may change the direction of analysis are not included. The legislative and environmental constraints were discussed in Chapter 9 and will be considered together with economic considerations in this chapter to yield a compromise strategy for federal, state, and local governments that face multi-goals and multi-interest groups.

Therefore, the purpose of this chapter is to provide an analytical tool that can take into consideration weights of various social goals and influences from different interest groups, for a governmental body to formulate compromise regulatory and fiscal strategies in achieving various social goals. The energy conservation goal is then used as an example. The essence of the tool is its methods for quantifying benefits and costs of various goals and influences from different interest groups, and for setting priorities among these goals and interest groups by pairwise comparison. The output of the analytical tool is a set of optimal strategies that will order the goals by weighting their costs and benefits and the influences of interest groups to achieve energy conservation targets or other regulatory targets (e.g., pollution reduction).

In order to consider all relevant information as completely as possible, the analytical tool consists of three basic simulations and a data management system:

- A products market simulation to represent market supply and demand for the products of the given industry, their substitutes, and complements
- A production simulation to represent input–output relationships of the given industry
- A prioritization simulation to represent the decision-making process of governmental bodies pursuing certain targets
- A data management system to handle data and formulas for linking the three simulations

The analytical tool works this way: The products market and production simulation models generate data concerning costs and benefits of alternative fiscal and regulatory scenarios designed by a governmental body to encourage or enforce a particular target such as a certain percentage of reduction of current energy consumption. Then the prioritization simulation model weighs all cost and benefits against various goals and influences of various interest groups to yield a compromise (optimal) strategy. This tool can be used to evaluate scenarios for energy conservation or it can be modified to be used to evaluate scenarios for other social goals like pollution reduction.

For convenience of our discussion, this chapter is divided into: the problem, the methodology, products market simulation, production simulation, prioritization simulation, data management system, energy management and policy—example of energy conservation in the primary copper production, and conclusions. Energy conservation will be used through the chapter to illustrate the use of this analytical tool.

THE PROBLEM

In pursuing energy conservation, federal, state, and local governments have available a number of regulatory and fiscal authorities to foster changes in the behavior of both industries and individuals. In general, there are two distinctive categories of questions to be asked:

- What can government do to encourage industry and business to perform in such a manner that a certain national goal can be achieved in a certain period of time frame? For example, if the government wishes to encourage a given industry to conserve a certain percentage of its current energy consumption in the next five years, what policy instruments or combination of policies can be employed to expedite the change of energy consumption pattern and achieve the goal?

- Who will be affected by these policy instruments? What will be their costs and benefits? What would be the compromise to satisfy most interest groups and other goals competing for the same resources?

Clearly, these questions originate from the government's desire to encourage a given industry to achieve a certain goal. Possible policy instruments can be fiscal policies (e.g., government subsidies, loans, energy tax credits, accelerated depletion or depreciation allowance), or regulatory measures (e.g., mandatory conservation targets for given industries and appliance efficiency standards). A variety of organizations and people would be affected if these instruments were implemented. They include industrialists, environmentalists, tax collectors, politicians, and consumers. An analytical tool is therefore needed to take all of these factors into consideration before any policy is implemented.

THE METHODOLOGY

The analytical tool proposed in this chapter is called an energy compromise model. As shown in Fig. 11–1, the model consists of three types of simulation and a data management system. The energy compromise model was outlined in the introduction and is explained as follows:

- A products market simulation which represents the relation of market supply and demand for the products of the given industry, their substitutes and complements. The basic function of this simulation is to predict the trend of the product for a given industry. Taking data from the data management system, the products market simulation generates as output prices and quantities for main products, substitutes, and complements under alternative scenarios.
- A production simulation which represents the production of individual firms in a given industry. It represents resource, conversion, and transportation of these firms. Taking data from the data management system and product price and quantity data from the products market simulation, it produces information on benefits and costs of production under alternate scenarios.
- A prioritization simulation which represents the decision-making process of a given governmental body. This simulation takes into account conflicting goals and competing interest groups in government decisions. Taking information of interest groups and goals from the data management system, data about product prices and quantities from products marked simulation, and data about benefits and costs of production from the production simulation, the prioritization simulation will either

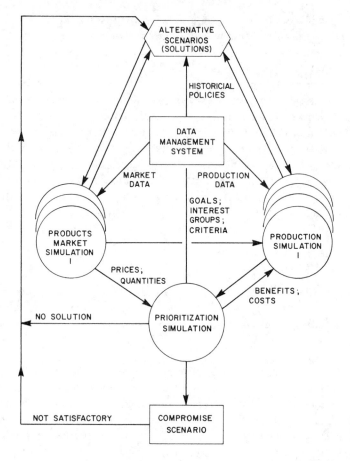

Fig. 11-1. The energy compromise model: flow of multi-goal, multi-interest group decision-making process.

result in no solution or a compromise scenario. This sceniario balances various goals and different interest groups or rejects all alternative scenarios selected. A new set of alternative scenarios must be chosen if all alternatives are rejected. The compromise scenario can be reexamined in the production simulation to yield more detailed benefit and cost information.

• A data management system which systematically contains information and formulas for linking the three simulations as described above. It also includes historical scenario data as references in constructing new scenarios to achieve certain goals.

Logic of Model Operation

The operation of the energy compromise model can best be described by the example of energy conservation in the refined copper production. In this example, a copper econometric model ("econometrics") is used for the products simulation, a copper resource assessment model ("resoumetrics") for the production simulation, a hierarchical analysis for the prioritization simulation, and a data management system for the linkage among the above three components.

By synthesizing three quantitative submodels—econometrics, resoumetrics, and hierarchical analysis—we attempt to produce a result that is sensitive and responsive to regulatory policies. The output of the analysis will be the assessment of the impacts of regulatory policy decisions by a given governmental body. These implications are identified in terms of supplies, costs, prices, material and energy requirements, water needs, environmental changes, land disturbances, labor requirements, international trade deficits, and so on. The construction of the methodology will be specified to the life cycle of refined copper.

The econometric model is discussed only where it is concerned with and related to refined copper. The resoumetric model is concerned with the flow of materials and energy in producing refined copper. The hierarchical analysis addresses the problems of decision-making in the social environment.

A decision by a regulatory agency interrelates not only with the responses and decisions of the same agency, but also with those of other public agencies, industry, labor, and international groups. This complex web of decisions and responses is translated into a flow of materials and energy within the physical environment, and monetary flows in economic systems, and possibly gives rise to changes in social values. The consequences of these flows are felt by all of the above-mentioned decision-makers (actors). In turn, each actor responds to changes he experiences and initiates a new set of feedback ripples which propagate anew throughout the systems.

A noteworthy point is that each actor responds to a stimulus of change according to his perception of interests in those changes. In addition, the actor reacts on the basis of available data no matter what the realities are. Therefore, any framework of analysis, if it is to be helpful to a decision-maker, must perceive the system from an overall standpoint. However, in the physical and economic worlds, events occur independently of the decision-makers' perceptions of reality. Therefore, the data that form the basis of the decision must originate independently of the decision-maker's biases—if the decision-maker is to have a chance to achieve the desired goal.

This state of affairs fortunately can be formulated into an assessment methodology through a scenario analysis with the help of the econometrics, resoumetrics, and hierarchical analysis models. A scenario is a portrayal of the state of affairs with a strong focus on a particular policy. An idea or interest is emphasized with an "adequate" account of its interaction with economic, sociopolitical, technological, resource, and environmental factors. Based on the decision-maker's information, interests and perceptions of other decision-making bodies' interests, and estimates of their behavior, he can create a number of scenarios allowing the dominance of specific social forces, or a combination of them.

The resoumetric and econometric models can then consider the set of policies corresponding to each scenario and provide the quantitative description of the "state of the commodity" in terms of quantitative parameters—such as level of production, demand, cost, price, water requirement, environmental disturbances, residual production, energy and other material needs, labor, and so forth—under the conditions set by each specific scenario.

At this point, the hierarchical analysis permits a decision-maker to view the whole set of scenarios and their resoumetric and econometric implications, make assumptions as to the behavior of other actors, and construct a "composite (compromise) scenario" of the state of the commodity that would result from the interactions of decisions made by various actors. Resoumetrics and econometrics will then estimate the costs and benefits associated with this composite scenario, thus enabling the decision-maker to assess the consequences of the decision. These consequences, as perceived by the actor, will be evaluated against the predetermined goals and, if desired, new decisions will be made that in turn will set the wheels in motion again.

The research procedure is outlined in the energy conservation flow chart of Fig. 11–2 and discussed below.

Initially, the econometric model generates information about market prices, demand, supply, and generation of scrap of various copper products for the whole industry. Second, a historical average profit in percentages is estimated for each company based on published or other available financial information. The differences between the prices and this profit establish the first approximation of products' cost for each company. Third, the cost function developed in the resoumetric model is used to estimate the quantity of production for each company at the calculated cost level. The sum of these productions, estimated in the above manner for all of the companies active in the industry, should be equal to or close to the production level estimated by the econometric model. If there is an unacceptable level of discrepancy, a systematic adjustment of profit level has to be made to bring these two quantities close to one another. Fifth, production functions developed in the resoumetric model are used to estimate the amount of energy needed by each company. The summation of this estimated energy yields the total energy need in the industry. This is the quantity that governmental policy attempts to reduce.

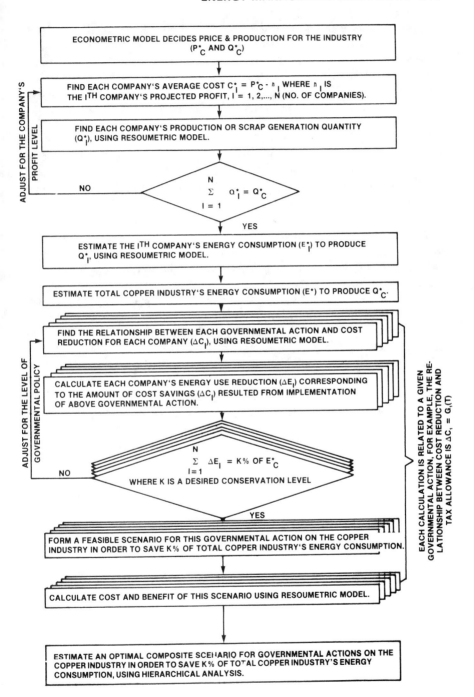

Fig. 11-2.   Energy conservation flow chart.

Since most measures to conserve energy will change production costs, it may be convincingly argued that the government, under the free enterprise system, will only be successful in bringing about the realization of this policy if it somehow compensates industry for any additional costs incurred by introducing the energy conservation. The following are some of the alternative governmental actions that can be used for this purpose:

- Additional tax credit for energy-saving equipment
- Corporation income tax rebate or tax amortization for the installation of energy-saving devices
- Subsidies or aids in the form of governmental contracts and floor price purchase agreements, etc.

There are many other examples of alternative policies available for governmental action.

It is obvious, however, that the best acceptable scenario will probably entail a composition of these various alternatives. The energy compromise model must have the capability of finding this composite scenario, a task that will be accomplished by initially examining each policy alternative separately and using the technique of the hierarchical analysis to create the desired course of action (or composite scenario).

Again referring to the flow chart of Fig. 11–2, consider the possibility of a tax credit to compensate for the expected cost increase incurred by the industry in its attempt to conserve energy. It can be argued that, provided the tax credit is sufficient to offset increased costs, the industry would not resist the conservation measure. The problem then reduces to estimating the percentage of tax credit that should be given to cause the desired energy conservation.

Using the resoumetric model, we can input a certain percentage of tax credit and calculate the benefit that would result for each company. Also, through the resoumetric model, we can calculate the amount of energy saved through the expenditure of this accumulated tax credit by each company on energy-conserving equipment. The sum of these energy savings for all the companies active in the industry can be calculated and compared with the desired level of conservation. If this does not meet the desired goal, then the size of the tax incentive will be reassessed, and the calculation will be repeated until an acceptable overall energy saving is achieved.

In general, such an iterative procedure can be used to estimate quantitatively the level of changes required for each governmental policy option (some of which are enumerated above) to produce the desired energy savings. In the real world, however, the best decision normally will not be associated with change in a single option. The most desired action will probably be a composite of various measures. The problem remains, then, of determining the combination of measures that will satisfy the interest of the decision-maker while producing the desired level of energy conservation in a given industry. This is accomplished by using the technique of the hierarchical analysis.

Thus, the individual econometrics, resoumetrics, hierarchical analysis, and data system will be discussed, and then the example of refined copper production will be used to demonstrate their interaction.

PRODUCTS MARKET SIMULATION—ECONOMETRICS*

Econometrics is one of the better approaches to products market simulation. In general, econometrics observes historical data to yield products' industry-wide supply-and-demand relationship, and then uses the observed relationship to project the future supply and demand for the products. Its structure and function are discussed here, and it is used in a later section to predict refined copper supply and demand.

There are three copper econometric models in the open literature: Arthur D. Little (ADL) model,[1] Fisher et al. (FCB) model,[2] and Charles River Associates (CRA) model.[3-5] The ADL** model was developed to evaluate the economic impact of environmental regulations on the primary copper industry for the U.S. Environmental Protection Agency. The FCB model was motivated by the interest of the Chilean government in determining the behavior of the world copper market as a prerequisite to governmental planning. The CRA model was developed to predict copper demand and supply for the General Services Administration; it has been revised three times.†

In this section we will explain these models; reestimate their coefficients using identical data (for the period 1957–1973); synthesize a copper econometric model by tailoring various elements of the models into a composite model; and compare several of the outputs of this composite model with actual past data.

Explanation of the Three Models

The major function of the existing copper econometric model is to determine industry-level copper prices, supply, and aggregate copper demand‡ in each desired time interval.

---

* A version of this section appeared in *Engineering Costs and Production Economics.*[13]
** The ADL model examined in this chapter is the Market Clearance Module.[1]
† The CRA model examined in this chapter is the third revision.[5]
‡ The existing models do not disaggregate copper demand for electrical and electronic needs, building construction, industrial machinery and equipment, or transportation needs.

Four prices are important: (1) London Metals Exchange (LME) price, (2) U.S. EMJ price (prices published in the *Engineering and Mining Journal*), (3) U.S. producers' price, and (4) scrap price.

Copper supply for the industry can be divided into three categories: new copper produced through mining, beneficiating, smelting, and refining; copper scrap generation consisting of secondary refined copper and nonrefined scrap; and inventory stock adjustment including imports in the form of ore, blister, or refined copper.

Model Coverage. The three models vary in the extent of their coverage. In general, the ADL model concentrates on the U.S. refined copper market; the FCB model covers the international copper primary market more thoroughly than the U.S. market; and the CRA model is evenly distributed between the international and U.S. primary market. Table 11–1 shows the coverage and overlap of the models. Note that all monetary units are deflated into 1974 dollars.

All three models use almost the same exogenous variables for the input information as shown in Table 11–2. Table 11–3 shows lagged endogenous variables used in each model. The FCB and CRA models also use some lagged exogenous variables, which are shown in Table 11–4. However, because the ADL model uses lagged endogenous variables, the information carried by lagged exogenous variables is indirectly used as well.

Model Structure. The relationships comprising the ADL, FCB, and CRA models are reported in Tables E-4, E-5, and E-6 (in Appendix E), respectively. All equations in the ADL model are simultaneous and in linear form. Except for two equations for secondary copper supply, the equations in the FCB model are also linear. However, most equations in the CRA model are in log–linear form.

In order to achieve each model's objectives as identified in the introduction, the ADL model simultaneously produces information for the U.S. EMJ price, scrap price, future price, secondary refined copper price, copper consumption, refined copper production, scrap collection, stocks, and net refined copper exports. The FCB model generates the LME copper price, world major producing countries' mine production, and major consuming countries' refined copper consumption, stocks, scrap collection, and exports. The CRA model also generates the above economic activity information for the world as well as U.S. markets in terms of a greater aggregation.

Re-estimated Coefficients

For the purpose of equally comparing the three models, based on the 1957–1973 time series data, the coefficients of equations for the three models were

Variable Description	Units	Composite[a]	ADL	FCB	CRA[c]
		Symbols			
EMJ copper price	¢/lb[b]	Y_0	PEMJ	PEMJ	PCUC
Future copper price	¢/lb	Y_1	FTUT		
Scrap price (no. 2 heavy scrap)	¢/lb	Y_2	PS		PCUSC
Secondary refined copper price	¢/lb	Y_3	PSR		
Copper consumption	1000 short-tons	Y_4	QD	USC	CCUT
Total production of refined copper	1000 short-tons	Y_5	QR		
Refined copper production from primary sources	1000 short-tons	Y_6	QPR		
Mine production	1000 short-tons	$Y_{6.1}$		USMP	QCU
Secondary refined copper production	1000 short-tons	Y_7	QSR		
Collection of nonrefined scrap copper	1000 short-tons	Y_8	QSNR		
Fabricator stocks	1000 short-tons	Y_9	IF		SCUQ
Refined copper stocks	1000 short-tons	Y_{10}	IRR	USS	SCU
Scrap stocks held by refiners	1000 short-tons	Y_{11}	IRS		
Net exports (refined copper)	1000 short-tons	Y_{12}	NE		MCURN
Net imports	1000 short-tons	Y_{13}			MCUTN
Change in U.S. government stocks	1000 short-tons	Y_{14}			SCUGD
Collection of new scrap	1000 short-tons	Y_{15}		USNS	QCUSN
Collection of old scrap	1000 short-tons	Y_{16}		USOS	QCUSO
Chilean mine production	1000 short-tons	Y_{17}		CHMP	QCCH
Canadian mine production	1000 short-tons	Y_{18}		CANMP	
Zambia mine production	1000 short-tons	Y_{19}		ZMP	QCXX QCW
Rest-of-world mine production	1000 short-tons	Y_{20}		RWMP	
European copper consumption	1000 short-tons	Y_{21}		EURC	
Japanese copper consumption	1000 short-tons	Y_{22}		JC	CCXT
Rest-of-world copper consumption	1000 short-tons	Y_{23}		RWC	
Rest-of-world secondary supply	1000 short-tons	Y_{24}		RWS	QCXS
Rest-of-world copper stocks	1000 short-tons	Y_{25}		RWSS	SCX
Foreign refined stock	1000 short-tons	Y_{26}			SCXR
Net exports from the rest-of-world to the U.S.	1000 short-tons	Y_{27}		RWX	
LME copper price	¢/lb	Y_{28}		PLME	
Copper producer price	¢/lb	Y_{29}		PPROT	

[a] These symbols were used in the following tables and related tables in Appendix E to exhibit the equations in various models. [b] In 1974 dollars for all monetary units used in this table, following tables, and related tables in [c] Appendix E.

Table 11-2. Input (Exogenous Variables).

Variable Description	Unit	Source	Composite[f]	Symbols ADL	Symbols FCB	Symbols CRA
Refining capacity	1000 sh. tons	American Bureau of Metal Statistics (ABMS)	X_1	KAPP		KCUP
Federal Reserve Board index of manufacturing production	%[e]	Current Business Survey (CBS)	X_2	YUD	USIP	YUD
LME copper price	¢/lb	ABMS	X_3	PLME		PLMEC
Price of scrap aluminum clippings	¢/lb	American Metal Market (AMM)	X_4	PAUSC	GALP[b]	PAUSC
Wholesale price index of durable manufacturing	%[e]	CBS	X_5	PUWD	USWI	PUWD
Producer variable operating cost	¢/lb	Composite Value[g]	X_6	VARCOS		FEU
Weekly wages of production workers in copper refining	$/week	Employment and Earnings and Monthly Report on Labor Force	X_7			FWUMC
Weekly wages of production workers in copper refining	$/week	Monthly Report on Labor Force	$X_{7.1}$			FWURC
Change of inventories of durable goods	billion $	CBS	X_8		WUSID	
Supply of domestic copper products		Copper Development Assoc. (CDA)	X_9	QFAB		SQUDC

				IGOV	WUSGS	
Government stock change	1000 sh. tons	CDA	X_{10}			
Time	1964 = 1	Calculated from CDA	X_{11}	T	T	T
Change in available scrap supply	1000 sh. ton		X_{12}		K	
West German index of manufacturing production	%[e]	International Financial Statistics (IFS)	X_{13}	YGR		
Japanese industrial production index	%[e]	IFS	X_{14}		JIP	
OECD European industrial production index[c]	%[e]	OECD, Main Economic Indicators (MEI)	X_{15}		EURIP	
Rest-of-world industrial production index[d]	%[e]	U.N. Statistical Yearbook (UNSY)	X_{16}		RWIP	
Canadian wholesale price index[a]	%[e]	MEI	X_{17}		CANWI	
Japanese wholesale price index[a]	%[e]	MEI	X_{18}		JWI	OXX
OECD European price index[a,c]	%[e]	MEI	X_{19}		EURWI	
Unit value of exports	%[e]	U.N. Statistical Yearbook	X_{20}			PDE

[a] Exchange rates for respective currencies are obtained from *Federal Reserve Board Bulletin.* [b] German aluminum price was used. [c] These figures are calculated by using the following weights in the OECD countries: Germany = 0.375, France = 0.212, Italy = 0.158, and England = 0.225. [d] These figures are computed by using the following weights for the rest of world: Canada = 0.394, South Africa = 0.098, Brazil = 0.237, Australia = 0.176, and Mexico = 0.095. [e] 1974 = 1. [f] These symbols were used in the following tables and related tables in Appendix E to exhibit the equations in various models. [g] Exogenously calculated for original ADL model, formula not available.

Table 11–3. Input (Lagged Endogenous Variables).

Lagged Variable Description	Unit	Composite	ADL	FCR	CRA
				Symbols	
EMJ copper price	¢/lb	$Y_0(-1)$		PEMJ(-1)	PCUC(-1)
Future copper price	¢/lb	$Y_1(-1)$			PCUSC(-1)
Scrap price (no. 2 heavy scrap)	¢/lb	$Y_2(-1)$			
Secondary refined copper price	¢/lb	$Y_3(-1)$			
Copper consumption	1000 short-tons	$Y_4(-1)$	QD(-1)	USC(-1)	CCUT(-1)
Total production of refined copper	1000 short-tons	$Y_5(-1)$			
Refined copper production from primary sources	1000 short-tons	$Y_6(-1)$			
Mine production	1000 short-tons	$Y_{6,1}(-1)$		USMP(-1)	QCU(-1)
Secondary refined copper production	1000 short-tons	$Y_7(-1)$			
Collection of nonrefined scrap copper	1000 short-tons	$Y_8(-1)$			
Fabricator stock	1000 short-tons	$Y_9(-1)$	IF(-1)		
Refined copper stocks	1000 short-tons	$Y_{10}(-1)$	IRR(-1)	USS(-1)	SCU(-1)
Scrap stocks held by refiners	1000 short-tons	$Y_{11}(-1)$	IRS(-1)		
Net exports	1000 short-tons	$Y_{12}(-1)$	NA		
Net imports	1000 short-tons	Y_{13}			MCUTN(-1)
Change in U.S. government stocks	1000 short-tons	$Y_{14}(-1)$			
Collections of new scrap	1000 short-tons	$Y_{15}(-1)$		USOS(-1)	QCUSN(-1)
Collections of old scrap	1000 short-tons	$Y_{16}(-1)$			QCUSO(-1)
Chilean mine production	1000 short-tons	$Y_{17}(-1)$		CHMP(-1)	QCCH(-1), (-2), (-3)
Canadian mine production	1000 short-tons	$Y_{18}(-1)$		CANMP(-1)	
Zambia mine production	1000 short-tons	$Y_{19}(-1)$			QCXX(-1)
Rest-of-world mine production	1000 short-tons	$Y_{20}(-1)$		RWMP(-1)	
European copper consumption	1000 short-tons	$Y_{21}(-1)$		EURC(-1)	CCXT(-1)
Japanese copper consumption	1000 short-tons	$Y_{22}(-1)$			
Rest-of-world copper consumption	1000 short-tons	$Y_{23}(-1)$		RWC(-1)	QCXS(-1)
Rest-of-world secondary supply	1000 short-tons	$Y_{24}(-1)$		RWS(-1)	SCX(-1)
Rest-of-world copper stocks	1000 short-tons	$Y_{25}(-1)$		RWSS(-1)	SCXR(-1)
Foreign refined stocks	1000 short-tons	$Y_{26}(-1)$			
Net exports from the rest-of-world to the U.S.	1000 short-tons	$Y_{27}(-1)$			
LME copper price	¢/lb	$Y_{28}(-1)$		PLME(-1)	

Table 11-4. Input (Lagged Exogenous Variables).

Lagged Variable Description	Composite	ADL	FCB	CRA
			Symbols	
Refining capacity	X_1 (−1)	NA		
Federal Reserve Board index of manufacturing production	X_2 (−1)			
LME copper price	X_3 (−1)			
Price of scrap aluminum clippings	X_4 (−1)		GALP (−1)	PUWD (−1)
Wholesale price index of durable manufacturing	X_5 (−1)		USWI (−1)	FEU (−1)
Operating costs of primary producers	X_6 (−1)			FWUMC (−1)
Weekly wages of production workers in copper mining	X_7 (−1)			FWURC (−1)
Weekly wages of production workers in copper refining	$X_{7,1}$ (−1)			SQUDC (−1)
Change of inventories of durable goods	X_8 (−1)		DUSID (−1)	
Supply of domestic copper products	X_9 (−1)			
Government stock change	X_{10}(−1)			
Time	X_{11}(−1)			
Change in available scrap supply	X_{12}(−1)		K (−1)	
West German index of manufacturing production	X_{13}(−1)			
Japanese industrial production index	X_{14}(−1)			
OECD European industrial production index	X_{15}(−1)			
Rest-of-world industrial production index	x_{16}(−1)			
Canadian wholesale price index	X_{17}(−1)			
Japanese wholesale price index	X_{18}(−1)		JWI (−1)	
OECD European price index	X_{19}(−1)		EURWI (−1)	

The necessary notes are the same as those in Table 11-2.

re-estimated (the coefficients given in this chapter, therefore, differ from the original estimates). In order to obtain the best results, some necessary proxies were used during re-estimation to replace the original explanatory variables; thus, a few minor changes were made. They are explained in Tables E-7, E-8, and E-9 in Appendix E. Because the samples were small, an instrumental variables scheme was used instead of a two-stage least-squares scheme to correct simultaneous errors (as the two-stage least-squares scheme was used in the original estimations for FCB and CRA models).

The re-estimated equations for the ADL model as well as their statistical significance are reported in Table E-7 in Appendix E. Comparing the statistical measures, such as R^2 and D.W.,* we may conclude that, except for scrap price, which may not be fully explained by the collection of nonrefined secondary copper as suggested by the ADL original estimation, all equations estimated fit the historical data better than the original estimation.[1]

Different samples were used to test the suitability of the functional type of equations for FCB** and CRA models, and satisfactory results are reported in Tables E-8 and E-9 in Appendix E, respectively. Re-estimation equations, as well as statistical criteria and sample periods, are all listed. An important aspect of the FCB model is that it treats the LME copper price as a function of the rest-of-world copper stocks.

The following points are worthy of note.

EMJ Copper Price. EMJ price is explained as a function of average variable operating costs of primary producers† in the ADL model; LME and U.S. producer copper prices in the FCB model; and copper scrap, mining wage difference, and lagged EMJ price in the CRA model. The functional types of these relationships for the three models are reported (as Equation 1) in Tables E-4, E-5, and E-6, respectively. The re-estimated results for the three models are shown (as Equation 1) in Tables E-7, E-8, and E-9 as well. Since the coefficients of determination (R^2) are very high for all equations, these equations are satisfactory. However, the re-estimated equation in the FCB model is better because it fits the historical data almost completely $R^2 \approx 1$).

U.S. Copper Consumption. The U.S. copper consumption has been related to deflated copper and aluminum prices, and the Federal Reserve Board

---

* R^2 = coefficient of determination; D.W. = Durbin-Watson test.
** Re-estimation of this model with some modifications by an ordinary least-squares scheme was made in 1975.[6] The re-estimation was not very satisfactory.
† The original formula used to calculate average variable operating costs of primary producers was not available; therefore, weekly wages of production workers in mining were substituted.

Production Index in all three models. Additionally, the one-year lagged fabricator stock is used in the ADL model, while inventory changes of durable goods for copper users are used in the FCB and CRA models. The functional types of relationships for the ADL, FCB, and CRA models are reported as Equations 5, 2, and 3, respectively, in Tables E-4, E-5, and E-6 (identified as Y_4). The re-estimated coefficients of these equations in the three models are listed in Tables E-7, E-8, and E-9. The results show a point common to the three models; that is, copper consumption was not discouraged when the price of aluminum (a copper substitute) decreased in the last 15 years. This may be due to the technological limit of substitution. From the statistical point of view, the equation for the ADL model is the most satisfactory in the sample period.

U.S. Copper Production. Because of different focuses in the three models, the ADL model predicts U.S. refined copper production, while the FCB and CRA models provide information for the U.S. mine production. The U.S. refined copper production in the ADL model is an equilibrium of other quantities such as consumption, scrap generation, stocks, and exports. The mine production is a function of the EMJ price and lagged mine production in the FCB model, and it is explained by the EMJ price, capacity index, and lagged mine production in the CRA model. The functional forms of these production equations are shown as Equations 7, 3, and 4, respectively, in Tables E-4, E-5, and E-6 (indicated as Y_6 and $Y_{6.1}$). The re-estimated coefficients are listed in Tables E-7, E-8, and E-9, respectively. Both re-estimated equations for the FCB and CRA models are satisfactory for the data in the sample period ($R^2 = 0.90$). The performance of the refined copper production equation is poor because the absolute mean value of difference between the ADL model's projection and actual values is over 20% (shown below in Table 11–6).

U.S. Copper Stocks. The ADL model estimates the change of fabricator stock, refined copper stock, and refiner scrap stocks. Based on the EMJ price and other factors (Equations 10, 11, and 12 in Table E-4), the FCB model uses an equilibrium of the change of various U.S. copper stocks (Equation 4 in Table E-5). However, the CRA model consists of an identity for scrap stocks (Equation 5 in Table E-6), and an equation for fabricator and refined copper stocks (Equation 6 in Table E-6). Re-estimated coefficients for the ADL, FCB, and CRA models are reported in Tables E-7, E-8, and E-9, respectively. The re-estimated equations for the predictions of fabricator stock and refined copper stock in the ADL model are satisfactory with respect to the statistical criteria. However, the equation for estimating scrap stock in the ADL model and those in either the FCB or CRA models do not appear to be satisfactory.

Penn Composite Model (PCM)

Examination of the three models indicates that the major difference among them is their coverage of the copper markets. In terms of the structure of the equations, little variation is observed. In fact, equations common to the three models use similar exogenous variables. The re-estimated coefficients for the three models show common strengths and weaknesses in representing the trend of the historical data. Results of estimation using equations common to all three models suggest that ADL or FCB models are better fits than CRA models. In addition, they are more readily usable because the relationships are in linear form. A weakness associated with the ADL model is due to its price equations such as Equations 1 and 2 in Table E-8—the explanatory variables may not explain the full cause of price changes. Thus, replacing Equation 1 by its equivalent from the FCB model (Equation 1 in Table E-8) improves the quality of the ADL model's prediction. This can be done because Equation 1 in the ADL model is independent of the remaining equations. These two models can also be combined further to create a more complete model. This has been attempted by generating the Penn Composite Model (PCM).

Structure of PCM. While the ADL model covers the U.S. market and assumes the LME copper price as exogenous, the FCB model predicts LME price and world copper stocks. Therefore, judicious combination of these two models will provide a more extensive coverage. Difficulty may arise due to a possible simultaneity of the equations in the ADL model. Nevertheless, the combining process is feasible if equations in the two models are arranged properly.

The task can be accomplished by combining all equations but Equations 2 and 4* in the FCB model and Equations 2–13 in the ADL model** to create the PCM for the copper commodity. Table 11–5 shows that the PCM consists of 26 equations, of which Equations 1–4, 25, and 26 are used to estimate EMJ, refined copper, refined scrap, nonrefined secondary copper, LME, and producer prices, respectively; Equations 5–13 are used to estimate U.S. refined copper demand, production, stocks, and net exports; and Equations 14–24 are used to estimate major countries' refined copper production,

---

* These equations are duplicates of equivalent equations in the ADL model.
** Equation 1 is independent of the rest of the equations in the ADL model.

consumption, imports, and stocks. The coefficients of equations shown in Table 11–5 were calculated from exogenous variables shown in Table E-10 in Appendix E and for endogenous variables shown in Table E-11 in Appendix E.

The underlying assumption for the equations related to major producing countries' mining production is that the production motivation is generated by copper price, and the refined copper consumption of the major consuming

Table 11–5. The Penn Composite Model.

1. Y_0/X_5 $= -0.224 + 0.995\ Y_{29}/X_5 - 0.0007\ Y_{28}/X_5$

2. Y_1/X_5 $= 28.197 + 4.225\ X_3/X_5 + 2.633\ D_1$

3. Y_2/X_5 $= 21.73 + 0.074\ Y_8 - 7.433\ D_3$

4. Y_3/X_5 $= 73.795 + 1.498\ Y_2/X_5 - 26.62\ Y_5/X_1 - 0.21112\ Y_7$

5. Y_4 $= 2088.13 - 25.62\ Y_0/X_5 + 15.15\ Y_2 - 8.73\ X_4/X_5 -$
$0.76\ Y_9(-1) + 0.57\ Y_4(-1) - 119.9\ DD_1$

6. Y_5 $= Y_6 + Y_7$

7. Y_6 $= Y_7 - Y_8 + Y_4 + Y_{12} + Y_9 + Y_{10} + Y_{11} + Y_{10}$

8. $(Y_2 - Y_0)/X_5$ $= -16.20 - 9.56\ Y_5/X_1 + 0.684\ (X_3 - Y_0)/X_5$

9. $(Y_3 - Y_0)/X_5$ $= 10.24 - 4.97\ Y_5/X_1 + 0.625\ (X_3 - Y_0)/X_5$

10. ΔY_9 $= 525.83 - 0.133\ X_9 - 0.605\ Y_9(-1) - 17.22\ D_1 +$
$16.33\ D_2$

11. ΔY_{10} $= 0.044\ Y_6 + 0.762\ Y_{10}/X_5 - 0.61\ X_3/X_5 - 1.27\ Y_{10}(-1) +$
$78.34\ D_3$

12. ΔY_{11} $= 5.966 + 0.301\ X_3/X_5 - 0.187\ Y_1/X_5 + 0.087\ Y_7 -$
$0.894\ Y_{11}(-1) + 12.37\ D_3$

13. Y_{12} $= 1461.73 - 1.16\ Y_0/X_5 + 5.06\ X_3/X_5 - 1430.97\ X_2/X_{13} -$
$36.69\ X_{11} - 33.35\ D_3$

14. $Y_{6.1}$ $= 180.73 + 18.27\ Y_0/X_5 - 80.63\ D_1 + 0.309\ Y_{6.1}(-1)$

15. Y_{17} $= 172.74 + 6.20\ Y_0/X_5 + 0.27\ Y_{17}(-1)$

16. Y_{18} $= -52.766 - 37.925\ Y_0/X_{11} + 1.196\ Y_{18}(-1)$

17. Y_{19} $= 581.64 + 12.71\ X_{11}$

18. Y_{20} $= -366.05 + 1.81\ Y_{28}/Y_5 + 1.26\ Y_{20}(-1)$

19. Y_{21} $= 1405.3 + 21.93\ X_4(-1)/X_5(-1) + 36.05\ X_{15} -$
$4.44\ Y_{28}(-1)/X_{19}(-1) - 0.26\ Y_{21}(-1)$

20. Y_{22} $= 443.40 + 14.55\ X_{24} - 1.96\ Y_{28}(-1)/X_{18}(-1)$

21. Y_{23} $= 560.97 - 1.04\ Y_{28}(-1)/X_5(-1) - 15.63\ X_4(-1)/X_5(-1) +$
$1.65\ X_{16} + 0.74\ Y_{23}(-1)$

22. $\ln[Y_{24}/(X_{12} +$
$140.000)]$ $= 0.26 + 0.42\ \ln[Y_{23}/(X_{12} + 140.000)] +$
$0.241\ \ln[Y_{24}(-1)/(X_{12}(-1) + 140.000)]$

23. ΔY_{25} $= Y_{17} + Y_{18} + Y_{19} + Y_{20} + Y_{24} - Y_{21} - Y_{22} + Y_{23} - Y_{27}$

24. Y_{27} $= -272.1 + 2.9\ Y_{29}/X_5 - Y_{28}/X_5 + 0.403\ Y_4 - Y_{6.1}$

25. Y_{28}/X_5 $= 28.50 + 0.14\ [\Delta Y_{25}/Y_{23} - \Delta Y_{25}(-1)/Y_{23}(-1)] \times 100 +$
$0.59\ Y_{28}(-1)/X_5(-1)$

26. Y_{29}/X_5 $= 29.44 - 0.075\ [Y_{28}(-1) - Y_0(-1)]/X_5(-1) -$
$155.47\ [Y_9(-1) + Y_{10}(-1)]/Y_4(-1) +$
$0.83\ Y_{29}(-1)/X_5(-1)$

Footnotes are as in Tables 11–1, 11–2, 11–3, 11–4; not all Y's are used.

countries is mainly affected by copper price, aluminum price, and their industrial activity level. Prices are assumed to be determined mostly by stock level and other related product prices; and stock change is assumed to be impacted by price and production level.

Operation of PCM. In order to test the reasonableness of the PCM, backcasting was used to simulate results for the years 1964 to 1973, and forecasting is used to simulate results for 1974.*

In the course of simulation, the outputs generated for the current year were used as lagged endogenous variables for the next year of operation** (see Fig. 11–3). Therefore, the input data for the simulation of the PCM are those of exogenous variables in each year and prior to the time of the model "starts." For an estimation of values shown in Fig. 11–4, the beginning year was 1964. Therefore, the necessary data are 1963's endogenous variables.

Data generated using the PCM include EMJ copper price, producer price, scrap price, refined copper production, and collection of nonrefined scrap.

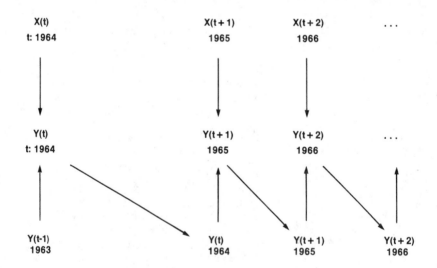

X(t) : vector of exogenous variables
Y(t) : vector of endogenous variables

Fig. 11–3. Operation of PCM.

---

* In order to compare the results with those generated from the ADL model, simulation was made for the U.S. market only.

** There are other ways to simulate models, but the method used here is a stricter one; it permits us to examine the tract of the model.

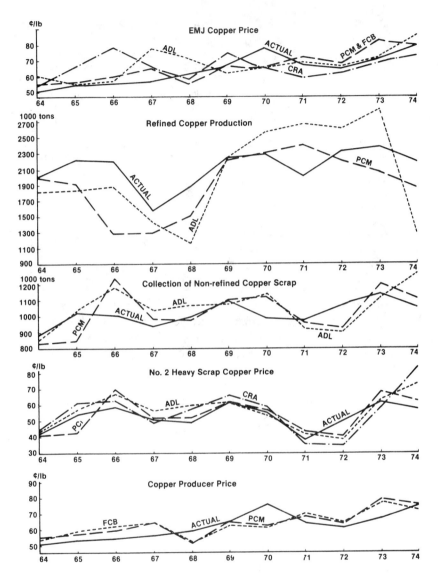

Fig. 11–4. Actual and various predicted values for five variables. Not all models predict these variables. If model name is not shown in the chart, then the model does not estimate the variable.

Table 11-6. Comparison of the Performance Between the PCM and the ADL Model.

Item	Absolute Mean Value of PCM's Difference Between the Projection and Actual Value (%)	Absolute Mean of Difference Between ADL Model's Projection and Actual Value (%)
EMJ copper price	8.97	11.19
Copper producer price	8.97	Not available
Scrap price	8.97	10.15
Secondary refined copper price	6.99	9.09
Total refined copper production	12.49	19.20
Refined copper production from primary sources	14.55	22.89
Secondary refined copper production	5.87	6.82
Collection of nonrefined scrap copper	9.75	9.75

PCM estimates for the years 1964–1974 are compared with estimates of the three existing models whenever applicable in Fig. 11–4.

The absolute mean values of difference between predicted values of the PCM and the actual values are reported in Table 11–6. In general, the average estimation error* of the simulation is between 6 and 15%. The PCM performs better for the production. For extraordinary events such as strikes (1966) and dramatic price changes (1970s), the outputs of the PCM appear behind the trend, and they pick up slowly. When compared with the ADL model, PCM performs much better, as shown in Table 11–6.

Summary of the Models

The three models examined show almost identical strengths and weaknesses in simulating copper markets. However, because of its objectives, each model emphasizes the different section of the copper markets (e.g., the world, U.S. copper mine production, or refined copper markets). Based on historical copper market relationships, the three models apparently are not able to predict future markets efficiently. For example, none of the three models could have predicted the dramatic copper price increases in 1973.

The Penn Composite Model, properly combined to contain only the strengths of the ADL and FCB models, has improved model predictability, but it is still far from perfect. However, we will later use the Penn Composite

---

* These values were not adjusted for variances; however, they should be smaller after adjustments for variances.

Model to predict future refined copper supply and demand in conjunction with production and prioritization simulations before we can make the model perfect.

PRODUCTION SIMULATION—RESOUMETRICS

As described in the section on methodology, resoumetrics will be used to simulate industrial production in this chapter. The resoumetric model, coined by Zandi et al.,[9] is a resource or production measurement tool. In a broad sense, it is a detailed process modeling that attempts to project production relationships on the basis of technological production structure. Originally, the resoumetric model was used in assessing copper resource and environment.[9]

The resoumetric model describes, for industrial production in engineering terms, the detailed material and energy flows and overall cost implications, including labor requirement, residual production, energy requirement, and so on, associated with alternative governmental policies. In this respect, resoumetric models differ from, but complement, econometric models frequently used in formulating national policy. The difference is that resoumetric models provide detailed impact information of individual industrial companies' production but the econometric models provide overall industry-wide and nationwide aggregate market information concerning prices, demand, and supply.

Copper Resoumetric Model for Energy Conservation

As an illustration of the type of models that can be developed, this chapter considers the case of energy conservation in the refined copper production. Copper, as a commodity, is characterized by highly volatile markets both nationally and internationally, and by energy inefficiency and pressing environmental problems at both extraction and smelting stages. The copper econometric model was discussed in the last section. The method of the copper resoumetric model for energy conservation was discussed in detail in Chapter 8 and is summarized below.

Method. The structure of the copper resoumetric model for energy conservation is as follows:

Minimize: Cost $= f$ (cost of production factors, production
            quantity)

Subject to: 1. Production quantity $= \delta$ (capacity, utilization
            rate, conservation technologies, sales
            commitments, transportation, and
            environmental constraints)
         2. Energy conservation target

3. Market penetration limits for conservation
   technologies

where f is a cost function, and δ is a set of production functions for capacity and other related constraints.

Results. Three representative leading companies (A, B, and C) were analyzed in detail. The first firm primarily mines, smelts, refines, and sells products by itself. The second one is a partially custom-smelting and refining company that does not produce sufficient ores for itself and takes large portions of its input from scrap or other firms. The third company has large deposits of rich ores and produces more ore than it uses. In 1974, these three companies owned 50% of the estimated recoverable copper reserves, contributed 40% of the total mine production, held 51% of the total U.S. smelting works, owned 51% of the total copper refinery capacity, and shared 54% of the total refined copper sales.[3,12]

Relation between Energy Savings and Cost Change. The capacity data (shown in Table E-1, E-2, and E-3), production data, and energy conservation method of Chapter 8 are used to estimate relationships between energy reduction and cost change. The resultant details, which were shown in Table 10–1 (in Chapter 10), are summarized in the following equations:*

$$\Delta C_A = 1.41 \ \Delta E_A \tag{11–1}$$

$$\Delta C_B = 1.63 \ \Delta E_B \tag{11–2}$$

$$\Delta C_C = 1.42 \ \Delta E_C \tag{11–3}$$

where ΔC_A, ΔC_B, and ΔC_C are conservation costs (in thousands of 1974 dollars) incurred by the installation of energy-saving equipment; and ΔE_A, ΔE_B, and ΔE_C are energy savings in 10^9 Btu for companies A, B, and C, respectively.

Relation between Pollution Emission and Production. Major items for the resoumetric model for evaluating the impacts of the government's policies are air emissions (mainly SO_2), waste water discharged, solid waste, and land disturbance. The relationships among these products and copper production level are estimated as follows:**

$$SO_2 = q^{1.03}/5.68 \text{ tons/day} \tag{11–4}$$

$$\text{Water} = 30 \ q^{.94} \text{ barrels/day} \tag{11–5}$$

---

* Detailed discussion is in Chapter 10.
** Rough estimates for demonstration purpose only.

$$\text{Waste} = 100 \ q \ \text{tons/day} \qquad (11\text{–}6)$$

$$\text{Land} = q \ \text{acres/day} \qquad (11\text{–}7)$$

where SO_2 is air emissions, in tons; q is refined copper production, in thousands tons; water is waste water discharged, in barrels; waste is solid waste in tons; and land is land disturbance in acres.

PRIORITIZATION SIMULATION—HIERARCHICAL ANALYSIS

The hierarchical analysis method is used to simulate prioritization processes for governmental bodies. Government's decision regarding certain fiscal regulatory policies for industry concerns many interest groups. Among them are environmentalists, government itself (the Administration as well as Congress), and industrialists. Each group has multiple considerations when lobbying to shape the final decision to its own benefit. Some of the principal concerns may be environmental (for clean air or undisturbed land), economic (for high net production or low prices), or political (for public support or ease in passing necessary legislation). The structure of the decision-making process is illustrated in Fig. 11–5.

Perhaps the most convenient and powerful way to construct a collection of the above policy-making goals and interests so that priorities can be established is to use a hierarchical structure (originally denoted a vertical authority structure in human organizations, and extensively discussed by Saaty[10]). This structure may be a complete (or partial) ordering or tree that represents some order of dominance. The final collective policy-making forms the top level of the hierarchy, while the criteria (objectives) of each interest group form lower levels of the hierarchy. Priorities of elements in each level are thus transferred from the lower level to the higher levels through the hierarchical structure (i.e., each interest group's judgments about future policies will then be transferred to affect the outcome of the final policy-making).

In determining the priorities of the elements in each level of the hierarchy, Kahn,[11] in investigating theories of social planning, observes that:

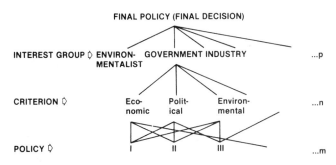

Fig. 11–5.   Simulated hierarchy of the federal government's decision-making.

Objectives and their order of priority are determined largely by value judgments rather than by analysis. These judgments should be made explicitly, and they should be based upon examination of the consistency, social consequences, and feasibility of objectives.

Saaty has developed a theory of hierarchies and priorities to capture Kahn's observation. The hierarchical approach involves a theory of pairwise comparisons between activities that derives a scale of priorities for these activities. Activities involved in the decision process may be grouped on different levels that form a hierarchy according to the structure of the process, whereby lower-level elements can be compared pairwise with respect to higher-level components, and a method of weighting yields the overall priorities for the activities on the lowest level. These overall priorities indicate the order of preference among feasible policies.

Hierarchical analysis can be applied to simulate the government's decision-making concerning the implementation of some fiscal policies in encouraging energy conservation.

The application of hierarchical analysis here is much more pragmatic than theoretical. Its theoretical frameworks have been largely completed by Saaty. While implementing the analysis, we simply invite all representatives from all interest groups and form a jury session. We first inform them of all the calculatable impacts of each possible government policy through the previous two simulations (PCM and the resoumetric model), and then ask them to compare the following items pairwise according to the cardinal scale shown in Table 11–7:

Table 11–7. The Scale of Pairwise Comparisons.

Intensity of Importance	Definition	Explanation
1	Equal importance	Two activities contribute equally to the objective.
3	Weak importance of one over another	Experience and judgment slightly favor one activity over another.
5	Essential or strong importance	Experience and judgment strongly favor one activity over another.
7	Demonstrated importance	An activity is strongly favored, and its dominance is demonstrated in practice.
9	Absolute importance	The evidence favoring one activity over another is of the highest possible order of affirmation.
2, 4, 6, 8	Intermediate values between the two adjacent judgments	Are used when compromise is needed.

Source: Saaty[10]

- Relative importance between policies with respect to various comparison criteria.*
- Relative importance between criteria with respect to various interest groups.*

Note that, in measuring importance, the difference between two items compared is much more crucial than their absolute values.

We then (acting as the decision-maker) compare pairwise the influence of interest groups.

The following rules should be followed in pairwise comparison:

- Comparison of each item with itself produces numeral one, as shown in the diagonal of the matrix in Table 11–8.
- The importance of I over II is the reciprocal of the importance of II over I, as demonstrated in Table 11–8.

The result of pairwise comparison of the policies by each interest group with respect to a given criterion is a matrix similar to the one shown in Table 11–8. For all possible policies (m), all interest groups (p), and with respect to all criteria (n), we generate $[(p + 1)n] + 1$ matrices. These are as follows:

- $p \cdot n$ matrices of comparisons for the importance between policies; each matrix is $m \times m$.
- p matrices of comparisons for the weights between criteria; each matrix is $n \times n$.
- One matrix of comparisons for the degree of influence between interest groups; the matrix is $p \times p$.

Table 11–8. An Example of Pairwise Comparisons by Industry with Respect to an Economic Criterion.

POLICY		POLICY		
		I	II	III
	I	1	5	3
	II	⅕	1	2
	III	⅓	½	1

---

*Saaty has proved that the number of each level of the hierarchy should most desirably be around seven or eight. This is because human beings are able to manage and distinguish differences better in this range, according to psychological and managerial experiments (by control span theory).

In general, these matrices can be written in the following format:

$$
\text{Matrix } A \equiv
\begin{array}{c|cccc}
 & A_1 & A_2 & \cdots & A_\eta \\
\hline
A_1 & W_1/W_2 & W_1/W_2 & \cdots & W_1/W_2 \\
A_2 & W_2/W_1 & & & W_2/W_\eta \\
\vdots & \vdots & \vdots & & \vdots \\
A_\eta & W_\eta/W_1 & & & W_\eta/W_\eta
\end{array}
$$

and $\eta \leqslant 8$.

W_i/W_j is the judgment of the importance of the ith item over the jth item and is a positive value. However, we do not know the value of W_i or W_j separately, comparing it with the rest of the elements in vector W. Thus, in order to obtain the weights (W_i) for all η items separately, Matrix A has to be transformed into a vector: $W = (W_i \ldots, W_\eta)$.

Saaty proved that if the judgments of importance in Matrix A are consistent—that is, Matrix A's maximum eigenvalue is close to its own rank (its consistency index† approaching zero)—then the dominant eigenvector (W^*) of Matrix A at its maximum eigenvalue ($\lambda^{\max}$) would represent the positive weights for the η items as in the following:

$$A \, W^* = \lambda^{\max} \, W^* \tag{11–8}$$

If the vector W^* is normalized, then it can be treated as a series of probability distribution among the above η items, since the summation of its elements is equal to one.†

Transforming all of the importance matrices by using Equation 11–8, we obtain:

- $p \cdot n$ dominant eigenvectors for the order of importance of policies, each vector containing m items.
- p dominant eigenvectors for the order of importance of criteria, each vector containing n items.
- One $p \times 1$ dominant eigenvector, P, for the degree of influence of interest groups.

We then concatenate the $p \cdot n$ vectors into p matrices M_i, for $i = 1$, $2, \ldots, p$, which have m rows, n columns. Let V_i^c represent the ith dominant eigenvector of the importance of criteria; then:

---

† See Appendix F for more information concerning definition and computation.

$$N_i = M_i \times V_i^c \tag{11-9}$$

where N_i is the weighting vector of the policies in the point of view of the ith interest group. The purpose of Equation 11-9 is to transfer the priorities of the lowest rank to the next lowest rank. This method of transfer of priorities will be applied for other ranks as well. Concatenation of N_i's into an $m \times p$ matrix, G, and multiplication by vector P (degree of influence), yields the vector H, which is the overall weighting of future policies (the composite policy), based on the hierarchical weighting of interest groups'influence, criteria's intensity, and policies' importance.

DATA MANAGEMENT SYSTEM

The data management system in Fig. 11-1 consists of data and formulas needed to link the three simulation elements in the energy compromise model. The most important elements in the system are formulas to allocate industry's total production among individual companies, and to indicate a cause–effect relationship between policies and induced energy savings for individual firms. Because the data management system is industry-specific, the primary copper industry is used to illustrate its basic functions.

Market Shares for Individual Firms

The determination of the market share for individual firms depends on the types of market in the industry: pure competition, monopoly, or something between. Historical data and economic theories show that the copper market is between pure competition and monopoly. In this type of market, there are two kinds of marketers, price leaders and price takers. The price leaders usually are major companies that have integrated their production from mining to refining and even to marketing. They decide the market prices for the whole industry to follow. The price takers are usually smaller independent producers who take market prices that have been determined by the price leaders.

The relationship of these two groups is shown in Fig. 11-6. Assuming that Group II is the set of price-taking firms and Group I is the set of price-leading companies, the relationship can be illustrated as follows:

When copper demand is smaller than Q_p, major firms do not produce, and only small competitive (price takers) produce copper for the market. Competitive firms in Group II produce Q_p at price P_p, when copper demand is Q_p. After the demand has increased over Q_p, the oligopolistic firms begin to produce. When copper demand is reaching Q_b, the price-leading firms set their production at the point where their marginal revenue equals their

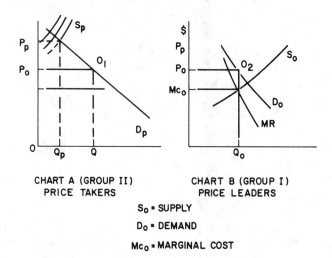

CHART A (GROUP II)
PRICE TAKERS

CHART B (GROUP I)
PRICE LEADERS

S_0 = SUPPLY

D_0 = DEMAND

Mc_0 = MARGINAL COST

Fig. 11–6.   Production of price takers and price leaders in the copper industry.

marginal cost, and post the price at P_0. The whole industry will produce Q at price P_0, and then the price takers in Group II will produce the difference between Q and Q_0 (the quantity produced by the price leaders).

The above discussion indicates that we will only have to determine the production for major producing firms (probably three or four price-leading companies in the copper industry), and the rest of the products will be produced by price takers at the price set by these major firms.

Theoretically, a firm's production is a function of its marginal cost of production as shown in Equation 11–10 and in Fig. 11–6:

$$q = f(Mc) \qquad (11\text{--}10)$$

where q is production, f is a symbol for function, and Mc is marginal cost of production. Because Mc is difficult to estimate, and because the competition among price leaders is substantially strong, its marginal cost would be very close to their average cost of production. Therefore, Equation 11–10 is estimated by Equation 11–11:

$$q = f(c) \qquad (11\text{--}11)$$

where c is average cost of production.

A historical relationship between production and average cost is thus estimated for major firms, and this relationship will be used to break down industry-wide production for individual major firms. Consequently,---- the price-taking independent producers will produce the difference between the industry-wide production (projected by using the PCM) and the sum of the four major firms, estimated through Equation 11–11.

Conventionally, the average cost of production is defined as

$$c = P - \pi \tag{11-12}$$

where c is average cost of production, P is price, and π is profit of production. Historically, major firms tend to keep their profit patterns. For example, when a price is low, the major copper companies are willing and able to arrange for production cuts in order to maintain their profit. This was evident in 1975's annual statements, which described how major companies shut down their plants from weeks to months to implement their production reduction agreements.

We will thus estimate equations for profit patterns and production functions by observing historical trends, and then use these equations to estimate production for the four companies.

Profit Patterns for the Four Major Companies. Four leading major companies (A, B, C, and D) were chosen and analyzed in detail. These four companies represent various types of copper companies. The first firm mines, smelts, refines, and sells products by itself. The second one is a partially custom-smelting and custom-refining company that does not produce sufficient ores for itself and uses large portions of its input from scrap or other firms. The third one has large deposits of rich ores and produces more ores than it uses. The last one consists of a large number of small mines and has a large capacity for smelting and refining as well as sales capability. In 1974, these four companies composed 70% of estimated---- recoverable copper reserves, contributed 60% of total mine production, held 72% of total U.S. smelting works, owned 72% of total copper refinery capacity, and shared 74% of total refined copper sales.

Annual profit, π, in percentage of annual revenue, can be expressed as:

$$\pi = \left(\frac{R - C}{R} \right) 100 \tag{11-13}$$

where R is annual revenue, and C is the incurred annual cost of a given company. The value of R and C for four companies, A, B, C,* and D, were estimated from the annual reports of these companies during 1966 through 1975. Table 11–9 shows these results.

---

* Some necessary adjustments of data for Company C are indicated in Table 11–9.

Table 11–9. Annual Profits for the Four Companies.

Year	Price (¢/lb Refined Copper in 1974 Dollars)	Index (I)	A Copper Concentrate in Ore in Percentage (× 10³)	(π^1)	Profit in Percentage of Revenue (π) Company B (π^2)	C (π^3)	D (π^4)
1966	55.45	792	792	37.87	20.33	20.49	22.23
1967	57.37	778	778	29.17	15.66	19.59	11.56
1968	60.69	747	747	27.30	13.51	18.00	14.72
1969	66.22	749	749	27.77	18.54	19.19	17.59
1970	77.33	773	773	31.95	15.94	*	21.79
1971	65.97	796	796	20.89	7.51	*	14.79
1973	69.08	766	766	25.27	12.48	*	16.36
1974	76.64	714	714	24.01	12.42	23.05*†	12.08
1975	57.51	716	716	1.13	1.53	*	2.16

* Because accounting practices were changes in the annual reports, we could not disaggregate copper revenue and cost from other products.
† This is an estimated value using a mixed-integer program (MIP). The MIP method is discussed in Chapter 8.

Source: Hu[8]

Examination of these historical annual profits shows that each company's profit has generally fluctuated along with producer copper prices. For example, profits in 1973 and 1974 were high, but they went down dramatically in 1975 because the prices in 1973 and 1974 were high. Profits in 1975 were very low owing to the slack in copper demand that resulted from worldwide economic recessions.

On the other hand, it is obvious that cost, at least for mining and milling, is a function of the ore grades. Therefore, the annual profit may be formulated in the following manner:*

$$\pi = f(Y_{29}, I) \qquad (11\text{-}14)$$

where Y_{29} is producer price in cents per pound of refined copper and I is percent of copper available in the ore. Table 11-9 shows that only Company A reported annual ore grade continuously from 1966 to 1975. Therefore, as an approximation in the present study, it is assumed that other companies exploit the same type of ore and calculate an ore index shown in Column 3 of Table 11-9. Using information in the table(except data for 1966 and 1967 because of strikes, and 1975 because of recessions), the set of equations shown in Table 11-10 is generated through regression analysis. It should be noted that better fits were achieved when $100 - \pi$ were regressed against Y_{29} and I.

Function for Each Company. The following relationship is used to estimate the average cost of production for each company:

$$c^i = Y_{29} \cdot (100 - \pi^i)/100 \qquad (11\text{-}15)$$

where c^i is the average cost in cents per pound of refined copper produced

Table 11-10. Profit Pattern.

Company A	$\pi^1 = 100 + .64\ Y_{29} - .160\ I$
	standard deviation $= 3\%$ of mean
Company B	$\pi^2 = 100 + .44\ Y_{29} - .160\ I$
	standard deviation $= 2\%$ of mean
Company C	$\pi^3 = 100 - .167\ Y_{29} - .09\ I$
	standard deviation $= 4\%$ of mean
Company D	$\pi^4 = 100 + .37\ Y_{29} - .14\ I$
	standard deviation $= 3\%$ of mean

---

* In his article, Lanzillotti indicated that the first objective of a large firm's pricing is to achieve certain profit rate (return on investment).[14]

Table 11-11. Average Costs* (c_i) and Productions (q_i) for Four Companies.

Year	Company A		Company B		Company C		Company D	
	c^1	q^1	c^2	q^2	c^3	q^3	c^4	q^4
1966	34.45	454	44.17	517	44.08	161	43.12	272
1967	40.63	289	48.38	334	46.13	86	50.74	157
1968	44.11	378	52.48	400	49.76	106	51.75	213
1969	47.82	496	53.94	582	53.50	156	54.56	283
1970	52.61	519	65.00	574	+	142	60.47	314
1971	52.19	456	61.01	473	+	182	56.34	281
1972	52.20	461	58.36	556	+	234	52.86	305
1973	51.61	472	60.45	488	58.98+	201	57.77	320
1974	58.23	402	67.12	437	+	190	67.38	281
1975	56.86	288	62.56	388	+	150	62.13	249

* Cost is in ¢/lb of 1974 dollars. Production is in one thousand short tons/year.
+ See note in Table 11-9.
Source: Hu[8]

Table 11-12. Production Functions for Four Companies.

Company A $q^1 = 150.0 \; e^{.023\,c^1}$
      standard deviation = 1% of mean

Company B $q^2 = 194.0 \; e^{.014\,c^2}$
      standard deviation = 2% of mean

Company C $q^3 = 67 \; e^{.017\,c^3}$
      standard deviation = 5% of mean

Company D $q^4 = 49 \; e^{.032\,c^4}$
      standard deviation = 2% of mean

by the ith company, Y_{29} is the copper producer price in cents per pound of refined copper, and π^i is the value of π for the ith company. The average cost has been calculated according to Equation 11-15 and is shown with production information (q^i) in Table 11-11.

Through regression analysis (except for data for 1966–1967 and 1975, for the reasons cited above), the relationships between production (q^i) and cost (c^i) for the four companies were estimated. These relationships are shown in Table 11-12.

Relationship Between Governmental Policy and Individual Firms' Savings

This subsection presents formulas and data for linking individual firms' production simulation and government's prioritization simulation.

The main function of this linkage is shown in Fig. 11-7, which indicates the effect of given energy tax credits on individual firms' energy savings. By using Fig. 11-7, a governmental body can predict the magnitude of any fiscal policy (e.g., energy tax credit) for the amount of energy savings it can induce. For example, Fig. 11-7 shows that a T_1 percent of energy tax credit increase will compensate ΔC_1 of costs increased because of installation of energy conservation devices (as shown in Chart D). This ΔC_1, in turn will induce firms to save $E\%$ of their energy consumption as shown in Chart C. The relationships of Chart C for primary producers A, B, and C are shown in Equations 11-1, 11-2, and 11-3, respectively.*

Chart D in Fig. 11-7 varies for individual producers, depending on their financial condition. We will first discuss the background and then the financial impact of alternative governmental fiscal policies on individual companies' energy savings.

---

* Only three companies (A, B, and C) have been analyzed because Company D consists of numerous small mines that are difficult to identify and model.

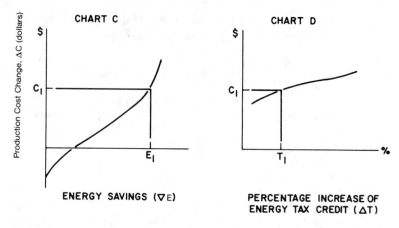

Fig. 11–7.   Effect of government's energy tax credit on individual firms' energy savings.

Historical Fiscal Policies.   Changes in government spending and taxes on the copper industry may significantly affect the industry's gain or loss in a given tax period. Among these policies, there have been aids in the form of purchase contracts and loans (via government spending), and depletion allowance, tax amortization, and import duty (via taxes), which were granted to individual copper companies in order to effectively ensure sufficient copper supply in the 1950s and 1960s.

The historical backgrounds of the above policies are described below.

Energy Tax Credits.   Energy tax credits were enacted by the National Energy Act and expanded by the Windfall Profits Tax legislation. Because a detailed discussion of energy tax credits was presented in Chapter 9, it is not repeated here.

Aids in the Form of Contracts and Loans.   Federal government aids were granted by the Defense Minerals Exploration Administration (DMEA) through purchase contracts with individual copper producers in exploration projects for copper. These contracts or loans were given to improve the long-term resource base. Almost all of the exploration loans were made between 1951 and 1958 with the exception of the last participation in 1969. Under the Defense Production Act of 1950, as amended, contracts were awarded for expansion and maintenance of the supply of copper. These contracts were accompanied by purchase commitments according to floor price agreements. All contracts and agreements for this assistance program were put into effect between 1951 and 1956 except for an $83 million contract in 1967 to the Duval Sierrita Corp. This contract was used to develop a copper supply to alleviate a copper shortage.

Depletion Allowance. For years the depletion allowance has been 15% of the net income from the property during the tax period for domestic ore deposits and 14% for foreign ore deposits. This allowance is provided by the current income tax laws for the depletion of copper ore deposits in each productive property. The provision of the allowance has effectively encouraged the copper industry to search for massive low-grade ore deposits in surface mining in recent years.

Tax Amortization. The tax amortization for copper companies' investment in facilities was awarded in the 1950s and 1960s. It was in terms of five years, at 75% of the total amount of costs incurred.

Import Duty. The import duty was 1.7 cents per pound of copper in the 1960s and 0.8 cent per pound in the 1970s with spontaneous suspensions (shown in Table 11–13). The regressional analysis regarding import duty and quantity of ore or blister imports, from the table, indicates that impacts of an import tariff on the ore and blister imports were not significant over the two decades shown. Therefore, changes of import duty may not be able to effectively influence copper companies' financial gains or loss.

The above analysis of the historical background of various fiscal policies suggests that energy tax credits, income tax rebate (or amortization), and aids in the form of contracts and loans can be effectively used by the federal government to compensate individual copper companies for costs incurred in energy conservation.*

The relationship between energy savings and alternative policies is given below. These relationships are estimated on the basis of the results of the resoumetric models for Companies A, B, and C which were previously detailed.

Relationship Between an Additional Energy Tax Credit and Energy Savings

The original energy tax credit provided in the Nation Energy Act for energy-saving equipment is 10%. In order to estimate the relationship, the energy conservation model and data discussed in Chapter 8 are used to simulate production for Companies A, B, and C, and then their results are analyzed

---

* Depletion allowance was effective historically; however, in light of concerns for energy conservation, this option is not considered here.

Table 11-13. Net Ore and Blister Imports and Import Duty (Thousand Short Tons).

	Net Ore Imports*	Net Blister Imports	Duty
1951	80.12	129.66	0.0
1952	103.59	173.42	0.0
1953	108.05	279.23	0.0
1954	123.56	257.38	0.0
1955	114.20	252.09	0.0
1956	109.93	274.59	0.0
1957	109.91	300.34	0.0
1958	82.67	266.62	1.7**
1959	81.11	266.53	1.7
1960	78.24	289.57	1.7
1961	47.03	335.08	1.7
1962	43.25	330.06	1.7
1963	48.32	367.85	1.7
1964	51.46	384.74	1.7
1965	26.25	327.71	1.7†
1966	40.52	349.17	0.0
1967	−5.73	248.33	0.0
1968	−37.42	254.96	0.0
1969	37.87	233.60	0.0
1970	−27.69	216.60	0.0
1971	23.09	128.03	0.0
1972	35.44	148.85	0.80#
1973	19.37	146.62	0.80
1974	36.04	205.15	0.0

* Includes concentrate and matte.
** Started on July 1, 1958.
† Suspended from 2/2/66 to 6/30/72.
\# Suspended from 7/1/73 to 6/30/75.

for the purpose of representing the industry. The relationship between additional energy tax credit and energy savings is thus estimated as follows:

$$\nabla E_T = 0.719 \; \Delta T$$
$$\text{for } 0 \leqslant \Delta T \leqslant 27.82 \tag{11-16}$$

where ∇E_T is the percentage of the copper industry's total energy consumption saved through conservation efforts induced by energy tax credits, and ΔT is the additional percentage of energy tax credits. For example, if the government desires to encourage 10% energy savings, the additional energy tax credit should be about 13.9%. Note that the realization of energy tax credits depends on whether individual companies have sufficient tax payments to cover the credits.

Relationship Between Income Tax Rebate and Energy Savings

Temporary or permanent income tax rebates or amortization can range from 0.0 to 50.0%. The realization of tax rebates depends on an individual company's yearly profit. In general, if it is assumed that each company makes sufficient profit to enjoy tax rebates, the relationship between tax rebate and energy savings for the industry, based on the results of Companies A, B, and C, is as follows:

$$\nabla E_R = 5.615R$$
$$\text{for } 0 \leqslant R \leqslant 3.6 \tag{11-17}$$

where ∇E_R is the percentage of energy reduction due to conservation efforts induced by tax rebates, and R is the percentage of income tax rebates. For example, if the government wants to encourage 10% energy conservation, the income tax rebate should be about 1.8%.

Relationship Between Government Aids and Energy Savings

The federal government can aid copper companies by granting them a certain number of cents per pound of copper produced, in order to compensate for cost increments incurred during implementation of energy conservation measures. Both taxes and spending are government fiscal policies that have different impacts on the economy. (This will be discussed later in this chapter.)

The relationship between government aids and energy savings is estimated, based on the results of Companies A, B, and C, as follows:

$$\nabla E_s = 24.67S$$
$$\text{for } 0 \leqslant S \leqslant .81 \tag{11-18}$$

where ∇E_s is the percentage of energy reduction due to conservation efforts induced by government aids, and S is the amount of aids in cents (1974 dollars) per pound of refined copper production. For example, in order to encourage conserving 10% of the energy used by the copper industry, the government can grant 0.4¢/pound of refined copper production in order to compensate any cost increase caused by energy conservation efforts.

ENERGY MANAGEMENT AND POLICY—EXAMPLE OF CONSERVATION IN THE PRIMARY COPPER PRODUCTION

Evaluation Procedure

We discussed the structure and function of the four elements of the energy compromise model in the preceding sections. In this section, we will discuss national energy management and policy by using the energy compromise model to evaluate energy conservation in the primary copper production. The energy conservation flow chart of Fig. 11–2 will be followed closely to demonstrate how to use the model.

The steps in the flow chart can be basically condensed as follows:

1. Determine refined copper market prices and production for the copper industry by use of econometrics.
2. Allocate the industry's production to major companies by using the data management system.
3. Estimate energy consumption for the above production by use of resoumetrics.
4. Estimate alternative scenarios to achieve a given conservation target using the data management system.
5. Determine benefits/costs of these scenarios using econometrics and resoumetrics.
6. Prioritize these scenarios by weighing their benefits/costs against interest groups and goals by using hierarchical analysis.

Determine Refined Copper Market Prices and Production. The PCM discussed in the previous sections is used to generate copper market information on the copper industry level as an input for the resoumetric model. Essentially, data for the producer's price, refined copper production, and nonrefined copper scrap recovery are simulated under given economic conditions.

The following assumptions are made for the copper markets:*

- Lagged fabricator and refined copper stock levels are assumed to be 110 and 50 thousand short tons, respectively.
- The U.S. industrial index is 110 (1974 = 100).
- The German industrial index is 120 (1974 = 100).

The PCM was operated, and a set of simulated results in respect to the above economic conditions was generated as follows:

- Refined copper price is 80.00¢/lb
- No. 2 heavy scrap price is 65.35¢/lb.

---

* These economic conditions are based on subjective judgments for illustrative purposes in the present version. In later versions, some well-known national econometric models such as the Wharton Model may be interconnected to obtain necessary data.

- Total refined copper production is 2200 thousand short tons.
- Nonrefined scrap generation is 1180 thousand short tons.

Allocate the Industry's Production to Major Companies. Because it is only necessary to allocate production among major producers, the equations shown in Tables 11–10 and 11–12 (in the data management system) are used to estimate four major companies' production. Other companies in the industry will make up the remaining production. Of course, one may argue that there are more than four leading companies in the copper industry. We only use them as an example; the number of major companies can increase if more are needed.

Recall the producer copper price (Y_{29}) and total industry production (Y_5) generated by the PCM as follows:

Producer copper price: $Y_{29} = 80 \text{¢/lb}$

Total industry production: $Y_5 = 2200$ thousand tons

Assume also that the index of ore grade will be .716%. Substituting the above information into the equations in Tables 11–10 and 11–12 yields the copper production for each company as shown in Fig. 11–8.

Each company's profit pattern must be considered to ensure that its magnitude is within a reasonable range. If it is not, equations shown in Tables 11–10 and 11–12 must be reevaluated.

Estimate Energy Requirement. The resoumetrics developed in the previous sections is used to estimate energy requirements for the above production allocated for the four major companies. Because we lack Company D's data, only three companies' energy requirements are estimated. The energy needed

Fig. 11–8.   Computation of Copper Production for Selected Companies.

Company A	$\pi^1 = 100 + 64\ Y_{29} - .160\ I = 36.64$
	$c^1 = (100 - \pi_1)\ Y_{29}/100 = 50.69$
	$q^1 = 150\ C^{.003\,C^1} = 150\ e^{.023 \times 50.69} = \underline{481}$
Company B	$\pi^2 = 100 + 44\ Y_{29} - .160\ I = 20.64$
	$C^2 = (100 - \pi)\ Y_{29}/100 = 63.48$
	$q^2 = 194\ e^{.014\,e^2} = 194\ e^{.014 \times 63.48} = \underline{472}$
Company C	$\pi^3 = 100 - .167\ Y_{29} - .09\ I = 22.2$
	$C^3 = (100 - \pi_3)\ Y_{29}/100 = 62.24$
	$q^3 = 67\ e^{.017\,c^3} = 67\ e^{.017 \times 62.24} = \underline{193}$
Company D	$\pi^4 = 100 + .37\ Y_{29} - .140\ I = 29.36$
	$C^4 = (100 - \pi_3) \times Y_{29}/100 = 56.5$
	$q^4 = 49\ e^{.032\,C^4} = 49\ e^{.032 \times 56.5} = \underline{299}$

Table 11-14. Energy Requirements for the Production Quantities Shown in Fig. 11-8.

Company	Energy Consumption (Billion Btu)
A	27529.4
B	18624.9
C	15269.0
Total	61423.3

for the production quantities shown in Fig. 11-7 is estimated in Table 11-14.

Estimate Alternative Scenarios to Achieve a Given Conservation Target. Equations 11-16, 11-17, and 11-18 (relationships between alternative government fiscal policies and energy savings) are used to estimate alternative scenarios to achieve a certain conservation target. If the conservation target is to reduce energy consumption for the refined copper production by 10%, the following three alternative scenarios are derived from Equation 11-16, 11-17, and 11-18, respectively:

- Scenario I: Additional 13.9% energy tax credit
- Scenario II: 1.8% income tax rebate
- Scenario III: 0.4 cent aid per pound of refined copper production

Determine Benefits/Costs of Alternative Scenarios. The PCM and resoumetrics are used to estimate benefits/costs of alternative scenarios.

Penn Composite Model. Sample items from the PCM for examining the impacts of the three policies are copper producer price, LME price, copper demand, and copper production. Net increases of Federal Reserve Board Index of Manufacturing Production for the above three alternative policies are estimated in Table 11-15. The increment of copper mine production

Table 11-15. Net Increase of Manufacturing Production.

Scenario	Net Increase of Index of Manufacturing Production (%)
I. Energy Tax Credit	2
II. Tax Rebate	2
III. Aid	3

Table 11-16. Economic Impacts of the Three Policies Via the Penn Composite Model.

| | Economic Impact* (Yearly) | | | | |
Policy	Col. 1 Copper Producer Price ($\cancel{c}$/lb)	Col. 2 LME Price ($\cancel{c}$/lb)	Col. 3 Copper Demand (1000 Tons)	Col. 4 Refined Copper Production (1000 Tons)	Col. 5 Copper Mine Production (1000 Tons)
I	+.0045	+.005	+30.3	+30.3	+78
II	+.0045	+.005	+30.3	+30.3	Undetermined
III	+.00747	+.0075	+45.45	+45.45	Undetermined

* In this table, "+" indicates increase.

$(Y_{6.1})$ in tons per year with respect to additional energy tax credits is assumed to be

$$\frac{\partial Y_{6.1}}{\partial D} = 3500 \, D \qquad (11-19)$$

where ∂ represents partial differentiation, and D is the percent of energy tax credit.

Substitution of Table 11-15 into the PCM with the aid of Equation 11-19 yields major economic impacts estimated for the three policies as shown in Table 11-16.

The Resoumetric Model. Sample items from the resoumetric model for evaluating the impact of the federal government's three policies are air emissions (mainly SO_2), waste water discharged, solid waste, and land disturbance. The relationships among these products and copper production level were estimated through the copper resoumetrics and Equations 11-4 through 11-7. These are used to evaluate environmental impacts for the three scenarios, which are shown in Table 11-17.

Prioritize Alternative Scenarios. To simulate the government's decision-making in choosing any of the three policies by hierarchical analysis, we are concerned about the following sample elements, for purposes of demonstration:

- Three interest groups: environmentalists, federal government, and the copper industry
- Three comparison criteria: economic, political, and environmental
- Three fiscal policies as discussed before

Table 11-17. Environmental Impacts of the Three Policies Via the Resoumetric Model.

	Environmental Impacts* Per Day†			
Policy	SO₂ Emissions (Tons)	Waste Water Discharged (Barrels)	Solid Waste (1000 Tons)	Land Disturbance (Acres)
I	+16.6	+2000	+8.5	+78
II	+16.6	+2000	+8.5	Undetermined
III	+25.1	+2883	+12.7	Undetermined

* In this table, "+" indicates increase.
† Assumed 1 year = 357 working days.

Instead of inviting representatives from the three interest groups to form a jury session and appraise all pairwise comparisons concerning the importance of policies and criteria, for the purpose of methodological demonstration we make all assessments based on our best judgments about the behaviors of various interest groups.

The ratio scale shown in Table 11-7 is applied for pairwise comparison. It is understood that the comparisons are subjective with respect to each interest group because each group weighs the scale value differently according to its value judgment on the information in Tables 11-16 and 11-17.

In the rest of the chapter, we shall appraise all values of pairwise comparisons and demonstrate the methodology from the bottom level of the hierarchy to the top level, that is, first policy pairwise comparisons with respect to each criterion, then criterion comparisons with respect to each interest group, and, finally, the interest group comparisons with respect to influence on shaping the future policy (final decision). According to Fig. 11-5, the following pairwise comparisons are made:

A. Policy pairwise comparisons
   1. From the point of view of environmentalists
     (a) With respect to economic criterion: Table F-1*
     (b) With respect to political criterion: Table F-2
     (c) With respect to environmental criterion: Table F-3
   2. From the point of view of the federal government
     (a) With respect to economic criterion: Table F-5
     (b) With respect to political criterion: Table F-6
     (c) With respect to environmental criterion: Table F-7

---

* These tables are shown in Appendix F. Values in the tables are subjective estimates.

3. From the point of view of the copper industry
    (a) With respect to economic criterion: Table F-9
    (b) With respect to political criterion: Table F-10
    (c) With respect to environmental criterion: Table F-11
B. Criterion pairwise comparisons
    1. With respect to environmentalists: Table F-4
    2. With respect to the government: Table F-8
    3. With respect to the copper industry: Table F-12
C. Interest group influence pairwise comparisons
    1. Table F-13

Maximum eigenvalue (λ^{max}), dominant eigenvector (W^*), and consistency index (Index) are calculated for each matrix and shown in the corresponding tables. The consistency index is a value indicating the consistency of value judgments among various comparisons.

The consistency index of each matrix is, in most cases, very close to zero; therefore, we may conclude that the dominant eigenvector is the representative of the priorities of the activities compared pairwise in the matrix.[10] Accordingly, we transfer lower-level priorities by concatenation of dominant eigenvectors in Tables F-1, F-2, and F-3 into a matrix and multiplication by the eigenvector in Table F-4, and yield a vector that is a weighting vector of policies based on the relative importance of each criterion for the environmentalists. In the same manner, we concatenate eigenvectors in Tables F-5, F-6, and F-7 with multiplication by the eigenvector in Table F-8 to yield a weighting vector of policies for the government; and repeat the process (using Tables F-9 through F-12) to yield the vector for the copper industry.

The above three weighting vectors of policies for the three interest groups are listed in Table 11–18. The weights in each vector in Table 11–18 clearly indicate each interest group's preference among the three policies. For example, the environmentalist strongly prefers the implementation of government's aids in the form of contracts or loans.

Table 11–18. Overall Importance of Policies with Respect to Various Interest Groups.

	Environmentalists	Federal Government	Industry
I (Energy tax credits)	.1720	.5249	.1612
II (Tax rebate)	.3278	.1451	.4822
III (Aid)	.5002	.3300	.3566

Concatenation of the above-yielded vectors into a matrix as shown in Table 11–18 and multiplication by the dominant eigenvector in Table F-13 yields vector H, which is the overall weighting of the possible future policies (the composite policy) based on the hierarchical weighting of interest groups' influence, criteria's magnitude, and policies' importance as the following vector:

$$H = (.4, .225, .375) \tag{11–20}$$

Equation 11–20 indicates that, in this particular case for an overall consideration by those interest groups, the implementation of granting energy tax credit or direct government aids in the form of contracts or loans is much more favorable than granting corporate income tax rebates.

Equation 11–20 also represents a composite (compromise) policy with the implication that, in order to meet each interest group's expectation as well as possible, the following policies may be preferred for the purpose of conserving 10% of the present energy consumption in copper production:

- Additional 5.56% of energy tax credits (multiply 13.9% by .4)
- 0.4% of income tax to be rebated (multiply 1.8 by .225)
- .15 cent aid per pound copper production (multiply .4 cent by .375)

Finally, we must be aware that these results are determined largely by our judgment in pairwise comparisons throughout the simulation of the decision-making process. They are mostly subjective and subject to personal biases.

Final Comment on the Hierarchical Analysis. As discussed in the last section, most judgments of importance in pairwise comparisons in Tables F-1 through F-13 are consistent because all consistency indices in the tables are close to zero.* We may thus conclude that the normalized weights in eigenvectors are representative of the priorities of activities compared pairwise in all matrices. Unfortunately, consistency is not equal to correctness of value judgments. All values in the matrices of Tables F-1 through F-13 are subjective. Any misjudgments in the matrices will be transferred to their corresponding eigenvectors and affect their resulting overall importance to decision-makers. On the other hand, misjudgments may reflect the nature of a human being with respect to his understanding of the decision environments and his evaluation system as it relates to his social and educational background.

---

* If a consistency index is close to zero, then the comparisons in its correspondent table are consistent.

CONCLUSIONS

The discussion in this chapter has had two purposes: (1) to present an analytical tool for governmental bodies to use to search for compromise policies involving multi-participants (interest groups) and multi-objectives; and (2) to use this tool to examine the nation's energy policies and evaluate various fiscal scenarios inducing energy conservation. The primary copper production was used as an example in analyzing the nation's energy policies.

In the discussion, we have proved that it is manageable to incorporate data concerning marketing, production, and governmental decision-making in evaluating energy policies from the nation's point of view. We are convinced that any comprehensive energy policies must be so evaluated before they are implemented. The purpose of the example of copper production is to illustrate that all participants and information involved in marketing, production, and governmental decision-making will eventually affect the outcome of the compromise scenarios. Absence of any interest group or information of the above three aspects is likely to lead to biased energy policies.

In this chapter, we identified energy tax credits as a preferred fiscal policy for governmental bodies to use for inducing conservation after having estimated responses of environmentalists, industrialists, and the federal government (representing the rest of the socioeconomic factors). However, the above result is based on the assumption that energy conservation is the nation's primary goal and can be achieved by using fiscal policies.* Also, the numbers computed in this chapter are only significant when they are used to indicate relative importance or relative magnitude in the demonstration of the methodology. We do not intend to attach any absolute accuracy to these figures.

REFERENCES

1. Arthur D. Little, Inc. "Economic Simulation and Impact Analysis Model of the U.S. Copper Industry," draft report to the U.S. Environmental Protection Agency, October 1976.
2. Fischer, F., Costner, P. and Baily, M. "An Econometric Model of the World Copper Industry," *The Bell Journal of Economics and Management Science*, Vol. 3, No. 2, 568–609, Autumn 1972.
3. Charles River Associates, Inc. "Economic Analysis of the Copper Industry," report to the Property Management and Disposal Service, General Services Administration, March 1970.
4. Charles River Associates. "An Econometric Model of the Copper Industry,"

---

* One may argue that we can sacrifice environmental protection for energy production. We exclude this option because it is not acceptable, owing to increasing awareness of environmental protection.

report to the Property Management and Disposal Service, General Services Administration, September 1970.

5. Charles River Associates. "The Effects of Pollution Control on the Nonferrous Metals Industries, Copper. Part III. The Economic Impact of Pollution Abatement of the Industry," NTIS, PB-207 163, December 1971.

6. Synergy, Inc. "A Forecasting System for Critical Imported Minerals," final report to Bureau of Mines, U.S. Department of the Interior, May 30, 1975.

7. Barry Silverman. "Resoumetric Modelling: Copper—A Case Study," Ph.D. dissertation, Department of Civil and Urban Engineering, University of Pennsylvania, 1977.

8. Hu, S. "The Copper Commodity Model and Energy Issues," Ph.D. dissertation, University of Pennsylvania, 1978.

9. Zandi, I. et al. "The Development of Resoumetric Methodology for Utilization of MAS in Policy Assessment," semiannual report, U.S. Bureau of Mines, December 1976.

10. Saaty, T. "A Scaling Method for priorities in Hierarchial Structures," *Journal of Mathematical Psychology,* Vol. 15, 111–159, 1977.

11. Kahn, A. *Theory and Practice of Social Planning.* New York: Russell Sage Foundation, 1969.

12. American Bureau of Metal Statistics. *Yearbook of the American Bureau of Metal Statistics.* New York: The Bureau, Annual, 1975.

13. Hu, S. and Zandi, I. "Copper Econometric Models." *Engineering Costs and Production Economics,* Vol. 5, 53–70, 1980.

14. Lanzillotti, R. F. "Pricing Objectives in Large Companies," *American Economic Review* 48(5), pp. 921–940, December 1958.

PART III
ENERGY CONSERVATION TECHNOLOGIES

This section discusses major generic conservation technologies commonly practiced in industry. A summary of specific conservation technologies is presented in Appendix C.

12

Waste Heat Recovery and Utilization

INTRODUCTION

Waste heat can include industrial process waste steam and heat, power plant reject heat, and heat generated from various other streams such as agricultural crops, food process waste, waste tires, and animal manure. Because the properties and recovery methods of the three types of waste heat are different, general methods for industrial process waste steam and heat are discussed in this chapter, in-plant electricity cogeneration (a special method of industrial waste heat recovery and utilization) is presented in Chapter 13, and power plant reject heat utilization in Chapter 14. Since the economic feasibility of heat generated from various waste streams is site- and waste-specific, only their potential is highlighted in this chapter.

Approximately two-thirds of industrial energy is used in process steam and heat, and this is in the form of thermal energy, rather than in the form of power. Consequently, the opportunities for waste heat recovery are plentiful. However, in establishing the opportunities, process energy requirements and waste streams need to be evaluated, and technologies of recovery and their costs and energy savings are vital to the determination of the economic viability of waste heat.

In this chapter an overview of waste heat is followed by a discussion of technologies for waste heat recovery and utilization, cost and energy savings of waste heat recovery and utilization, and research and development in waste heat recovery and utilization.

OVERVIEW OF WASTE HEAT

Definition of Waste Heat

Waste heat is defined as heat to be rejected from a process at a temperature enough above the ambient temperature to permit the engineer to extract additional value from it. Three temperature ranges are used to classify waste heat: (1) the high-temperature range—above 1200°F; (2) the medium-temperature range—between 450°F and 1200°F; and (3) the low-temperature range—below 450°F.

High- and medium-temperature waste heat can be used to produce process steam. Gas or steam turbines should be used to perform work before this waste heat is extracted. Low-temperature waste heat can be applied for mechanical work through heat pumps, an interesting application being in petroleum distillation, where the working fluid of the heat pump can be the liquid being distilled.*

Sources of Waste Heat

Depending on its temperature, waste heat comes from various fuel-fired processes. The temperature level and the location of the waste heat streams relative to the user's need are significant in waste heat recovery.

High-Temperature Waste Heat. The maximum theoretical temperature for combustors is about 3500°F, and most practical combustors are operated under 3000°F. Table 12–1 shows the location and temperature level of the high-temperature waste heat streams. All of the temperature ranges result from direct fuel-fired processes.

Medium-Temperature Waste Heat. Table 12–2 shows the location and temperature level of medium-temperature waste heat streams. Most of the waste heat in this temperature range comes from the exhausts of directly fired process units. Medium-temperature waste heat is still hot enough to allow consideration of the extraction of mechanical work from the waste heat, by a steam or gas turbine.

Low-Temperature Waste Heat. Table 12–3 lists the sources and temperature levels for some low-temperature waste heat streams. In this temperature range, it is usually not practical to extract work from the waste heat source, although steam production may not be completely excluded if there

---

* This application was developed by the British Petroleum Co.[1]

Table 12–1. High-Temperature Waste Heat.

Type of Device	Temperature (°F)
Nickel refining furnace	2500–3000
Aluminum refining furnace	1200–1400
Zinc refining furnace	1400–2000
Copper refining furnace	1400–1500
Steel heating furnaces	1700–1900
Copper reverberatory furnace	1650–2000
Open hearth furnace	1200–1300
Cement kiln (dry process)	1150–1350
Glass melting furnace	1800–2800
Hydrogen plants	1200–1800
Solid waste incinerators	1200–1800
Fume incinerators	1200–2600

Source: National Bureau of Standards.[1]

Table 12–2. Medium-Temperature Waste Heat.

Type of Device	Temperature (°F)
Steam boiler exhausts	450–900
Gas turbine exhausts	700–1000
Reciprocating engine exhausts	600–1100
Reciprocating engine exhausts (turbocharged)	450–700
Heating furnaces	800–1200
Drying and baking ovens	450–1100
Catalytic crackers	800–1200
Annealing furnace	800–1200

Source: National Bureau of Standards.[1]

is a need for low pressure stream. Low-temperature waste heat may be used supplementarily for preheating. For example, it is possible to use economically the energy from an air conditioning condenser operating at about 90°F to heat the domestic water supply. Since the hot water must be heated to about 160°F, obviously the air conditioner waste heat is not hot enough. However, because the cold water enters the domestic water system at about 50°F, energy interchange can take place, raising the water to something less than 90°F.

Table 12–3.   Low-Temperature Waste Heat.

Source	Temperature (°F)
Process steam condensate	130–190
Cooling water from:	
Furnace doors	90–130
Bearings	90–190
Weldings	90–190
Injection molding machines	90–190
Annealing furnaces	150–450
Forming dies	80–190
Air compressors	80–120
Pumps	80–190
Internal combustion engines	150–250
Air conditioning and	
refrigeration condensers	90–110
Liquid still condensers	90–190
Drying, baking, and curing ovens	200–450
Hot processed liquids	90–450
Hot processed solids	200–450

Source: National Bureau of Standards.[1]

Quantity of Waste Heat

It is difficult to estimate U.S. total industrial waste heat that is worth recovering. The following eight industries are commonly believed to have the most potential for waste heat recovery:

- Petroleum refining
- Steel
- Aluminum
- Paper
- Olefins and derivatives
- Textiles
- Cement
- Glass

As shown in Table 12–4, the total waste heat in the eight industries is estimated at 8.5 quadrillion (Q) Btu/year, an equivalent of 4.2 million barrels of oil/day.[2] Table 12–5 further indicates the concentration of waste heat in six processes: kraft pulping, atmospheric distribution, electrolytic reduction, iron products, coal products, and ethylene production.

Waste heat in these six processes is estimated at 4.25 Q/year, an equivalent

Table 12–4. Estimated Waste Heat in Eight Industries.

Industry	Total Energy Consumed (10^{15} Btu/Year)	Estimated Waste Heat (10^{15}/Year)
Petroleum refining	2.96	2.23
	(plus 23.0 for feedstock)	
Steel	3.8	1.82
Aluminum	1.39	1.15
Paper	1.5	1.72
	(plus 2.5 for feedstock)	
Olefins		
and derivatives	0.99	0.46
	(plus 0.2 for polymerization)	
Textiles	0.54	0.37
Cement	0.53	0.39
Glass	0.33	0.32
Total		8.46

Source: ERDA.[2]

Table 12–5. Estimated Waste Heat in Six Processes.

Process	Industry	Estimated Waste Heat (10^{15} Btu/Year)
Kraft pulping	Paper	1.29
Atmospheric distribution	Petroleum	0.87
Electrolytic reduction	Aluminum	0.71
Iron products	Steel	0.61
Coal products	Steel	0.44
Ethylene products	Olefin	0.34
Total		4.25

Source: ERDA.[2]

of 2.1 million barrels of oil/day. Note that U.S. crude oil imports are about 6.5 million barrels/day.

Potential for Heat Recovery from Various Waste Streams

There is some potential for heat recovery from industrial, agricultural, and municipal waste streams, and there is much literature in this area. This subsection is intended to be a commentary discussion on this subject. The interested reader is referred to other publications.[3-6]

Table 12–6 shows the heat content, quantity, and energy potential for the seven most common waste streams. It is interesting to note that the heat content in most of these waste streams is high, and that a substantial amount of energy could be recovered. This potential is even higher than that of waste heat recovery in the eight most promising industries discussed in Table 12–4. Like industrial waste heat recovery, heat recovery from waste streams can reduce environmental pollution. However, unlike industrial waste heat recovery, heat recovery from various waste streams is still facing major obstacles because of economic, institutional, and technical problems. For this reason, only the technologies of industrial waste heat recovery are presented in this chapter.

TECHNOLOGIES FOR WASTE HEAT RECOVERY AND UTILIZATION

There are two basic ways to recover heat from the sources previously discussed: (1) using heat exchangers to transfer heat in one fluid stream to another (e.g., from flue gas to feedwater or combustion air); and (2) cogeneration, which transfers waste thermal power to electric power. Heat exchangers are discussed in the rest of this chapter, and cogeneration is discussed in Chapter 13.

A heat exchanger is a system that separates the stream containing waste heat and the medium that is to absorb it, but allows the flow of heat across the separation boundaries. Industrial heat exchangers have many pseudonyms. They are sometimes called recuperators, regenerators, waste heat steam generators, condensers, heat wheels, temperature and moisture exchangers, and so forth. Whatever name they may have, they all perform one basic function: the transfer of heat.

Overview of Heat Exchangers*

In industrial plants, heat exchangers are used to reclaim waste heat. Most exchangers can be retrofitted in existing plants.

Definition. Heat exchangers are characterized as single or multipass, gas to gas, liquid to gas, liquid to liquid, evaporator, condenser, parallel flow, counter flow, or cross flow. We are defining these terms as follows:

Single or Multipass. The terms "single" or "multipass" refer to the heating

---

* Most information in this section was obtained from NBS.[1]

Table 12-6. Estimates for Heat Recovered from Various Waste Streams in 1980.

Waste Stream Type	Waste Stream* (Millions of Dry Tons)	Specific Heat Content (Million Btu/Dry Ton)	Total Energy Potential (Quads)
Agricultural crops and food processing waste	390	11†	4.29
Animal manure	266	7	1.86
Urban refuse	222	9**	2.00
Logging and wood manufacturing residues	59	12	0.71
Miscellaneous industrial wastes	55	20	1.00
Miscellaneous organic wastes	60	14	0.84
Municipal sewage solids	14	4	0.06
Totals	1066	(Average 10.1)	10.76

* Data obtained from Anderson.[4]
† Data calculated from Combustion.[5]
** Data calculated from Schultz.[3]

or cooling media passing over the heat transfer surface once or a number of times. Multipass flow involves the use of internal baffles.

Gas to Gas, Liquid to Gas, Liquid to Liquid. These three terms refer to the two fluids between which heat is transferred in the heat exchanger, and imply that no phase changes occur in those fluids. Here the term "fluid" is used in the most general sense.

Evaporator. An evaporator is a heat exchanger in which heat is transferred to an evaporating (boiling) liquid.

Condenser. A condenser is a heat exchanger in which heat is removed from a condensing vapor.

Parallel Flow. A parallel-flow heat exchanger is one in which both fluids flow in approximately the same direction.

Counter Flow. A counter-flow heat exchanger is one in which the two fluids move in opposite directions.

Cross Flow. When the two fluids move at right angles to each other, the heat exchanger is considered to be of the cross-flow type.

Applications. Some of the practical applications for various types of heat exchangers (with different names) are listed as follows:

- Preheat the combustion air by medium- to high-temperature exhaust gases for
  - Boilers using *air preheaters*
  - Furnaces using *recuperators*
  - Ovens using *recuperators*
  - Gas turbines using *regenerators*
- Preheat liquid or solid feedstocks by low- to medium-temperature exhaust gases using *heat exchangers*
- Preheat boiler feedwater or makeup water using *economizers,* which are gas to liquid water-heating exchangers
- Generate steam in *waste heat boilers*
- Transfer waste heat to liquid or gaseous process unit directly through pipe and ducts or indirectly through a secondary fluid such as steam or oil

Classifications. Although the functions of various heat exchangers are similar (i.e., the transfer of heat), their structures and principal modes of heat transfer are different.

The specification of an industrial heat exchanger must include the heat exchange capacity, the temperatures of the fluids, the allowable pressure drop in each fluid path, and the properties and volumetric flow of the fluids entering the exchanger. These specifications will determine construction parameters and, thus, the cost of the heat exchanger. The final design will be a compromise between pressure drop, heat exchanger effectiveness, and cost. Accordingly, the essential parameters to be considered for an optimum choice of waste heat recovery devices are as follows:

- Temperature of waste heat fluid
- Flow rate of waste heat fluid
- Chemical composition of waste heat fluid
- Minimum allowable temperature of waste heat fluid
- Temperature of heated fluid
- Chemical composition of heated fluid
- Maximum allowable temperature of heated fluid
- Control temperature, if control is required
- Effectiveness of heat transfer
- Capital, operation, and maintenance costs

Commercial heat exchangers are commonly classified in the following nine categories with respect to the above specifications:

- Gas-to-gas heat exchangers
  - Recuperators
  - Heat wheels
  - Air preheaters (passive regenerator)
  - Heat pipes
- Gas-to-liquid regenerators
  - Waste heat boilers
  - Heat pumps
  - Finned-tube heat exchangers (economizers)
  - Shell and tube heat exchangers
  - Gas and vapor expanders
- Liquid-to-liquid exchangers
  - Shell and tube heat exchangers (same as above under gas-to-liquid regenerators)

All of the above heat exchangers can be retrofitted, and package units are available. Different types of heat exchangers are used to recover various-temperature waste heat streams.

Gas-to-Gas Heat Exchangers

A gas-to-gas heat exchanger is often used to preheat the waste gas before oxidation and to lower the fuel consumed when a gaseous waste stream requires thermal oxidation to meet pollution standards. A temperature profile and dew point for sulfur-bearing residual fuel fired in the heater or oxidizer must be considered. Payout on heat exchangers is attractive, sometimes less than one year. This type of heat exchanger includes recuperators, heat wheels, air preheaters, and heat pipes.

Recuperators. Recuperators are the simplest form of heat exchangers. Generally there are two types of recuperators, radiation and convective (tube). The ceramic recuperator, vertical tube-within-tube recuperator, and combined radiation and convective type recuperator are examples of derived types of recuperators.

Radiation Recuperators. A radiation recuperator consists of two concentric lengths of metal tubing as shown in Fig. 12–1. The heat is transferred from the hot waste gases to the surface of the inner tube by radiative heat transfer. The inner tube carries the hot exhaust gases, while the external annulus carries the combustion air from the atmosphere to the air inlets of the furnace burners. The hot gases are cooled by the incoming combustion air, which now carries additional energy into the combustion chamber. This is energy that does not have to be supplied by the fuel; consequently, less fuel is burned for a given furnace loading. The savings in fuel also means a decrease in combustion air; thus stack losses are decreased not only by lowering the stack gas temperatures, but also by discharging smaller quantities of exhaust gas.

As shown in the diagram, the two gas flows are usually parallel because recuperators frequently serve the additional function of cooling the duct carrying away the exhaust gases and consequently extending its service life.

Convective Recuperators. A convective recuperator consists of a shell surrounding a number of parallel small-diameter tubes as shown in Fig. 12–2. The incoming air to be heated enters the shell and passes over the hot tube one or more times. If the tubes are baffled to allow the gas to pass over them twice, the heat exchanger is termed a two-pass recuperator; if two baffles are used, a three-pass recuperator; and so on.

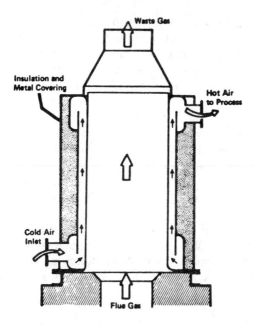

Fig. 12–1. Diagram of a metallic radiation recuperator. Source: NBS.[1]

Shell-and-tube-type recuperators are generally more compact and have a higher effectiveness than radiation recuperators, because of the larger heat transfer area made possible through the use of multiple tubes and multiple passes of the gases.

The principal limitation on the heat recovery of metal recuperators is the reduced life of the liner at inlet temperatures exceeding 2000°F.

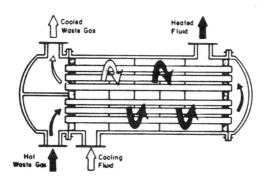

Fig. 12–2. Diagram of a convective recuperator. Source: NBS.[1]

Ceramic Tube Recuperators. In order to overcome the temperature limitations of metal recuperators, ceramic tube recuperators have been developed. Ceramic tube recuperator materials allow operation on the gas side to 2800°F and on the preheated air side to 2200°F on an experimental basis, and to 1500°F on a more or less practical basis. Figure 12–3 illustrates the structural difference.

Tube-within-tube Recuperators. An alternative arrangement for the convective recuperator, in which the cold combustion air is heated in a bank of parallel vertical tubes that extend into the flue gas stream, is shown schematically in Fig. 12–4. The advantage of this arrangement is the ease of replacing individual tubes, which can be done during full-capacity furnace operation.

Combined Radiation and Convective Recuperators. For maximum effectiveness of heat transfer, combinations of radiation and convective recuperators are used, with the convective type always following the high-temperature radiation recuperator. A schematic diagram of this arrangement is seen in Fig. 12–5.

Applications. Recuperators are used for recovering heat from exhaust gases to heat other gases in the medium- to high-temperature range. Some typical applications are in soaking ovens, annealing ovens, melting furnaces, afterburners and gas incinerators, radiant-tube burners, and reheat furnaces.

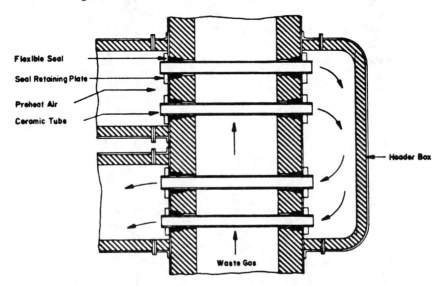

Fig. 12–3. Ceramic recuperator. Source: NBS.[1]

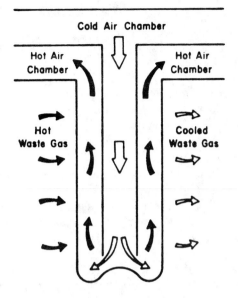

Fig. 12–4. Diagram of a vertical tube-within-tube recuperator. Source: NBS.[1]

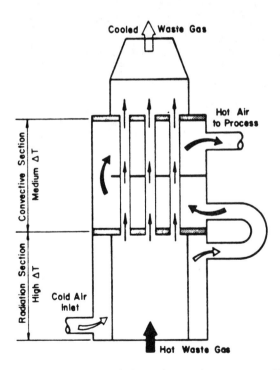

Fig. 12–5. Diagram of a combined radiation and convective recuperator. Source: NBS.[1]

Heat Wheels. A heat wheel is a rotary regenerator (another form of heat exchanger) used to recover low- to medium-temperature heat. As shown in Fig. 12–6, a heat wheel consists of a sizable porous disk, fabricated from some material having a fairly high heat capacity, rotating between two side-by-side ducts: one a cold gas duct, the other a hot gas duct. The axis of the disk is located parallel to, and on the partition between, the two ducts. As the disk slowly rotates, sensible heat (and, in some cases, moisture containing latent heat) is transferred to the disk by the hot air and, as the disk rotates, from the disk to the cold air. The overall efficiency of sensible heat transfer for this kind of regenerator can be as high as 85%. Heat wheels have been built as large as 70 feet in diameter with air capacities up to 40,000 ft³/min. Multiple units can be used in parallel.

The temperature range of the heat wheel is limited primarily by mechanical difficulties introduced by uneven expansion of the rotating wheel when the temperature differences mean large differential expansion, causing excessive deformations of the wheel and thus difficulties in maintaining adequate air seals between duct and wheel.

Heat wheels are available in four types. The first consists of a metal frame packed with a core of knitted mesh stainless steel or aluminum wire. The second, called a laminar wheel, is fabricated from corrugated metal and is composed of many parallel flow passages. The third variety is also a laminar wheel but is constructed from a ceramic matrix of honeycomb configuration. This type is used for higher-temperature applications with a present-day limit of about 1600°F. The fourth variety is of laminar construction, but the flow

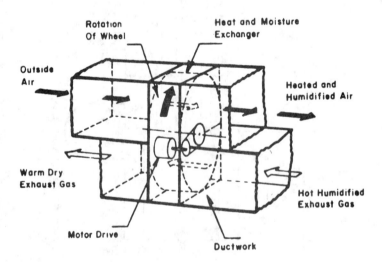

Fig. 12–6.   Heat and moisture recovery using a heat wheel type regenerator. Source: NBS.[1]

passages are coated with a hydroscopic material so that latent heat may be recovered. The packing material of the hygroscopic wheel may be any of a number of materials.

Applications. One application of heat wheels is in space-heating situations where unusually large quantities of ventilation air are required for health or safety reasons. In the summer season, the heat wheel can be used to cool the incoming air with cold exhaust air, reducing the air conditioning load by as much as 50%.

Heat wheels are finding increasing use in process heat recovery in low- and moderate-temperature environments. Typical applications would be curing or drying ovens and air preheaters in all sizes for industrial and utility boilers.

Air Preheaters (Passive Regenerators). Passive gas-to-gas regenerators, sometimes called air preheaters, are available for applications that cannot tolerate any cross contamination. They are constructed of alternate channels (see Fig. 12–7) that put the flows of the heating and the heated gases in close contact with each other, separated only by a thin wall of conductive metal. They occupy more volume and are more expensive to construct than heat wheels, since a much greater heat transfer surface area is required for the same efficiency. An advantage, besides the absence of cross contamination, is the decreased mechanical complexity, since no drive mechanism is required. However, it becomes more difficult to achieve temperature control with passive regeneration, and, if this is a requirement, some of the advantages of its basic simplicity are lost.

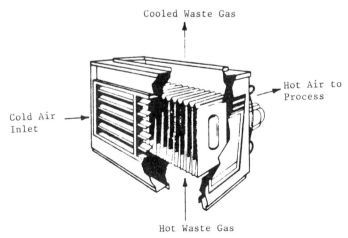

Fig. 12–7. A passive gas-to-gas regenerator. Source: NBS.[1]

Applications. Gas-to-gas regenerators are used for recovering heat from exhaust gases to heat other gases in the low- to medium-temperature range. A list of typical applications follows:

- Heat and moisture recovery from building heating and ventilation systems
- Heat and moisture recovery from moist rooms and swimming pools
- Reduction of building air conditioner loads
- Recovery of heat and water from wet industrial processes
- Heat recovery from steam boiler exhaust gases
- Heat recovery from gas and vapor incinerators
- Heat recovery from baking, drying, and curing ovens
- Heat recovery from gas turbine exhausts
- Heat recovery from other gas-to-gas applications in the low- through high-temperature range.

Heat Pipes. Heat pipes have only recently become commercially available, but they show promise as an industrial waste heat recovery option because of their high efficiency and compact size. In use, the pipe operates as a passive gas-to-gas finned-tube regenerator. As can be seen in Fig. 12–8, the elements form a bundle of heat pipes that extend through the exhaust and inlet ducts in a pattern resembling the structure of finned coil heat exchangers. Each pipe, however, is a separate sealed element consisting of an annular wick on the inside of the full length of the tube, in which an appropriate heat transfer fluid is entrained.

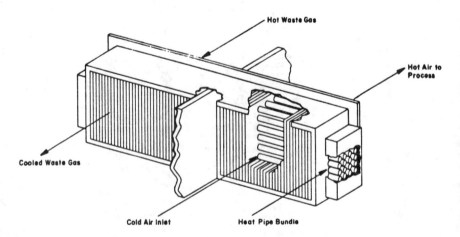

Fig. 12–8. Heat pipe bundle incorporated in gas-to-gas regenerator. Source: NBS.[1]

The heat pipe is compact and efficient because: (1) the finned-tube bundle is inherently a good configuration for convective heat transfer in both gas ducts, and (2) the evaporative-condensing cycle within the heat tubes is a highly efficient way of transferring the heat internally. It is also free from cross contamination.

Applications. Heat pipes are usually used for recovering heat from low- to medium-temperature heat sources. Possible applications include the following:

- Drying, curing, and baking ovens
- Waste steam reclamation
- Air preheaters in steam boilers
- Air dryers
- Brick kilns (secondary recovery)
- Reverberatory furnaces (secondary recovery)
- Heating, ventilating, and air conditioning systems

Gas- and Liquid-to-Liquid Regenerators

Gas-to-liquid regenerators include five types of heat exchangers: waste heat boiler, heat pump, finned-tube heat exchanger (economizer), shell and tube heat exchanger, and gas and vapor expanders. The shell and tube heat exchanger is the only type that can also transfer heat from liquid to liquid.

Waste Heat Boilers. Recovery of waste heat through the use of boilers is becoming increasingly accepted in most industrial processes, as a way to maintain maximum plant efficiency and simultaneously solve the problem of pollution control. Waste heat boilers are ordinarily water tube boilers in which the hot exhaust gases from gas turbines, incinerators, and so on, pass over a number of parallel tubes containing water. The water is vaporized in the tubes and collected in a stream drum, from which it is drawn off for use as heating or processing steam. Figure 12–9 indicates one arrangement that is used, in which the exhaust gases pass over the water tubes twice before they are exhausted to the air.

The exhaust gases are usually in the medium-temperature range. Thus, in order to conserve space, a more compact boiler can be produced by finning the water tubes to increase the effective heat transfer area on the gas side. The diagram shows a mud drum, a set of tubes over which the hot gases make a double pass, and a steam drum that collects the steam generated above the water surface. The pressure at which the steam is generated and the rate of steam production depend on the temperature of the hot gases

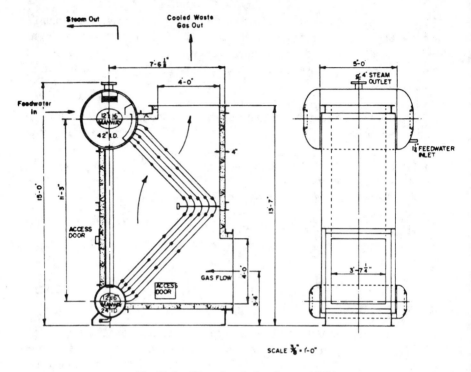

Fig. 12–9. Waste heat boiler. Source: NBS.[1]

entering the boiler, the flow rate of the hot gases, and the efficiency of the boiler.

Applications. Waste heat boilers are built in capacities ranging from less than a thousand to almost a million ft³/min of exhaust gas. In application, special designs or modified techniques with waste heat boilers may be necessary to also recover heat available from waste products and particulates.

Typical applications of waste heat boilers are to recover energy from the exhausts of gas turbines, reciprocating engines, incinerators, and furnaces.

Heat Pumps. In thermodynamics, heat must flow spontaneously from a system at high temperature to one at a lower temperature. However, a heat pump reverses the direction of spontaneous energy flow, producing higher energy efficiency. This device consists of two heat exchangers, a compressor, and an expansion device. Figure 12–10 shows the arrangement schematically.

The logic of a heat pump is described as follows. A liquid or a mixture of liquid and vapor of a pure chemical species flows through the evaporator,

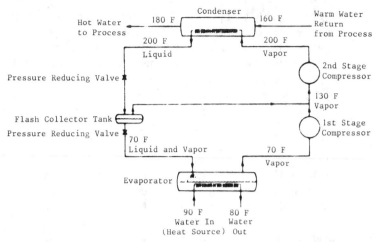

Fig. 12–10. Heat pump schematic diagram, two-stage compression, 90°F heat source, 180°F hot water delivered. Source: NBS.[1]

where it absorbs heat at low temperature and, in doing so, is completely vaporized. The low-temperature vapor is compressed by a compressor, which requires external work. The work done on the vapor raises its pressure and temperature to a level where its energy becomes available for use. The vapor flows through a condenser where it gives up its energy as it condenses to a liquid. The liquid is then expanded through a device back to the evaporator, where the cycle is repeated.

The heat pump was developed as a space-heating system in which low-temperature energy from the ambient air, water, or earth is raised to heating-system temperatures by doing compression work with an electric-motor-driven compressor.

COP. The performance of the heat pump is odinarily described in terms of the coefficient of performance (COP), which is defined as:

$$COP = \frac{\text{Heat transferred in condenser}}{\text{Compressor work}} \qquad (12\text{–}1)$$

which in an ideal heat pump is found as:

$$COP = \frac{T_H}{T_H - T_L} \qquad (12\text{–}2)$$

where T_L is the absolute temperature at which waste heat is extracted from the low-temperature medium, and T_H is the high absolute temperature at which heat is given up by the pump as useful energy. The coefficient or performance expresses the economy of heat transfer.

Because a heat pump normally uses outside air as the heat source, its output is affected substantially by outside temperature. Generally, a heat pump will not be economical if the outside temperature is below 35°F. Figure 12–11 shows that greater energy savings are realized with higher heat source temperatures. Accordingly, when the heat source temperature is 0°F, the COP is less than 1.8; COP is close to 3.6 when the heat source is 70°F.

Applications. In the past, the heat pump has not been generally used for industrial applications. However, several manufacturers are now redeveloping their domestic heat pump systems as well as new equipment for industrial use. Some advantages of using heat pumps in industrial plants are as follows:

- No introduction of new pollutants inside or outside the plant
- No cross contamination, because of separate heat exchangers
- Possible incremental installation
- High coefficient of performance during the hot season

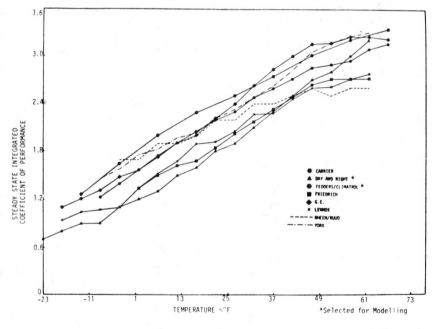

Fig. 12–11. Coefficient of performance (heating) for heat pumps with 3-ton nominal cooling capacity (data from manufacturers' specifications).

- Use for cooling and dehumidification in summer
- All-electric operation with ease of control

The heat pump is a low-temperature heat recovery device and can be applied in the following industrial processes:

- Washing, blanching, sterilizing, and clean-up operations in food processing
- Grain drying
- Metal cleaning and treating processes
- Recycling heat in distillation operations in the food and petrochemical industries
- Industrial space heating

Finned-Tube Heat Exchangers (Economizers). When waste heat in exhaust gases is recovered for purposes such as providing domestic hot water, heating the feedwater for steam boilers, or for hot water space heating, the finned-tube heat exchanger is generally used. Round tubes are connected together in bundles to contain the heated liquid, and fins are welded or otherwise attached to the outside of the tubes to provide additional surface area for removing the waste heat in the gases. Figure 12–12 shows the usual arrangement for the finned-tube exchanger positioned in a duct, and details of a typical finned-tube construction.

This particular type of application is more commonly known as an economizer. The tubes are often connected all in series but can also be arranged in series-parallel bundles to control the liquid side pressure drop. The air side pressure drop is controlled by the spacing of the tubes and the number of rows of tubes within the duct.

Finned-tube exchangers are available prepackaged in modular sizes or can be made from custom specifications very rapidly from standard components. Temperature control of the heated liquid is usually provided by a bypass duct arrangement which varies the flow rate of hot gases over the heat exchanger. Materials for the tubes and the fins can be selected to withstand corrosive liquids and/or corrosive exhaust gases.

Applications. Finned-tube heat exchangers are used to recover waste heat in the low- to medium-temperature range from exhaust gases for heating liquids. Typical applications are domestic hot water heating, heating boiler feedwater, hot water space heating, absorption-type refrigeration or air conditioning, and heating process liquids.

Shell and Tube Heat Exchangers. When the medium containing waste heat is a liquid or a vapor that heats another liquid, then the shell and

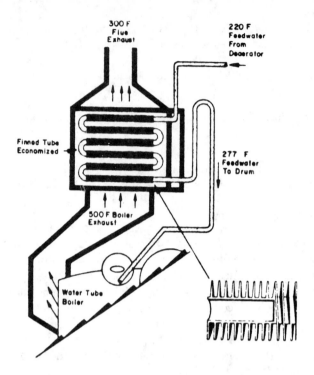

Fig. 12–12.   Finned-tube gas-to-liquid regenerator (economizer). Source: NBS.[1]

tube heat exchanger must be used, since both paths must be sealed to contain the pressures of their respective fluids. The shell contains the tube bundle, and usually internal baffles, to direct the fluid in the shell over the tubes in multiple passes. The shell is inherently weaker than the tubes so that the higher-pressure fluid is circulated in the tubes, while the lower-pressure fluid flows through the shell. When a vapor contains the waste heat, it usually condenses, giving up its latent heat to the liquid being heated.

In this application, the vapor is almost invariably contained within the shell. If the reverse is attempted, the condensation of vapors within small-diameter parallel tubes causes flow instabilities. Tube and shell heat exchangers are available in a wide range of standard sizes with many combinations of materials for the tubes and shells.

Applications.   Typical applications of shell and tube heat exchangers include heating liquids with the heat contained by condensates from refrigeration and air conditioning systems; condensates from process steam; coolants from furnace doors, grates, and pipe supports; coolants from engines, air compressors, bearings, and lubricants; and condensates from distillation processes.

Gas and Vapor Expanders. Industrial steam and gas turbines are in an advanced state of development and are commercially available. Special gas turbine designs for low pressure waste gases have recently become available; for example, a turbine is available for operation from the top gases of a blast furnace. In this case, as much as 20 MW of power could be generated, representing a recovery of 20 to 30% of the available energy of the furnace exhaust gas stream. Maximum top pressures are on the order of 40 lb/in²g.

Perhaps of greater applicability than the turbine of the last example are steam turbines used for producing mechanical work or for driving electrical generators. After removing the necessary energy for doing work, the steam turbine exhausts partially spent steam at a lower pressure than the inlet pressure. The energy in the turbine exhaust stream can then be used for process heat in the usual ways.

Steam turbines are classified as back-pressure turbines, available with allowable exit pressure operation above 500 lb/in²g, or condensing turbines, which operate below atmospheric exit pressures. The steam used for driving the turbines can be generated in direct-fired or waste heat boilers.

Applications. A list of typical applications for gas and vapor expanders follows:

- Electrical power generation
- Compressor drives
- Pump drives
- Fan drives

Liquid-to-Liquid Exchangers

Shell and tube heat exchangers are the only type of liquid-to-liquid exchangers commercially available. The functions and applications of the shell and tube heat exchangers have been discussed (under "Gas- and Liquid-to-Liquid Regenerators").

COST AND ENERGY SAVINGS OF WASTE HEAT RECOVERY AND UTILIZATION

Because technologies for waste heat recovery as described in the previous section are site- and plant-specific, it is unfeasible to generalize their costs and energy savings. The interested reader is directed to NBS[1] for some successful applications of waste heat recovery. Equipment required, costs and energy savings were examined in the reference for each of the following applications:

- Radiation recuperators
- Convective recuperators
- Heat wheels
- Air preheaters (passive regenerators)
- Heat pipes
- Waste heat boilers
- Heat pumps
- Finned-tube heat exchangers (economizers)

RESEARCH AND DEVELOPMENT IN WASTE HEAT RECOVERY AND UTILIZATION

In this section, ten major projects in the original Industrial R&D of Conservation and Renewable Energy, Department of Energy (DOE), for the period of fiscal year (FY) 1980–1984 are discussed.* These projects are as follows:[7]

- Ceramic recuperator
- Metallic counter-flow recuperator
- Reradiant recuperator
- High-temperature recuperators/burners/systems
- Stirling cycle heat pump
- Brayton heat pump
- Rankine cycle—steam recompressor heat pump
- Electric heat pumps
- Heat pump study
- Oxygen enrichment

Most of these projects are in recuperators and heat pumps; they are summarized below.** Note that the oxygen enrichment technique will increase fuel use and reduce waste heat (see Chapter 7 for details). These projects are subject to termination because of the Reagon Administration's intention to abolish DOE.

Ceramic Recuperator

Various stages of development and demonstrations were under way in FY 1980 on two types of ceramic recuperator. One was based on built-up, cross-

---

* Many projects in the waste heat recovery area have been developed by industries. Because of incomplete information, they are not included here. In addition, the Electric Power Research Institute has studied intensively the application of heat pumps.

** Because of differences of new administration's research priorities, the content of these projects is subject to change.

flow, thin-wall unit modules of nominal 850-cfm (cubic feet per minute) capacity. Five early demonstrations of this technique were being completed, having several thousand hours of test time under a variety of severe and corrosive conditions at inlet temperatures up to 2600°F. Improvements in materials and minimizing leakage were accomplished. A sixth demonstration, in a steel mill soaking pit, was installed. Preheating of inlet combustion air by these units decreased fuel usage by more than 30%. These modules are expected to provide "building blocks" for even greater fuel savings when used in very high-temperature inlet air burners. Some initial commercial sale of these recuperator modules were made.

A second ceramic recuperation development was to preheat combustion air in glass furnaces by installing counter-current special-geometry units in the exhaust stack. The exahust air is especially aggressive in its effect on materials. A major effort was required in selection of the ceramic material. These units should provide greater fuel savings than the presently used checker units because of their higher effective combustion air inlet temperature. Initial technical studies and material component tests were completed, and engineering work and a search for a demonstration site were under way.

Phase three design, construction, and test operation of a full-scale unit were initiated in FY 1980. Using these recuperators, cumulative projected fuel savings through the year 2000 was estimated by DOE to exceed 15 quad.

Metallic Counter-flow Recuperator

Highly efficient, high-temperature metallic recuperators represent a sizable energy conservation potential. DOE sponsored the development of a brazed construction, stainless steel recuperator that uses rectangular, offset extended heat transfer surfaces. The AiResearch Manufacturing Co., a division of Garrett Corporation, developed a metallic, counter-flow recuperator. It was demonstrated in commercial service in a gas-fired aluminum remelt furnace. This test unit was then installed to recover waste heat from the exhaust stream and used to heat inlet combustion air. More than 430 tons of aluminum were processed in this furnace with a resulting energy savings of more than 32%. During the demonstration period, no service or operational difficulties had been encountered. AiResearch has pursued commercialization of this technology with DOE's assistance. Consideration was given to demonstrating a heavy-duty unit in a steel mill.

It has been estimated by DOE that the annual energy savings for this generic conservation technology will exceed 0.3 quad through the year 2000. The technology has good economic potential, showing a projected return on investment in excess of 20%.

Reradiant Recuperator

The performance of existing annular-passage, counter-current flow recuperators (sometimes called Escher types) can be improved by providing a reradiant insert in the exhaust passage. A demonstration project to evaluate the magnitude of these possible energy savings accurately has progressed to the point that the first two test units have been installed on aluminum remelt furnaces in Alabama. The first unit was to evaluate the increased performance obtainable in a conventionally sized unit. The initial tests revealed the need for more flue gas chemical composition information and improved materials. The second test unit was then fabricated and installed. This second unit was of a smaller size and lower capital cost than a normal-size unit, but was projected to provide equivalent energy savings. No further results were reported.

High-Temperature Recuperators/Burners/Systems

This project would build upon the success achieved in the ongoing ceramic recuperator development program. High-temperature (up to 2600°F) inlet air would provide even further energy savings, but existing burners are unable to use it. The cost of the ducting for the very hot gas from the recuperator to the burner inlet is very high, approaching that of the recuperator alone.

Preliminary technical evaluation of the potential unit fuel savings had been completed in FY 1980, and confirmation of the actual fuel savings obtainable by test of a specially modified installation was under way. Projected market applications and corresponding national fuel savings were yet to be determined. No further results were reported.

Stirling Cycle Heat Pump

Several studies indicate that many specific industrial processes exist that use 250-psia and above process steam from fossil-fuel-fired boilers while at the same time rejecting considerable quantities of low-grade waste heat contained in streams such as nonreturned condensate and process cooling water.

The objective of this project was to evaluate the technical and economic feasibility of using the Stirling cycle operated as a heat pump for these types of industrial heat recovery applications at temperatures generally above 400°F. Other thermodynamic cycles (i.e., Rankine and Brayton) are limited to lower-temperature industrial processes because of the thermal stability of available working fluids and/or increasing complexity and decreasing efficiency of the system. Stirling cycle machines employ gaseous working mediums such as helium or argon, which are not limited by thermal stability, material compatibility, or toxicity considerations. Stirling cycles, in addition, offer the potential for achieving higher system performance. A detailed technical performance and economic assessment study of various Stirling cycle heat pump concepts was under way in FY 1980 to define configurations that offer potential for satisfying industrial process needs and for becoming economically viable commercial products.

Brayton Heat Pump

The objective of this project was to determine the technical feasibility and economic viability of using high-temperature industrial heat pumps to reclaim low-temperature waste heat from industrial processes. Because of the noncontaminating nature of the Brayton-cycle heat pump and process operating conditions, ideal applications were found in the food process industries. Milk drying was a representative process and selected for concept demonstration. The high-temperature Brayton-cycle heat pump is ideally suited for energy conservation in many kinds of dryers and ovens. The energy consumed by industrial dryers and baking ovens in the United States was estimated by DOE to exceed 2 quad per year. Drying conditions are generally similar, and most of the energy supplied (usually as direct-fired natural gas) is exhausted to the atmosphere.

The purpose of a Brayton heat pump is to recover the wasted energy from the process, and in particular from the exhaust gas leaving the evaporator and dryer, and use it to reduce system energy requirements. For the conditions normally used for milk and whey drying, a heating COP of between 3 and 4 can be obtained with the heat pump. Initial design studies were under way in FY 1980 to show that the technique can make substantial savings when installed on the evaporator as well as on the dryer, and that a variety of prime movers will be economical.

Rankine Cycle—Steam Recompressor Heat Pump

A great deal of energy is unavailable for further use in industry because of its low temperature. This project would take 190°F hot water (or subatmospheric steam) and put it through a heat exchanger to vaporize a fluorinated hydrocarbon working fluid that expands through a special turbine. This turbine will drive a conventional steam compressor to pressurize the subatmospheric vapor to 20 psig (pounds per square inch—gauge) where it can be further used as process steam. The demonstration unit was planned for installation and testing in an acetone recovery plant in a chemical complex in Virginia, although additional sites are being considered. The preliminary design and performance predictions had been completed in FY 1980. Design approval was given, and release of long-lead items for procurement requested. No further results were reported.

The commercial equipment developed from this demonstration would have the potential of providing sizable energy savings (estimated by DOE to exceed 1 quad per year average between now and the year 2000) because of the very large amount of process steam in use.

Electric Heat Pumps

This project was established to improve components and materials to be used in electrically driven, high-COP heat pumps and to conduct a demonstration of individual-scale units in the pulp and paper industry.

Development work on compressors, seals, and other components was under way in the laboratory in FY 1980. Especially good results were being achieved in developing and evaluating higher-temperature working fluids. The initially chosen demonstration installation in a Kraft paper plant was deferred; a number of more appropriate sites were investigated, and their individually desirable operating conditions were determined. No further progress was reported.

Heat Pump Study

This project was initiated with a university to study and prepare a report on all the significant parameters that affect industrial heat pump design and operations. The actual effect of these several parameters was checked by obtaining data from a laboratory model designed, constructed, and operated under the auspices of this project.

Oxygen Enrichment

The enrichment of combustion air by modest increases in the oxygen content has been demonstrated (on a non-cost-effective basis) to achieve significant savings in fuel requirements by increasing the flame temperature and by not heating additional nitrogen, which is subsequently exhausted.

Two separate techniques were selected for generating and using oxygen-enriched air in multiple types of applications. They are pressure swing absorption and spiral wound membranes. Design work was initiated in FY 1979, with fabrication, installation, and tests occurring in the subsequent two years. Projected payback periods for individual installations using these techniques, based on energy savings alone, are less than two years.

Table 12-7. Major Operation and Application Characteristics of Industrial Heat Exchangers.

Commercial Heat Transfer Equipment	Low Temperature Sub-Zero–250°F	Intermediate Temp. 250°F–1200°F	High Temperature 1200°F–2000°F	Recovers Moisture	Large Temperature Differentials Permitted	Packaged Units Available	Can Be Retrofit	No Cross-Contamination	Compact Size	Gas-to-Gas Heat Exchange	Gas-to-Liquid Heat Exchanger	Liquid-to-Liquid Heat Exchanger	Corrosive Gases Permitted With Special Construction
1. Radiation recuperator			x		x	1	x	x		x			x
2. Convective recuperator		x	x		x	x	x	x		x			x
3. Metallic heat wheel	x	x		2		x	x	3	x	x			x
4. Hygroscopic heat wheel	x	x		x		x	x	3	x	x			
5. Ceramic heat wheel		x	x		x	x	x		x	x			x
6. Air preheater (passive regenerator)	x	x			x	x	x	x		x			x
7. Heat pipe	x	x			4	x	x	x	x	x	x		x
8. Waste heat boiler	x	x				x	x	x	x		x		5
9. Heat pump	x	x	x		6	x	x	x	x		x		
10. Finned-tube heat exchanger (economizer)	x	x			x	x	x	x	x		x		x
11. Shell and tube heat exchanger	x	x			x	x	x	x	x		x	x	5

1. Off-the-shelf items available in small capacities only.
2. Controversial subject. Some authorities claim moisture recovery. Do not advise depending on it.
3. With a purge section added, cross-contamination can be limited to less than 1% by mass.
4. Allowable temperatures and temperature differential limited by the phase equilibrium properties of the internal fluid.
5. Can be constructed of corrosion-resistant materials, but consider possible extensive damage to equipment caused by leaks or tube ruptures.
6. Not efficient under large temperature differentials.

Source: Revised from NBS.[1]

CONCLUSIONS

We have discussed various technologies (heat exchangers) for waste heat recovery. Heat exchange is one of the two methods used for waste heat recovery. The other method, cogeneration, is discussed in Chapter 13.

Various types of heat exchangers for waste heat recovery can be retrofitted. Most of them have less than five-year payout periods, and they will become more economically attractive as fuel prices continue to rise rapidly. Because waste heat recovery technologies are retrofitted onto existing steam-producing systems, they are site- and technology-specific; therefore, their economics must be carefully examined. This examination includes engineering design, costs and savings, supervision, and future maintenance and pollution controls. Methods presented in Chapters 4–11 should be used to aid decision-making for investments in waste heat recovery equipment.

Table 12–7 summarizes major operation and application characteristics of 11 of the most common types of industrial heat exchangers. This table provides rapid comparisons among exchangers with regard to temperature range, moisture recovery, temperature differential allowance, ability as packaged units, retrofitability, compactness, heat transfer fluids, and corrosion resistance. It is worth noting that many of the heat exchangers operating in the low-temperature range may condense vapors from the cooled gas stream. Provisions must be made to remove those liquid condensates from the heat exchanger.

REFERENCES

1. National Bureau of Standards. "Waste Heat Management Guidebook," prepared for Federal Energy Administration (CG-04-75-031-00 FEA), February 1977.
2. Streb, A. "Priority Listing of Industrial Processes by Total Energy Consumption and Potential for Savings," ERDA Report CONS/5015-1, 1977.
3. Schultz, H. et al. "Resource Recovery Technology for Urban Decision-makers," Columbia University, New York, funded by National Science Foundation (RANN), January 1976.
4. Anderson, L. "Energy Potential from Organic Wastes: A Review of the Quantities and Sources," Bureau of Mines Information Circular 8549, 1972.
5. Combustion, *Special Resource Recovery Issue,* Combustion Publishing Co., Inc., New York, February 1977 (interpreted by A. R. Graham).
6. Sternlicht, B. "Capturing Energy from Industrial Waste Heat," *Mechanical Engineering,* August 1978, pp. 30–41.
7. Department of Energy, "Industrial Programs Multi-Year Plan FY 1980–84," Conservation and Solar Applications, Draft, October 25, 1979.

REFERENCE BOOKS FOR WASTE HEAT RECOVERY

8. *Industrial Gas Magazine*, one issue published annually as a product directory (209 Dunn Avenue, Stamford, Conn. 06905).

9. *ASHRAE Handbook and Product Directory—1973, Systems*, published every four years by American Society of Heating, Refrigerating, Ventilating & Air Conditioning Engineers, Inc. (345 East 47th Street, New York, N.Y. 10017).

10. *ASHRAE Guide and Data Book—1975, Equipment*, published every four years by American Society of Heating, Refrigerating, Ventilating & Air Conditioning Engineers, Inc. (345 East 47th Street, New York, N.Y. 10017).

11. *Thomas Register of American Manufacturers*, Thomas Publishing Company (461 Eighth Avenue, New York, N.Y. 10001).

12. *Heating/Piping/Air Conditioning*, Reinhold Publishing Corporation (10 S. LaSalle Street, Chicago, Ill. 60603).

13. *Industrial Heating*, National Industrial Publishing Company (Union Trust Building, Pittsburgh, Pa. 15219).

14. *Plant Engineering*, Technical Publishing Company (1301 S. Grove Avenue, Barrington, Ill. 60010).

15. *Heating/Combustion Equipment News*, Business Communications, Inc. (2800 Euclid Avenue, Cleveland, Ohio 44115).

16. *Power Magazine*, McGraw-Hill Publications Company (New York, N.Y. 10020).

13
Cogeneration: In-plant Power Generation Systems

INTRODUCTION

Cogeneration, a means of waste heat recovery that was cited as an important technique for conserving domestic energy resources in the National Energy Plan in April 1977, is the sequential production of electrical or mechanical power and process heat from the same primary fuel. It can contribute significantly to the nation's efforts in using fuel more efficiently.

Depending on the market sectors in which they are used, cogeneration may be called in-plant power generation systems, dual energy use systems, waste- (or reject-) heat utilization systems, district-heating systems (only with electricity generation), and total-energy systems (or integrated energy/utility systems).

Although cogeneration has been practiced by industry since about 1905, it has been the recent several-fold price increases and supply uncertainty of petroleum-based fuels that have made it an important policy issue, deserving careful study with regard to its potential and economics. In this chapter, cogeneration is discussed with respect to definition and historical development, technology review, potential benefits and costs, costing and pricing, legislative constraints, and applications and research and development efforts.

DEFINITION AND HISTORICAL DEVELOPMENT

Definition of Cogeneration

Cogeneration is broadly defined as the sequential production of electrical or mechanical power and process heat from the same primary fuel or energy source. The fundamental difference between a conventional energy (electric or steam) system and a cogeneration system is illustrated in Fig. 13–1. The conventional system produces either electricity (as shown in Chart A) or

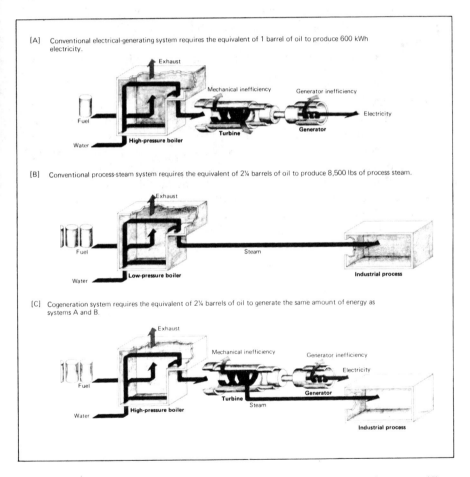

[A]   Conventional electrical-generating system requires the equivalent of 1 barrel of oil to produce 600 kWh electricity.

[B]   Conventional process-steam system requires the equivalent of 2¼ barrels of oil to produce 8,500 lbs of process steam.

[C]   Cogeneration system requires the equivalent of 2¼ barrels of oil to generate the same amount of energy as systems A and B.

Fig. 13–1.   Conventional electrical (A) and steam systems (B) and congeneration system (C). Conventional steam and electrical systems need more fuel to produce the same energy output than do cogeneration systems. Source: DOE.[1]

thermal energy (as shown in Chart B), whereas the cogeneration system produces both (as shown in Chart C) by using the same primary fuel.

In most market sectors, thermal energy is produced through the operation of equipment such as boilers and furnaces, and electricity is purchased from a utility. The utilities, which are in the business of selling electricity, seek to maximize their production of electricity and historically have had limited incentive to buy power from industrial plants. By recapturing and using some of the thermal energy that is normally discharged from an engine, a cogeneration system can reduce system fuel requirments by 10–30%. Cogeneration thus offers significant energy-saving potential. A cogeneration system can be installed in either a utility site or an industrial steam plant. If the system is installed in an industrial steam plant, it is termed an "in-plant (or by-product) power generation system"; it is called a "reject heat utilization system" if the cogeneration system is installed in a utility site. The major difference is explained as follows:

- In-plant power generation system: The configuration of this system is shown in Fig. 13–2. Chart A. Industrial steam is produced at a higher temperature and pressure than needed for process, and is brought down to the required state through a back-pressure turbine, thereby generating electricity before moving on to its intended end use. The by-product electricity can be used locally or sold to the neighboring utility.
- Reject heat utilization system: A schematic illustration of this system is shown in Fig. 13–2, Chart B. Some of the steam is extracted from the utility's turbine at the pressure required by an adjacent industry and supplied for its use, therefore eliminating or reducing the need for boilers in the industry's plants.

The in-plant system is discussed in the remainder of this chapter, and the reject heat utilization system in Chapter 14. Thus, in this chapter, cogeneration will mean in-plant power generation, except as otherwise noted.

Two Basic Cogeneration (In-plant Power Generation) Concepts

As shown in Fig. 13–3, there are two fundamental types of cogeneration systems, topping and bottoming cycles, differentiated on the basis of whether electrical or thermal energy is produced first.

In a topping cycle system, fuel is burned to produce high-temperature heat, which is expanded through a turbine to generate electrical or mechanical power. After passing through the turbine, the reject heat is then used in industrial applications as process heat. On the other hand, in a bottoming

CHART A: IN-PLANT POWER GENERATION SYSTEM

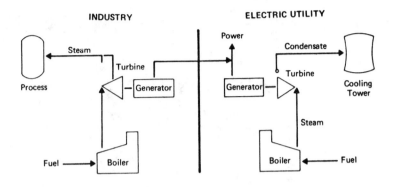

CHART B: REJECT HEAT UTILIZATION SYSTEM

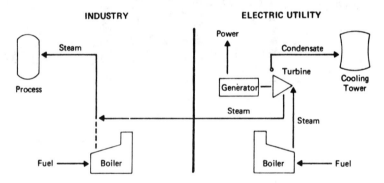

Fig. 13–2. Cogeneration systems: in-plant power generation and reject heat utilization systems. Source: Dow Chemical.[2]

cycle system, fuel is burned initially to produce process heat, with the reject heat to be used to generate either electrical or mechanical power.

Because of the energy required to generate the electrical or mechanical power, more fuel is consumed in a cogeneration system than in producing process heat alone. However, the total fuel required to produce both power and process heat in one system is less than the fuel required to produce power and heat in separate systems. For example, the overall efficiency of a steam turbine topping cycle cogeneration system is about 79%, compared with a combined efficiency of about 58% for two separate systems.

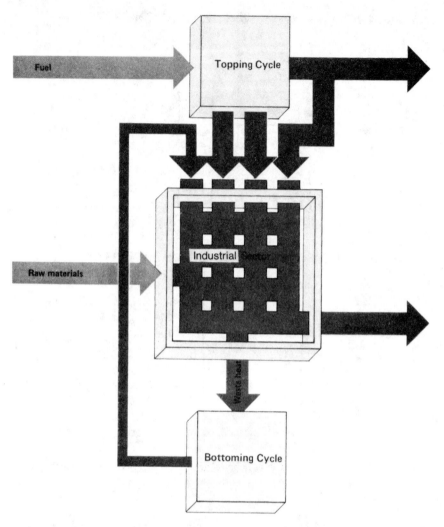

Fig. 13-3.   Two basic cogeneration concepts—topping and bottoming cycles. Source: DOE.[1]

Both cycles vary in size and hardware depending on the specific electrical and thermal needs of the particular user application. However, low-heat temperature requirements in many industrial processes and a lack of available high-temperature resistance components have limited the use of bottoming cycle systems. Consequently, the bottoming cycle may not make a major contribution in industrial energy conservation within the next ten years.

Thermodynamic Advantage of Congeneration

The thermodynamic advantage (theoretical energy savings) of cogeneration results from the use of cogeneration systems to utilize reject heat from power generation. Figure 13–4 shows an ideal steam turbine cogeneration system which maximizes the use of the reject heat of power generation for steam process heat. In the figure, the broken-line shaded area (including the double-shaded area) represents the waste heat resulting from a power-only plant, and this area minus the double-shaded area represents the waste heat from a steam-only heat boiler. The combination of power generation and steam production reduces the waste heat to approximately the area of the separate steam process. Therefore, cogeneration results in a savings proportional to the double-shaded area (approximately *agbcd*). The magnitude of the savings depends on the temperature and the cogeneration technologies.

Cogeneration Efficiency

Because the power portion of the topping cycle rejects no heat to the environment (i.e., all reject heat goes to the steam process), it has a theoretical efficiency of 100%. However, unlike the case of a power-only plant, the reject heat is not condensed for follow-up power production, and the cogenerators produce less electricity than does the power-only station, only 58% of the electricity produced by the latter. The efficiency for the process heat portion is the same as that of steam-only boilers.

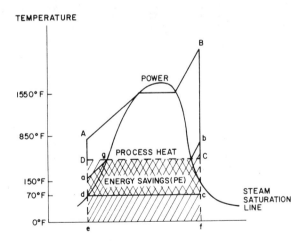

Fig. 13–4. Thermodynamic energy savings of an ideal topping cogeneration system, compared with a steam power cycle and a steam process heat cycle.

Historical Development of Cogeneration

At the turn of the century, most industrial plants generated their own electricity with coal-fired boilers and steam-turbine generators. The percentage of total power cogenerated in the industry has been estimated at 58% for the early 1900s although the actual cogeneration data are not available.

The historical trend of cogeneration shares in the U.S. electricity supply is shown in Fig. 13–5. Data for this figure were compiled by the Edison Electric Institute.[3] Cogeneration declined from an estimated 58% in the 1890s to about 20% in 1940 and to about 5% by 1974. The Department of Energy has cited the following eight reasons for this decline:[1]

- Availability of less expensive and more reliable electricity
- Increasing regulation over all forms of electrical generation at both the state and federal levels

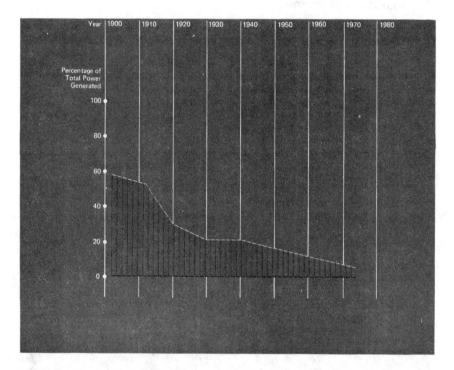

Fig. 13–5. Historical trend of percentage of total U.S. power for in-plant cogeneration. Since 1900, in-plant industrial power generation has declined in the United States relative to total generation. Source: EEI.[5]

- Long-standing policies on the part of most electric utilities that discouraged on-site generation of electricity
- Low energy costs that represented a gradually declining percentage of industry expenses
- Higher fuel costs relative to electricity rates
- Industry bias in favor of market-oriented rather than cost-cutting investments
- Changing corporate income-tax structures
- Advances in technology—specifically, the availability of the "package boiler"

Cogeneration systems have been installed primarily in chemical, steel, paper, petroleum refining, textile, and food industries because those industries use a bulk of steam and process heat. Examples of early cogeneration systems are Dow Chemical Company's Midland, Michigan plant in the 1920s and Republic Steel's Youngstown, Ohio plant in 1945. The latter was even upgraded in 1971.

Cogeneration in Europe

As shown in Fig. 13–6, industrial cogeneration has been five to six times more common in some parts of Europe than in the United States, according to the International Federation of Industrial Cogenerators. In 1972, 16% of West Germany's total power production was cogenerated by industries; in Italy, 18%; in France, 16%; and in the Netherlands, 10%. The two main reasons for the more extensive use of cogeneration in Europe are that:

- Because reliable utility grids became available about 25 years later in Europe (1945) than in the United States (1920s), European industries had to generate their own electricity.
- Most European countries have fewer regulations restricting the sale of cogenerated electricity. In France, for example, under a decree passed in 1955, the state-owned utility is required to buy surplus power from in-plant generators.

REVIEW OF COGENERATION TECHNOLOGIES

The following subsections discuss how to integrate cogeneration into industrial processes and review alternative cogeneration technologies.

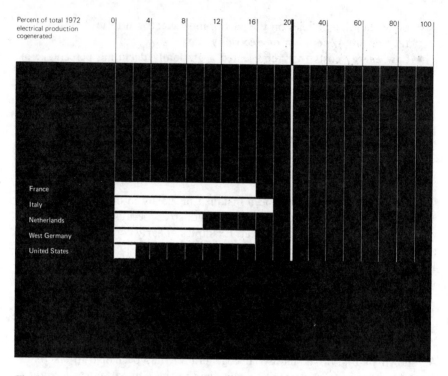

Fig. 13–6.   Percentage of total electricity cogenerated in the five nations. Industrial cogeneration is far more common in Europe than in the United States. Source: IFOIC.[4]

Integration of Industrial Processes with Cogeneration

As shown in Fig. 13–7, technologies for the topping cycle are installed before the low-temperature processes, and those for the bottoming cycle are installed after the high-temperature processes.

Sources of energy are traditionally industrial waste products and fossil fuels, although nuclear and solar energy may find direct industrial application in the future. Energy is first consumed to generate electrical or shaft power. Conventional diesel, Brayton (gas), Rankine (steam), and combined cycle equipment as it exists today may be used, with minor modification for topping-cycle cogenerators. Fluidized-bed combustion (FBC), fuel cells, MHD generators, and other advanced energy conversion techniques may find application in future industrial cogeneration systems. Heat rejected from the topping cycle may be directly applicable to the industrial process. If a process requires steam, it can be derived directly from a steam turbine by operating in a back-pressure or extraction mode or indirectly from a diesel or gas turbine

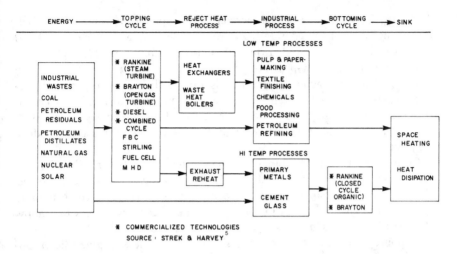

Fig. 13–7. How to integrate cogeneration technologies into industrial processes. Source: Streb and Harvey.[5]

via a waste heat boiler. Gas turbine exhaust can be employed directly as preheated air for process heaters, dryers, and conventionally fired boilers.

Although partial topping cycles may be possible for certain high-temperature processes, direct firing is generally more practical. The temperature of exhaust products leaving many high-temperature processes is sufficiently high to permit application of a bottoming cycle. Advanced Rankine and Brayton cycle equipment has been developed that operates efficiently at process exhaust temperatures of 800°F to 1100°F. Lack of available components and economic factors have limited the use of bottoming-cycle systems. Recent research and development efforts are expected to greatly enhance the prospects and availability of bottoming-cycle cogeneration systems.

Review of Technologies

Steam, gas (including Brayton), diesel, and combined gas/steam turbine systems are four commercialized topping cogeneration technologies. The steam turbine cycle is the most common device for topping industrial process steam. Rankine and Brayton cycles are the two commercialized bottoming systems. FBC, bi-phase turbine bottoming systems, and fuel cells are promising near-term technologies. Energy storage systems, Stirling engines, MHD generators, Lysholm expanders, and advanced heat pumps are future cogeneration technologies requiring further research and development work.

Topping Systems. For commercial steam topping systems, a fossil-fuel-fired boiler is used to generate high-pressure steam (e.g., 1000°F and 1,450 psig), which is then expanded through a steam turbine-generator set to produce electricity and low-pressure process steam.

The gas turbine (including Brayton) is a more direct way to generate electricity than the steam turbine. Fuel such as oil or natural gas, burned in the combustors of the turbine, heats air that has been previously compressed. The burned mixture expands through a turbine generator set to produce electricity.

Combined gas/steam turbines provide greater flexibility for industrial topping of process steam than either simple gas or steam turbines. However, their performance can only duplicate that of separate gas turbine topping or steam topping. They can include special heat recovery boilers that allow the gas turbine exhaust to be used as combustion air.

Topping of process steam or process heat with a diesel engine (diesel topping) provides the highest electricity-to-steam ratio and is available in a range of sizes from 100 kW to 25 MW. Basic thermal efficiencies are approximately 35%, compared to less than 30% for gas and steam turbines. Heat rates are in the 6000–7000 Btu/kWH range, less favorable than for the other prime movers. However, electricity generation per unit is twice that of gas turbines and more than ten times that of steam turbines. The diesel engine drives an electric generator, and process steam is made in heat recovery boilers that use the engine exhaust as heat sources.

Bottoming Systems. Organic Rankine and Brayton cycles are the two commercial bottoming systems that convert waste heat to electric and useful thermal energy. The Rankine system heats and vaporizes liquid working fluid in a vapor generator at constant pressure. The hot vapor is expanded to a lower pressure in an expander, producing electric and shaft power. The turbine is then cooled and condensed at constant pressure, and the condensed fluid is pressurized by a pump, preheated, and returned to the vapor generator. The Brayton bottoming system is similar to the Brayton topping system.

Potential of Technologies

Although diesel and gas turbine systems are more efficient and generate more electricity per unit of steam, as illustrated in Fig. 13–8, the greatest potential for cogeneration exists in steam turbine systems (see Table 13–1).

Choice of cogeneration systems will be determined by site-specific factors (e.g., the ratio of thermal-to-electrical loads, fuel availability, and environmental constraints). Cogeneration can be designed to use standard components. The success of future cogeneration largely depends on the development of coal-derived fuels and effective environmental control systems.

System Description	Heat Rate	Energy Out / Energy In	Energy Cog. / Energy Conv.	Percent Energy Saving
Boiler—Back Pressure Turbine 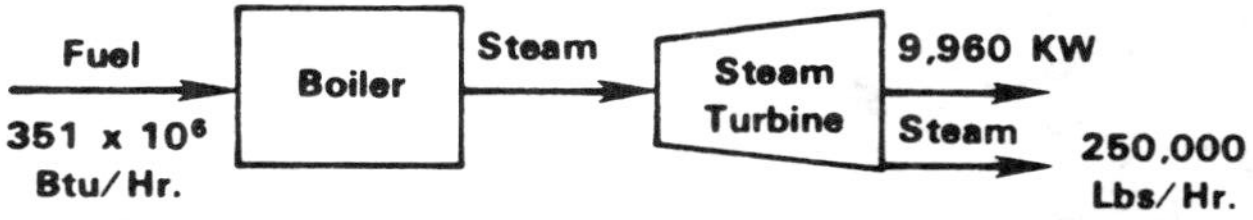	4,590 Btu/KW	.81	.86	14%
Gas Turbine—Waste Heat Boiler 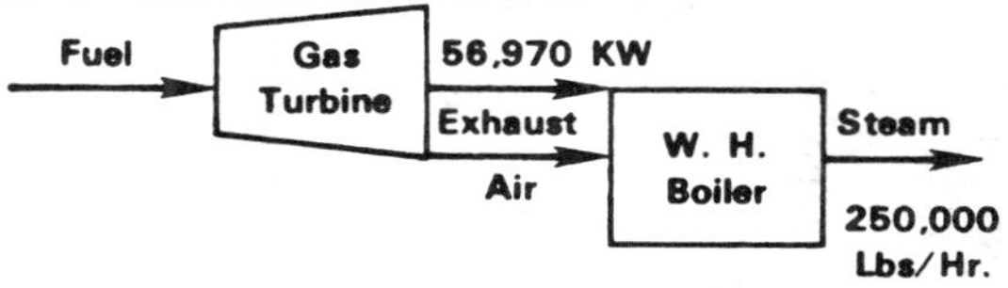	6,540 Btu/KW	.66	.75	25%
Combined Cycle 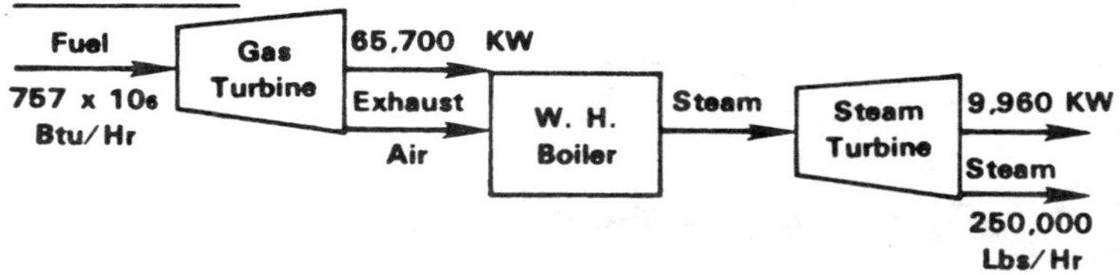	5,980 Btu/KW	.67	.69	31%

Boiler—Back Pressure Turbine: Fuel 351 x 10⁶ Btu/Hr. → Boiler → Steam → Steam Turbine 9,960 KW → Steam 250,000 Lbs/Hr.

Gas Turbine—Waste Heat Boiler: Fuel → Gas Turbine 56,970 KW, Air → Exhaust → W. H. Boiler → Steam 250,000 Lbs/Hr.

Combined Cycle: Fuel 757 x 10⁶ Btu/Hr → Gas Turbine 65,700 KW, Air → Exhaust → W. H. Boiler → Steam → Steam Turbine 9,960 KW → Steam 250,000 Lbs/Hr

Fig. 13–8.   Comparison of three industry cogeneration options. Source: Streb and Harvey.[5]

Table 13-1. Estimated Cogeneration by Plant Type in 1976.

Cogeneration Plant Type	Process Steam Cogenerated (Trillion Btu)	Electric Energy Cogenerated (Billion kWh)	Cogeneration Capacity (MW)
Coal-fired steam turbine	97	4.16	550
Residual-fired steam turbine	204	8.81	1,157
Distillate-fired gas turbine	20	3.51	446
Natural gas turbine	65	4.34	554
Waste fuel steam turbine	182	9.90	1,254
Heat recovery steam turbine	16	0.65	84
Total	584	31.37	4,045

Source: RPA.[6]

In the remainder of this section, we will review the four topping systems and one bottoming system (organic Rankine system), and the future cogeneration technologies discussed in Fig. 13–7 (fluidized-bed combustion, Stirling engine, fuel cells, and MHD generators), as well as the Lysholm expander and heat pumps. The review will emphasize technological characteristics, development potential, and benefit/cost.

Steam Turbine Topping System. The configuration of a steam turbine topping system is shown in Fig. 13–9. One of the most important features of a steam turbine is substantial fuel flexibility. Back-pressure steam turbines are generally used for topping because they exhaust low-pressure steam and are most efficient.

The boiler in a steam turbine system, which generates steam through the combustion of fuel, can be "fired" by oil, natural gas, coal and coal-derived liquids and gases, wood, or synthetic liquids and gases. Mechanical energy is produced as the high-pressure steam drives a turbine. Through the use of a generator, this mechanical energy is then converted to electricity. The low-pressure exhaust from the turbine can be used for industrial-process applications or to provide space heating and cooling.

Steam turbine technology has been available for several decades, and reliability is high. However, since almost all industrial plants use packaged boilers for steam generation, modifications are required for steam turbine topping

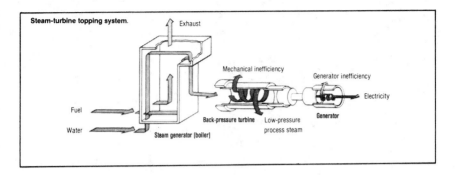

Fig. 13–9.   Steam turbine topping system with a back-pressure turbine. Source: DOE.[1]

systems because the pressure and temperature generated by these boilers are too low.

Increased use of steam turbines for the topping of process steam is not expected to have adverse effects on the environment except where coal is used as a fuel. Sulfur oxides are major pollutants from the burning of high-sulfur coal. Fly ash is also a problem. Special emission control systems such as exhaust stack scrubbers will be required in most areas of the country where coal is burned.

Gas Turbine Topping System.   A schematic illustration of a gas turbine topping system is shown in Fig. 13–10. A gas turbine topping system is a reversible thermodynamic cycle that describes the heat-to-work conversion process in an open gas turbine plant. The gas turbine burns fuel directly and uses the combustion gases to produce mechanical shaft power, which, in turn, is used to drive a generator to produce electricity.

The best performance and least emissions are achieved if natural gas is used as fuel. Clean light petroleum distillates (such as naphtha) also provide acceptable performance, and some gas turbines are currently available with fuel flexibility. These turbines can switch without interruption from natural gas to distillates. Dependence on these premium fuels has limited the widespread application of cogeneration systems using gas turbines and diesel engines, even though they offer potential energy savings of up to four times those possible with a steam turbine system.

The exhaust from gas turbine and diesel engine systems can be used for a variety of purposes, ranging from process heat to the production and subsequent use of low-pressure process steam. To generate steam, the hot exhaust gases are passed through a waste-heat-recovery boiler, where the thermal energy is transferred to water. The gases effectively serve as fuel for the boiler. The exhaust stream, which is relatively clean and dry, can also be

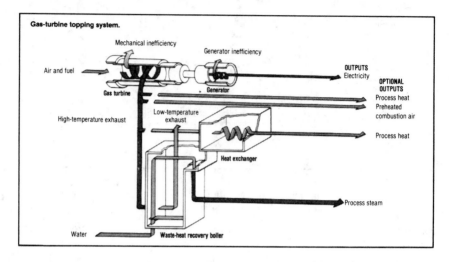

Fig. 13–10. Gas turbine topping system with a heat exchanger and waste-heat recovery boiler. Source: DOE.[1]

used directly for heating or drying, an important consideration in the food industry. If lower temperatures or increased volume is desired, the exhaust stream can be mixed with ambient air. If the back pressure of the process is low, a simple inductor is all that is required to provide the dilution air. At higher back pressures, a fan is required.

High-temperature, high-performance gas turbines are currently limited by the materials from which they are made. However, improved blade and nozzle materials, such as ceramics, can allow higher firing temperatures.

Because of the scarcity of petroleum products, unless gas turbine systems can operate efficiently on low-grade fuels, they will not sell well until coal-derived fuels are available.

Gas turbines require more frequent maintenance than steam turbines. The effects on the environment of increased industrial gas turbine topping cycles will less depend on the fuels they burn than steam turbines. EPA requirements will limit NO_x emissions and the sulfur content of fuels. Better combustors are needed to meet new emissions levels with all fuels except lean distillates and natural gas.

Combined Topping System. A schematic representation of a combined gas/steam turbine topping system is shown in Fig. 13–11. A combined cycle is the process of using the high-pressure steam produced in the waste-heat recovery boiler to feed a back-pressure steam turbine and produce additional electricity and low-pressure process steam. It is similar to gas and steam turbine topping systems but has greater flexibility for topping process steam.

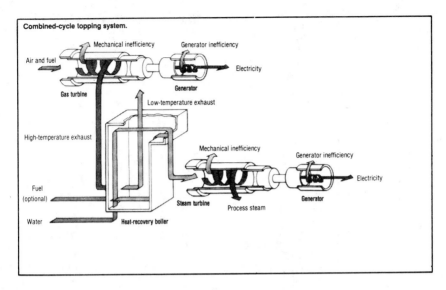

Fig. 13–11.   Combined gas/steam turbine topping system with a heat-recovery boiler. Source: DOE.[1]

Diesel Engine Topping System.

The configuration of a diesel engine topping system is shown in Fig. 13–12. The diesel engine drives an electric generator. Process steam is made in heat recovery boilers that use the engine exhaust and jacket cooling as heat sources. Diesel engines can be coupled to bi-phase bottoming cycles or Rankine bottoming cycles to generate additional electricity.

Diesel engine topping systems currently depend on petroleum-based fuels. They can be designed to use both high distillates and low-grade crude and residual oils.

The spread of diesel topping systems will depend on two major factors. First, the electricity-to-thermal ratio of diesel engines is significantly higher than most industrial plants need. Diesel engines will not be economically attractive unless excess electric power can be sold or stored. Second, they will also have to convert more abundant coal-derived fuels.

The environmental impact of diesel topping cycles is minimal if light distillate fuels are burned. If coal is used as a fuel, sulfur oxides must be removed from the stack gas to meet emission standards.

Rankin Bottoming System.*

The schematic of a Rankin bottoming system is shown in Fig. 13–13. Bottoming systems can be used to recover

---

* Another similar bottoming cycle is the bi-phase turbine, which has a gas-to-liquid heat exchanger instead of a boiler.

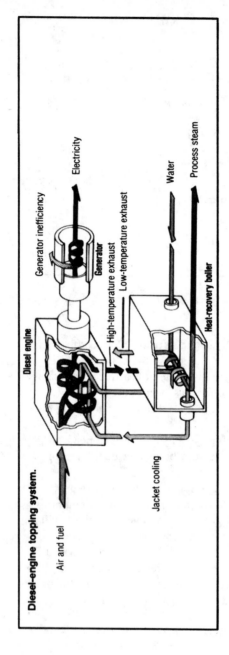

Fig. 13–12.   Diesel engine topping system with a heat-recovery boiler.   Source: DOE.[1]

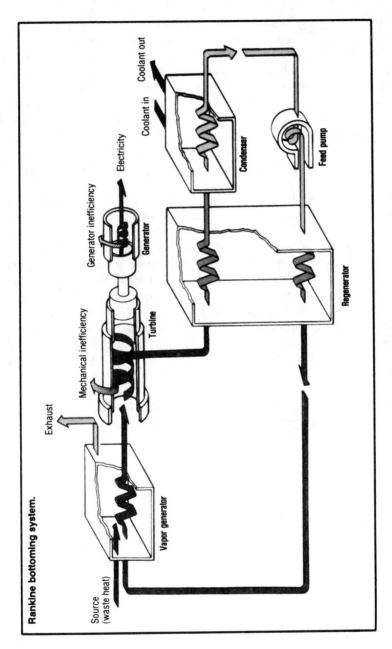

Fig. 13–13. Rankine bottoming system with a regenerator. Source: DOE.[1]

waste heat from many high-temperature industrial processes and from diesel engine exhaust and jacket cooling water. In a bottoming system, fuel is consumed to produce the high-temperature thermal energy needed in such applications as a steel-reheat furnace, a glass kiln, or an aluminum-remelt furnace. Heat is extracted from the hot exhaust waste stream and transferred to a fluid (generally through a waste-heat-recovery boiler), which is then vaporized. The vapor is used to drive a turbine, which produces electrical or mechanical energy. The closed-loop operation is completed when the stream is liquefied and pumped back to the waste-heat recovery boiler, where the squence is repeated. An application of a binary Rankine cycle system is shown in Fig. 13–14.

Two types of fluid have been used in bottoming systems: water (steam) or an organic fluid such as toluene, fluorinol, or Freon II. Organic fluids provide greater flexibility because their properties permit operation at temperatures lower than those at which water vaporizes. All of the Rankine cycle systems currently in commercial use employ steam as the working fliud. They are used for a wide variety of mechanical drive and electrical power generating purposes. These systems have thermal efficiencies from 14 to 36%, operate over a temperature range of 400°F to 1000°F, and are greater than 670 hp (approximately 500 kW) in size.

New Rankine cycle systems using organic working fluids are commercially available in a limited number of sizes, ranging from 3000 to 5000 hp (approximately 2000 to 3800 kW). These operate at much lower temperatures (200°F to 340°F).

The maintenance requirements and reliability of bottoming cycles are generally unknown because there is little commercial experience with these systems to date.

The effects of bottoming systems on the environment include the reduction of thermal pollution of the atmosphere. However, organic Rankine system working fluids are generally toxic and therefore pose an environmental risk.

Fluidized Bed Combustion System.   Figure 13–15 shows the configuration of a fluidized bed boiler. Unlike conventional boilers, which require costly scrubbers to remove sulfur from the combustion gases of coal, the FBC cleans the coal when it burns. The FBC will enhance industrial cogeneration by the following three topping methods (see Fig. 13–16):*

- An atmospheric FBC connected with a steam turbine cycle (Chart A)
- A pressurized FBC combined with a gas and a steam turbine cycle (Chart B)

---

* Variations other than these are also available, e.g., an atmospheric FBC with gas turbines and waste heat recovery boilers.

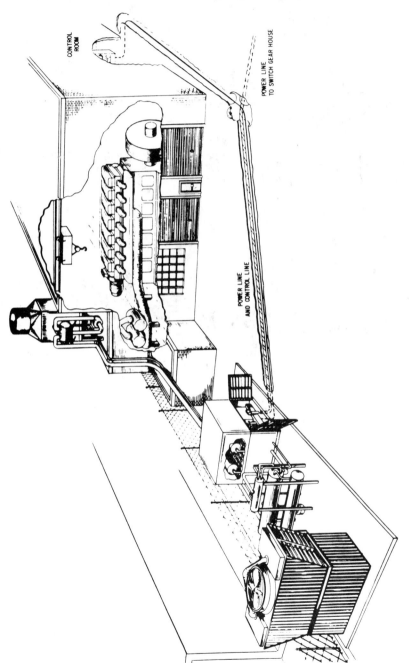

Fig. 13–14. General arrangement of MTI binary Rankine cycle system for waste heat recovery/electric power generation at Municipal Power Plant, Rockville Centre, New York. Source: DOE photo.

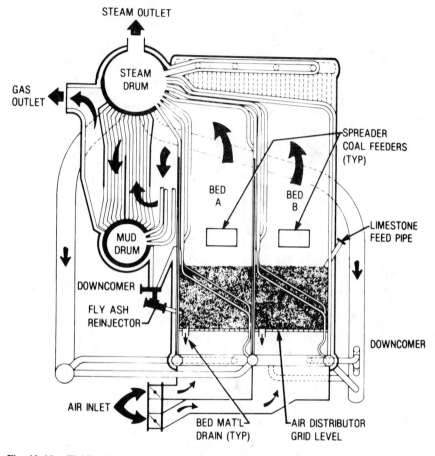

Fig. 13–15. Fluidized bed boiler at Georgetown University in Washington, D.C. The result of a $16 million DOE research and development program begun in 1975, it is the first commercial-scale unit of its type to operate in a metropolitan area. Source: DOE photo.

- A pressurized adiabatic FBC (no water tubes) combined with a gas turbine and a steam bottoming cycle (Chart C)

The FBC has significantly lower NO_x emissions than conventional combustion. It can be commercially available by the early 1980s.

Stirling Engine. The Stirling engine is an externally fired, multifuel, high-efficiency closed cycle heat engine. Large Stirling engines are expected to reach efficiencies in the range of 40 to 45%, to be clean and quiet, and to operate at low speeds. Without adjustment, interruption, or loss of power

CHART A: ATMOSPHERIC FLUIDIZED BED COMBUSTION

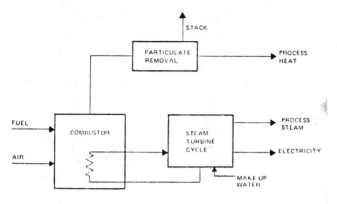

**CHART B: PRESSURIZED, COMBINED CYCLE
FLUIDIZED-BED COMBUSTION**

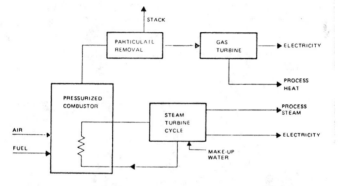

**CHART C: PRESSURIZED, COMBINED CYCLE ADIABATIC
FLUIDIZED-BED COMBUSTION**

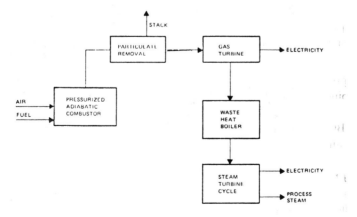

Fig. 13–16.   Three fluidized-bed combustion topping methods. Source: RPA.[6]

and efficiency, the Stirling engine can use a variety of fuels (or thermal sources), including petroleum distillates, shale or coal distillates, methane, low-Btu gas, powdered coal, waste heat, and solar energy. Figure 13–17 shows a derivation of a P-40 engine designed for a stationary power plant by United Sterling of Sweden. The engine was used in a test car produced by the U.S. Department of Energy–MTI, Inc. Stirling engines have a relatively pollution-free exhaust. The NO_x content of emissions can be further reduced by recirculating part of the exhaust gas.

Because of the long-term investment needed, the Stirling engine will not be commercially available before 1985. However, multifuel Stirline engines might be considered insurance against interruptions of petroleum supply.

Fuel Cells. Figure 13–18 shows a commercial application of a fuel cell in-plant integrated energy system. A fuel cell plant consists of three major subsystems: The fuel processor (reformer) converts a conventional, available utility fuel to a hydrogen-rich gas; the power section (composed of fuel cell stacks) converts hydrogen and oxygen (from ambient air) to water and electricity; and the power processor converts dc power to ac power compatible with the utility bus.

Fig. 13–17. Application of Stirling engine. A P-40 engine designed for a stationary power plant by United Stirling of Sweden was used in a test car produced by the U.S. Department of Energy–MTI, Inc. Source: DOE photo.

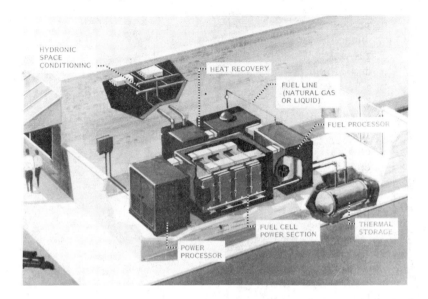

Fig. 13-18. A commercial application of a fuel cell on-site integrated energy system. The heart of a fuel cell energy system is in the power section, made up of a number of cell stacks. Each stack contains many individual cells; each cell is capable of electrochemically reacting fuel and air to produce a fractional kilowatt of dc power. Source: DOE photo.

The major advantages of fuel cells derive from their environmental compatibility and ready siting, and from their efficiency across a wide range of loads and unit sizes. The present fuel cell energy system can reach a maximum fuel efficiency of 80%. It will use any of several liquid or gaseous fuels derived from petroleum or coal. NO_x, SO_x, and particulate emission levels from fuel cell energy systems are lower than any projected requirements, and these systems are quiet and water-conservative.

Currently phosphoric acid and molten carbonate fuel cell technologies are being developed. Only the phosphoric acid fuel cells can be commercialized by 1990.[12]

MHD Generator. Figure 13-19 shows a schematic representation of a typical MHD energy system. The MHD generator uses the advanced energy conversion concept of magnetohydrodynamics. It converts the thermal energy of a high-temperature plasma directly into electrical power by letting the plasma flow through a strong magnetic field. As in a conventional generator, the motion of the conductor (the hot plasma) through the magnetic field generates an electric potential. The generated dc power must be converted to ac for utility applications. In its most common configuration for coal-

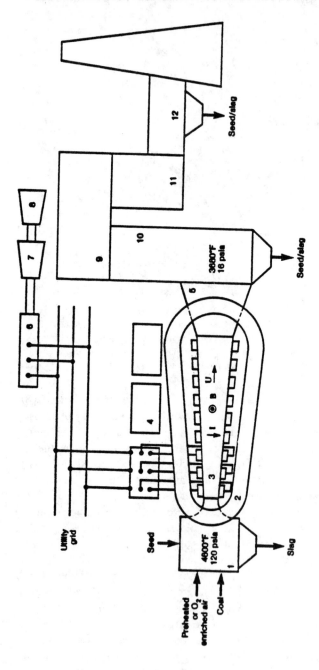

Fig. 13–19.   Schematic diagram of MHD power plant. Source: EPRI.[7]

SUBSYSTEMS

1)	Combustor	5)	Diffuser
2)	Magnet	6)	Alternator
3)	Channel	7)	Steam turbine
4)	Inverter	8)	Air compressor

9)	Super heater/reheater
10)	Radiant boiler
11)	Low temp. air heater
12)	Precipitator

I)	Current
B)	Magnetic field
U)	Gas velocity

NOT SHOWN: Oxygen plant (optional); high temp. air heater (optional); seed processing plant

Source: EPRI[7]

fired applications, the MHD generator is operated as a topping cycle to a steam plant. This cycle configuration, consisting of a combined Brayton–Rankine cycle, is thermodynamically similar to a conventional cumbustion turbine/steam turbine combined cycle.

The MHD generator can be designed to meet 1976 New Source Performance Standards for coal-fired plants. It can be further improved to meet future standards by using stack cleaning procedures or using the potassium seed method for sulfur control.

The MHD generator is not expected to be commercially available before 1990.[7]

Lysholm Expander. The Lysholm expander is a high-temperature (greater than 2500°F), helical expansion device made of ceramic materials. It will be developed for industrial cogeneration and topping cycles for central power plants.

Ceramic helical expanders are especially tolerant of a variety of fuels. They can use solid, liquid, or gaseous fuels that can be derived from coal or other sources. The commercial potential of Lysholm expanders presently is inhibited by the materials limitation on operating temperatures, pressures, steam kinetics, and so on. Lysholm expanders are not expected to be commercially available until the late 1980s.

Heat Pumps and Energy Storage Systems. These two technologies are not topping or bottoming cycles, but they will help the development of cogeneration.

Heat Pumps. Heat pumps can be used in industry to elevate the temperature of thermal energy in waste streams. Thus they can be used in numerous industrial applications to reduce thermal demands by recovering and transforming low-quality waste heat to usable levels.

Present systems have demonstrated a performance level up to 3.0 (COP). Advanced high-temperature heat pumps are expected to become commercially available in the 1980s and achieve a reasonable degree of penetration in the industrial, commercial, and residential markets by 1985. (For more detail, see Chapter 12.)

Energy Storage. Compressed air and pumped hydro are the cheapest forms of energy storage for large facilities (200–2000 MW). Lead acid batteries can be used by smaller facilities, and, as they improve (between now and 1985), they will be suitable in the 20–50 MW range. New designs and materials may make flywheels a good form of storage in the 20–50 MW range as well. Between 1985 and 2000, advanced batteries should raise efficiencies and lower costs considerably. Thermal storage has a longer expected life, but is more expensive. Production and storage of hydrogen will be the least efficient and most expensive technique.

POTENTIAL BENEFITS AND COSTS OF COGENERATION

Cogeneration results in energy conservation: its efficiency depends on the technologies implemented. Benefits of congeneration include energy savings, utility capital cost savings, and waste-heat reductions. Costs of cogeneration include increases in capital investment and industrial energy use, and an unfavorable fuel shift (likely from coal to oil or gas).

Potential Nationwide Benefits of Cogeneration

The potential of energy savings from industrial cogeneration depends on the market penetration of commercially available cogeneration technologies. Since industrial cogeneration is site-specific, it is important first to identify suitable sites where the thermal and electrical loads are matched for cogeneration systems before cogeneration energy savings can be determined. In addition, future expected electricity rates, capacity of electric utilities, availability of fuel for cogeneration, the prevailing environmental standards, and the anticipated federal policies concerning energy use in industry will all affect cogeneration potential.

Because the interaction of the above factors may prevent us from accurately projecting the future expansion of cogeneration, an alternative is to examine the performance of European cogeneration, which has successfully provided about 16% of total electricity generated (see Fig. 13-6). Unfortunately, details of cogeneration data for specific industries in Europe are not currently available.* On the other hand, the GAO and Resource Planning Associates recently studied cogeneration energy saving potential under four scenarios with respect to the above-mentioned factors. The four scenarios are status quo, economic maximum, technical maximum, and incentive cases. The industries that are most suitable for cogeneration are discussed first, followed by a consideration of the potential under these four scenarios.

The Industries Most Suitable for Cogeneration.
The paper and pulp, chemical, petroleum refining, iron and steel, and cement industries historically have used cogeneration systems to produce various percentages of electricity consumed (see Table 13-2). Among these five industries, the

---

* As shown in Fig. 13-6, only aggregate country-wide cogeneration data are available through the International Federation of Industrial Cogenerators.

Table 13–2. Estimated Cogeneration in Industries in 1977.

Industry	Electricity Consumption (10^9 kWh/Yr)	Estimated Percentage Cogenerated	Waste Heat Temperature*	Maximum Cogeneration Potential (10^9 kWh/Yr)
Paper and Pulp	59	49	<400°F	211.8
Chemical	106	24	300– 450°F	279.2
Petroleum refining	24.5	11	300–1000°F	90.8
Iron and steel	50	23	800–2000°F	> 50
Cement	9	5	1000–1100°F	> 10

* The data are only representative because large variations may occur in site-specific instances.
Sources: Icerman and Staples,[10] GAO.[9]

paper and pulp industry has cogenerated about 50% of its electric power. The paper and pulp, chemical, and petroleum refining industries have been studied extensively by DOE, the Thermo Electron Corporation, and Resource Planning Associates. In addition, the food processing and textile industries have been identified as major potential cogenerators.[5]

Potential of In-plant Cogenerated Electricity. The General Accounting Office recently studied the potential of cogeneration from the paper and pulp, chemical, and petroleum refining industries under the status quo, economic maximum, technical maximum, and incentives cases. The definitions of the four cases are summarized below:[9]

- Status quo case: some generation will be developed without government incentives or further disincentives.
- Economic maximum case: maximum electricity production under economic conditions is assumed; distillate-fired gas turbines will be used exclusively.
- Technical maximum case: generation of all steam identified as technically suitable is assumed; all economic conditions are ignored.
- Incentives case: a series of governmental incentives for cogeneration is assumed. This case, with the same type of technology mix as the no-action case, increases the amount of cogeneration, assuming the existence of the following incentives: a 30% investment tax credit for cogeneration equipment, marginal cost pricing of electricity (rate reform), and exemption from FERC and PUC regulations. The 30% investment tax credit consists of an existing 10% general investment tax credit with an additional 20% for cogeneration.

Tables 13–3 and 13–4 summarize GAO's projection of electricity cogenerated in 1985 for the three industries and various cogeneration systems. These

Table 13–3. Projections of Electricity Cogenerated in 1985 for the Three Industries.

	1977 Electricity Use, 10^9 kWh/Yr	Estimated 1977 Percentage of Cogeneration**	Status Quo Case† 10^9 kWh/Yr*	Status Quo Case† % of 1977 Electricity Consumption	Economic Maximum Case† 10^9 kWh/Yr*	Economic Maximum Case† % of 1977 Electricity Consumption	Technical Maximum Case† 10^9 kWh/yr*	Technical Maximum Case† % of 1977 Electricity Consumption
Paper and pulp	59	49%	38.33	65%	93.92	158%	211.8	359%
Chemical	106	24	21.12	20	92.53	87	279.2	263
Petroleum refining	24.5	11	6.02	25	22.61	92	90.8	371
Total	189.5	30%	65.47	35%	209.06	110%	581.8	307%

* Data taken from GAO.[9]

** 1977 data as in Table 13–2.

† See text for definitions of status quo, economic maximum, and technical maximum cases.

Table 13-4. Projections of Electricity Cogenerated in 1985 by Various Turbine Cycles for Paper and Pulp, Chemical, and Petroleum Refining Industries|

Cogeneration System Type	Status Quo Case†	Electricity Cogenerated (10⁹ kWh)	
		Economic Maximum Case†	Technical Maximum Case†
Coal-fired steam turbine	18.34	—	—
Residual-fired steam turbine	8.27	—	—
Distillate-fired gas turbine	4.03	167.86	581.8
Natural gas turbine	2.50	—	—
Waste fuel steam turbine	30.51	38.27	—
Heat recovery steam turbine	1.81	2.93	—
Total	65.46	209.06	581.8

† For definition, see text.
Source: GAO.⁹

cogeneration systems include coal-fired steam, residual-fired steam, distillate-fired gas, natural gas, waste fuel steam, and heat recovery steam turbines. Estimated current electricity cogenerated in the three industries is about 30% of the three industries' total electricity consumption. Cogeneration in the three industries will increase to 35% of their total electricity consumption in the status quo case, to 110% in the economic maximum case, and to 307% in the technical maximum case. This electricity will mostly be cogenerated in the paper and pulp and chemical industries, and part of this electricity will have to be sold to neighboring utilities or consumers.

Among the cogeneration systems shown in Table 13–4, coal-fired and waste fuel steam turbines will cogenerate the most electricity in the status quo case. Distillate-fired gas turbines will cogenerate the most electricity (80%) in the economic maximum case and all electricity (100%) in the technical maximum case. Therefore, in the economic and technical maximum cases, oil will be used extensively to cogenerate electricity, and coal and natural gas will be saved.

Impact of Cogeneration on Fuel Use Patterns

Fuel shift is an important factor to consider in cogeneration. Tables 13–5 and 13–6 show the impacts of cogeneration on U.S. fuel and industrial fuel use patterns. In the status quo case (i.e., cogeneration neither encouraged nor discouraged), national fuel consumption as well as crude oil imports

Table 13–5. Impacts of the Three Industries' Cogeneration on U.S. Fuel Use Pattern* in 1985.

	Changes in Consumption, 10^{15} Btu/Yr			
	Status Quo† Case	Incentives** Case	Economic Maximum† Case	Technical Maximum† Case
Total fuel consumption	−.26	−.48	−.93	−1.52
Specific fuel types:				
Petroleum	−.14	−.26	+.64	+3.20
Gas	−.05	−.08	−.12	−1.53
Coal	−.08	−.15	−1.39	−2.97
Nuclear	+.01	0.	−0.06	−0.23
Crude oil imports	−.06	−.1	+.36	+1.72

* The U.S. fuel use pattern is the one forecasted in the base case of the DOE's Mid-Range Energy Market Model (MEMM)

** A series of governmental incentives for cogeneration is assumed.[9]

† For definition, see text.

Source: GAO.[9]

Table 13-6. Impacts of the Three Industries' Cogeneration on Industrial Fuel Consumption* in 1985.

Type of Fuel**	Changes in Consumption, 10^{15} Btu/Yr			
	Status Quo Case	Incentives Case	Economic Maximum Case	Technical Maximum Case
Electricity	−.06	−.15	−5.6	−1.58
Natural gas	−.03	−.07	.14	−.80
Petroleum:				
Distillate	−.03	−.06	+1.52	+5.53
Residual	+.02	+.06	−.21	−.67
Liquified gas	−.02	−.05	−.09	−.19
Coal	+.05	+.15	−.44	−1.12

* See footnotes in Table 13-5.
** Not including refinery oil and gas.
Source: GAO.[9]

will decrease. However, if the incentives go too far, there is a danger that U.S. petroleum use and crude imports will increase, as in the economic and technical maximum cases. An increase of petroleum consumption is unfavorable in long-run energy planning.

Potential Costs and Economies of Scale

The annualized capital cost for cogeneration is between \$.92 and \$1.875 per million Btu (as shown in Table 13-7) or \$.93 to \$1.92 per kWh equivalent.* This cost is based on the capital recovery rate of a 15% interest rate for 20-year-life cogeneration systems, and does not include operation and maintenance (O&M) costs, which are higher than average utilities' power generation because natural gas and oil are usually used in cogeneration. Table 13-7 also indicates that cogeneration in the chemical industry needs the lowest capital among the three industries.

Cost of Cogeneration Policies. In order to compare costs of cogeneration policies, incremental energy cogenerated (sum of electricity and steam) and incremental capital requirements are compared in Table 13-8 for two cogeneration policies, incentive and economic maximum cases. Table 13-8 shows that significantly more capital must be spent to install cogeneration equipment when additional energy is cogenerated. The annualized marginal

---

* Assumed power generation efficiency is ⅓.

Table 13–7.   Annualized Capital Cost for Cogeneration in the Paper and Pulp, Chemical, and Petroleum Refining Industries (1977 dollars).

	Annualized Capital Cost* ($/MMBtu)		
	Status Quo Case	Incentives Case	Economic Maximum Case
Paper and pulp	1.81	1.84	1.875
Chemical	.92	.97	1.27
Petroleum refining	1.566	1.69	1.74

* Two steps are taken to calculate these data: (1) Annualize capital requirement data in Table III–13 through 15 in the General Accounting Office Report[9] by the capital recovery factor of .15976 (at a 15% interest rate for 20-year-life-cycle cogeneration systems). (2) Derive the above data by energy cogenerated in Table 13–8.

Table 13–8.   Incremental Energy Cogenerated and Capital Requirement for the Two Cogeneration Policies (1977 dollars).

		Cogeneration Policies			
		Incentives Case		Economic Maximum Case	
	The Base Energy Cogenerated* (10^{12} Btu)	Energy Increment* (10^{12} Btu)	Capital Cost Increment (10^6 Dollars)	Energy Increment* (10^{12} Btu)	Capital Cost Increment (10^6 Dollars)
Paper and pulp	406.9	60.15	775	166.85	2065
Chemical	429.7	105.5	751	224.2	2811
Petroleum refining	75.8	40.54	488	42.6	501
Total			$2014		$5377

* Includes steam and electricity.
Source: Calculated from GAO.[9]

capital cost per million Btu energy cogenerated is shown in Table 13–9 for the two cogeneration policies. The annual marginal cost at 15% interest rate for 20-year-life cogeneration systems is between $1.14 and $2.06 per million Btu of between 1.17¢ and 2.10¢ kWh equivalent.*

Economies of Scale.   Table 13–9 shows that cogeneration in the paper and pulp and petroleum refining industries may enjoy economies of scale if O&M costs are not considered, but this is not so in the chemical industry.

---

* Assumed power generation efficiency is ⅓.

Table 13–9. Incremental Annual Capital Cost* of Cogeneration Policies in 1980 for Paper and Pulp, Chemical, and Petroleum Refining (1977 dollars per million Btu).

	Cogeneration Policies	
	Incentives Case	Maximum Economic Case
Paper and pulp	2.06	1.98
Chemical	1.14	2.00
Petroleum refining	1.92	1.88

* Two steps are taken to calculate annual capital cost: (1) Annualize the incremental capital requirment in Table 13–8 by multiplying the capital recovery factor by .15976, which is computed at a 15% compound interest rate for 20-year-life cogeneration systems. (2) Divide the above results by the energy increment.

Because more oil has to be used for the distillate-fired gas turbines in the economic maximum case, cogeneration investment may show a diminishing return if all costs (capital and O&M costs) are considered.

Other Impacts of Cogeneration

Utility capital cost savings and national energy price change are discussed in this subsection.

Utility Capital Cost Savings. Table 13–10 shows the incremental utility capital cost savings in 1985 for the two cogeneration policies, calculated from the GAO study.[9]

Comparing Table 13–10 with Table 13–8, we find that the incremental utility capital savings are almost identical with the three industries' incremental capital increases. Cogeneration makes capital requirement shift from the utilities to industry. However, in this process of shifting, a large portion of reject heat is used in the industrial processes. Because a cogeneration plant has no transmission and distribution losses, additional energy savings (about 9% of all power generated by central plants) are achieved. Other side benefits include short construction lead (3 months to 1 year versus 5 years to 10 years), less pollution concentration, and higher availability (99% versus 70%) due to the shorter repair and maintenance period.

National Energy Price Changes Due to Cogeneration. According to the GAO study[9] for the status quo and incentive cases, energy prices show little movement. All fuels—coals, distillate, residual steam, and natural gas—show small reductions reflecting the overall decline in fuel demand.

In the economic and technical maximum cases, distillate oil prices increase slightly, reflecting the increased distillate demand with gas turbine cogenera-

Table 13–10. Incremental Utility Capital Cost Savings* in 1985 under Two Cogeneration Policies (millions of 1978 dollars).

Plant Type	Incentives Case	Economic Maximum Case
Residual steam	$ 0	$ 0
Coal high-sulfur	0	0
Coal low-sulfur	399	738
Subbituminous coal	814	2,627
Lignite	124	2,016
Distillate turbine	328	1,074
Combined cycle	175	1,355
Gas steam	0	0
Hydro	0	0
Pumped storage	0	0
Nuclear	47	295
	$1,887	$8,105

* See footnote in Table 13–5.

Source: Data for preparing the table were obtained from GAO.[9]

tion. Coal and natural gas prices decrease accordingly. For all four scenarios, electricity rates decline by as much as $.09 to $1.48 per MWh.

COSTING AND PRICING COGENERATED PRODUCTS

The cost of cogenerated products (electricity and steam) is affected by the ownership of cogeneration systems and the method of cost allocation; the price of cogenerated products is determined by the cogeneration cost and electric regulations. In other words, the price of cogenerated electricity may not be exactly its allocated cost (this cost includes a fair rate of return). Cogeneration has been advocated in many states (e.g., California) because of the recent high cost of generating electricity. Many incentives for cogeneration of electricity are included in rate-making. In order to cover all aspects of cogeneration pricing and costing, this section is divided into five subsections: (1) motivations for cogeneration, (2) costs of cogeneration facilities, (3) ownership of cogeneration facilities, (4) cost allocation between steam and power production and (5) pricing cogenerated products.

Motivations for Cogeneration

In general, there are three parties interested in cogeneration: industries, public utility commissions (PUCs), and electric utilities. The motivations of the three parties are substantially different.

Motivation of Industries. Because industries seek profits, possible reasons for an individual firm to invest in cogeneration equipment are as follows:

- Cogeneration provides a substantial savings in fuel, or produces sufficient electricity for sale to enable a competitive rate of return on the investment.
- Uncertainties exist regarding future electricity supply, and electricity is a major portion of the firm's fuel sources.

Cogeneration investments appear to be attractive options to satisfy one or both of these conditions. In the face of escalating power prices in the future, cogeneration may prove to be a justifiable long-term protection from these problems.

Motivation of Public Utility Commissions. Motivations and actions of public utility commissions vary from region to region and time to time. Many public utility commissions such as the California and Louisiana Public Utility Commissions are aggressively pursuing cogeneration because of opposition to construction of new coal or nuclear power plants. Other commissions are interested in cogeneration because of the requirements of PURPA.

However, owing to recent energy prices, inflation, and high interest rates, more and more public utility commissions will advocate cogeneration and conservation to alleviate the rapid increase of electricity rates.

Motivation of Electric Utilities. Ideally, cogeneration will reduce the need to construct new power plants and help alleviate financial difficulties of utilities. Cogeneration will also save their transmission and distribution costs because electricity cogenerated need not be transmitted. Two reasons for the utilities to support cogeneration are summarized as follows:

- To reduce power requirements, eliminate the high cost of new additional capacity, and save transmission and distribution costs for the cogenerated electricity.
- To fulfill the requirements of public utility commissions and PURPA.

Costs of Individual Cogeneration Facilities

The costs of cogeneration are site- and plant-specific. Generally, the costs include installed (capital) cost, O&M costs, and fuel costs. A general discussion about the range of these costs is presented below.

The Installed (Capital) Cost. The installed cost is the initial investment for cogeneration facilities. The primary cogeneration components are the boiler, steam turbine, pollution control, and fuel handling equipment for steam turbine topping; and the waste heat boiler, accessory equipment and piping, foundation, turbine building, instruments, insulation, painting, and installation labor for gas turbine generation. The system components for diesel cogeneration are similar to those for gas turbine generators, and a waste heat boiler and turbine cogeneration are needed for bottoming cycles.

A summary of capital costs per kilowatt installed is shown in Table 13–11 for cogeneration systems. This table allows comparison of various power-generation systems (cogeneration and conventional central station power-generation systems). Accordingly, gas, diesel, and 20-MW oil-fired steam turbine cogeneration systems are cheaper than nuclear ($716–$818/kW) or coal-fired (approximately $730–$830/kW) power plants.[13] However, attention must be paid to the fuel costs in cogeneration. Because the above cogeneration systems use either oil or gas, which is more expensive in the long run, the overall economics (including capital, O&M, and fuel costs) must be weighted carefully. O&M and fuel costs are discussed below.

O&M Costs. The O&M costs vary with the rate of production (or system load factor) and plant size. The range of these costs is estimated to be between 2 (oil plants) and 5 (coal plants) mills/kWh. Theoretically, O&M costs for cogeneration should be no higher than those of a peaking station because both systems use mostly oil or gas, and because cogeneration systems have short downtime for maintenance. In addition, cogeneration systems have better fuel efficiency, and their costs are shared by both steam and electricity.

Table 13–11. Capital Costs per kW Installed and Fuel Used for Cogeneration Systems (1978 dollars).

	Size of System			
	1 MW	**5 MW**	**20 MW**	
Type of System	———	(Cost Per kW)	———	**Fuel Used**
Gas turbine	$ 693	$ 533	$ 426	natural gas and distillate oil
Steam turbine:				
Coal-fired	2,664	1,332	1,065	Any fuel
Oil-fired	*	932	613	
Diesel	586	586	*	Natural gas and distillate oil

* Cost estimates unavailable.

Source: Values (in 1977 dollars) of the GAO report[9] adjusted by the producer price index of material and components for manufacturing (1977 = 195.5, 1978 = 208.3).

Fuel Costs. Fuel costs vary over production rate, type of fuel, and region. Given the current price rates, the estimated range of fuel cost is between 12 (for coal) and 50 (for oil) mills/kWh. The fuel costs should be less than those for utility peaking stations because cogeneration is more efficient, and its cost is shared by both steam and electricity.

Ownership of Cogeneration Facilities

Cost allocation between steam and electricity for cogeneration facilities largely depends on the ownership of cogeneration property, because utilities are regulated by public utilities commissions. The ownership affects the amortization schedule of the capital invested in cogeneration facilities. For example, the capital recovery factor used to amortize capital costs is about 18% for utilities, but it is between 20 and 30% for nonregulated industries.

There are five ownership arrangements for cogeneration facilities: (1) industrial ownership with firm sales of power, (2) industrial ownership without firm sales of power, (3) utility ownership and operation of the cogeneration system, (4) third-party ownership, (5) joint-venture agreements between an industry and a utility. Most present cogeneration ownership falls in the first two categories. This is particularly true in California.[8] A summary of the main characteristics of the above five types is as follows:

- Industrial ownership without firm sales of power
  - The industrial firm owns and operates the cogeneration system.
  - The cogeneration system partially satisfies the firm's electrical demand.
  - Additional power, if required, is purchased from an electric utility.
  - Excess by-product power, if and when available, is sold to the utility.
  - Standby power is purchased from the utility.
- Industrial ownership with firm sales of power
  - The industrial firm owns and operates the cogeneration system.
  - All cogenerated power or excess by-product power is sold to the utility as firm power.
  - The industrial firm purchases all its required power from the utility.
- Utility or third-party ownership
  - The utility company or a third party owns and operates the cogeneration system.
  - The industrial firm purchases both steam and electricity from the utility or third party.
- Joint ownership between the industrial firm and the utility
  - The utility owns and operates the electricity generation equipment and switchgear, while the industrial firm owns the land and the process heat systems.

– The industrial firm purchases both steam and electricity from the utility.

Joint ownership between industries and utilities has been favored by many advocates because they feel it is cheaper for utilities to own cogeneration facilities owing to their lower rate of return (compared to industries) and their capital recovery factors.

Methods of Joint Ownership Between Industry and Utility. The concept of this joint ownership is to let industries bear the burden of financing construction and utilities bear the responsibility of financing operation of the cogeneration facility. Three arrangements are possible:

- The industrial firm finances and owns the cogeneration facility during construction and then sells it to the utility.
- The utility sets up a trust to own the project financed by the industrial firm during construction. Once the cogeneration facility is completely functional, the utility takes title and repays the industrial firm.
- The utility and the industrial firm share construction costs, and then the industrial firm makes payments to the utility for its capital invested in construction-work-in-progress.

Two difficulties arise from this financing: industries' unwillingness to finance other companies' assets, and the question of assigning risk in constructing the cogeneration facility. Agreements must be made to overcome these difficulties in arranging joint ownership for cogeneration.

Cost Allocation Between Steam and Power Production

Because cogeneration is a process of sequentially producing steam and electricity, the principle of cost allocation is to distribute incurred costs due to cogeneration between steam and electricity. The cost allocation is affected by the thermodynamic efficiency of the steam and electricity production and by ownership of the cogeneration facilities. The thermodynamic efficiency determines theoretical shares of incremental costs for cogeneration between electricity and steam production, and the ownership determines how realistically the costs are distributed between electricity and steam.

Incremental Costs for Cogeneration. Figure 13–20 shows incremental costs for cogeneration. Additional annual capital, fuel, and O&M costs are needed for cogenerating steam and electricity sequentially. The magnitude of incremental cost for cogeneration depends on whether the original cost

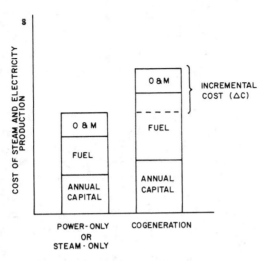

Fig. 13–20. Incremental costs of cogeneration compared with power-only or steam-only production.

base is power-only or steam-only as shown in Fig. 13–20. In other words, the determination of incremental cost for cogenerated electricity to an industrial firm depends on whether the firm treats the electricity as a by-product or as a joint-product. The distinction of by-product and joint-product is further complicated by the ownership of cogeneration facilities. For example, if an industrial firm owns the facilities and treats the electricity production as a by-product, then certain fixed costs are already burdened by steam production, and the incremental cost for the electricity generation would be less than it otherwise would have been. However, if a utility owns the facilities and treats steam production as a joint-product, then the electricity generation must share a large portion of fixed costs. Therefore, it is important to know the treatment of cogenerated electricity in order to determine its incremental costs.

Cost Treatments of In-plant Electricity Generation. For costing purposes, the method of treatment of in-plant electricity generation can be any one of the following:

- Joint-product: the benefit of energy savings is shared between electricity and steam production in proportion to their thermodynamic availability or other distribution criteria.
- Electricity as a by-product: the benefit of energy savings may be mostly enjoyed by electricity generation because steam production is the basis

for computing incremental costs. The ownership of cogeneration facilities is most likely to be industrial.

● Steam as a by-product: the benefit of energy savings may be enjoyed mostly by steam production. The owner of the cogeneration facilities is most likely to be a utility.

In general, the joint-product treatment of costing in-plant electricity production is the most complex of the three methods because cost items must be broken down in detail and distributed between steam and electricity according to thermodynamic availability as efficiency or other costing criteria (e.g., main turbine output). The treatment of electricity as a by-product or steam as a by-product somewhat simplifies the computation of cost, but it may fail to proportionate the costs of cogeneration to outputs of steam and electricity. A good example of the by-product cost treatment is the "OSW method," which was first introduced in the early 1960s by workers in the former Office of Saline Water for dealing with combined power in desalting plant production.

Steam is chosen as the primary product to illustrate the by-product treatment for allocating costs between steam and electricity production. First, a steam-only industrial plant is designed to a desired capacity as the base for computing incremental costs for the production of by-product electricity. Then a cogeneration plant with an identical steam capacity plus a quantity of electricity is designed and similarly costed. Note that because the primary product is steam, steam capacities for both steam-only and cogeneration plants are identical. This costing procedure is repeated for fuel and O&M costs. The costs for the steam-only and cogeneration plants are then compared (see Fig. 13–20). The cost increment is assigned to electricity production (as shown in the figure) and the base amount to steam production.

Pricing Cogenerated Products

Because excess electricity from cogeneration will be sold mostly to electric utilities regulated by PUCs, prices of cogenerated electricity are also regulated by PUCs. They are thus isolated from the free market's supply and demand mechanism.

PURPA requires that electric utilities offer to buy power from, and sell power to, qualifying cogenerators at fair rates. In establishing rules for enforcing this part of the act, FERC is required to ensure that the rates are just and reasonable, in the public interest, and not discriminatory against qualifying cogenerators.

There are two primary bases on which a just and reasonable electricity rate can be scheduled:

- Cost of cogeneration allocated for the in-plant electricity production as described in the last subsection.
- Avoided cost of a utility's power generation, i.e., the cost that a utility has to pay in order either to obtain the electricity from other utilities or to generate it itself.

Cost of Cogeneration for Electricity. As previously discussed, there are three ways to calculate costs of cogenerated electricity. If the rate of cogenerated electricity is computed on the basis of these costing methods, a fair rate of return must be added to the costs thus computed. This pricing method is fair because it is based on the cost of in-plant electricity generation, but it is difficult for the utility to manage because the number of the cogenerators may be too large, and because each has a different cost and price value. In short, this method is good for the products of competitive markets. The prices based on actual costs of cogeneration should be used as reference prices in evaluating cogeneration potential and establishing rates for cogenerated electricity.

Avoided Costs of Utility's Power Generation. FERC requires utilities to buy power from cogeneration systems or from small generation facilities using solar energy, wind, municipal waste, and other renewable fuels. It also requires utilities to purchase this power at rates equal to avoided cost of their generation costs (i.e., what it would cost for them to generate the electricity themselves or buy it elsewhere). Clearly, the rate set at the avoided cost of a utility power generation is not related to the power costs of cogeneration. Consequently, the rate thus established is not warranted to cover cogeneration costs. According to FERC, utilities do not have to purchase power from a cogenerator whose power generation costs are higher than the utilities' avoided costs. Appendix D shows an example of a power sales agreement (including rates) for cogeneration and small power production. This is prepared by the Pacific Gas and Electric Company in accordance with the California Public Utilities Commission requirements.

The concept of avoided cost is easy in principle and difficult in practice. There are a number of elements involved in this concept. Their definitions require classification before an avoided cost for a utility's power production can be computed. Major elements are discussed below.

Avoided Cost. FERC defines avoided cost as "costs to an electric utility of energy or capacity or both which, but for the purchase from a qualifying facility, the electric utility would generate or construct itself or purchase from another source." Under this definition, FERC recognizes two points. First, that both capacity and energy cost must be analyzed. Second, that the analysis of the energy cost must include the effect on energy cost and generation of the difference in load factor between that of the qualifying facility and that of the deferred capacity. Judgmental inputs must be made in determining load factors and deferred capacity.

Deferred Capacity. A number of points must be considered when defining deferred capacity. Three of them are explained here.

First, there is the question as to whether the deferred capacity should be the optimal deferred capacity, where optimal is defined as that which minimizes net avoidable utility system costs while maintaining the utility reliability. Second, should the deferred capacity necessarily correspond to a full generating unit, or should portions of a unit be considered? Finally, in calculating the effect of the deferred unit, should reliability be maintained each year or averaged over the entire study period?

Firm Power. Firm power can be interpreted as power from small self-generation facilities that are available to meet the load equivalent to any unit in the purchasing utility system. Maximum forced outage rates and maintenance requirement would be specified in the power purchasing agreement between the utility and cogenerators.

Incremental Cost. Section 210(d) of PURPA states that rates charged for required purchases should not exceed the incremental cost of alternative electric energy. Incremental cost is defined as the cost of energy that, but for the purchase, the utility would generate or purchase from another source. As we have stated previously, the incremental cost depends on the cost base from which it is calculated. The estimation of incremental costs for any electric energy can be very judgmental.

Transmission and Distribution (T&D) Cost. The T&D cost is a part of the problems raised with the avoided cost concept. Should it be included as a part of avoided costs? Theoretically, it should, but the allocation of the T&D costs avoided in purchasing cogenerated power can be very difficult.

Interconnection Cost. The interconnection cost is the cost for connecting cogeneration facilities into a utility's power grids. FERC has ruled that the cogenerators should pay these costs.

Capacity Credit. Capacity credit refers to the credit for the deferred capacity discussed previously. The capacity credit is also one of the debatable items in the avoided-cost concept. How much credit should be given to cogenerators because they defer the need of a utility's construction for new power plants? This should be specified upon the signing of power purchasing agreements.

Impacts of Pricing Cogenerated Electricity on Cogenerators and Rate Payers

Appropriate rates for cogenerated electricity will encourage cogeneration. Rates that are too low will not cover the cogeneration cost for electricity and, therefore, will reduce the interest of cogenerators. However, rates that are too high will, in effect, subsidize cogeneration and, indirectly, sacrifice the rate payers. This argument is illustrated in Fig. 13–21 and is discussed below. As shown in the figure, if a cogenerator has a total production cost, including fair rate of return, of c, and produces a units of electricity and b units of steam, the mathematical relationship for this joint production to recover the production cost is as follows:

$$a \cdot x + b \cdot y = c \qquad (13–1)$$

where x and y are prices for steam and electricity, respectively. If a, b, and c, in Equation 13–1 are kept constant, that is, production units (a, b) and cost (c) are kept constant, any given steam price (OS in Fig. 13–21) will determine a corresponding electricity price (OK in Fig. 13–21) required for fully recovering the total-production electricity price (steam and electricity cost).

The rate set by a utility to purchase cogenerated electricity will fall into

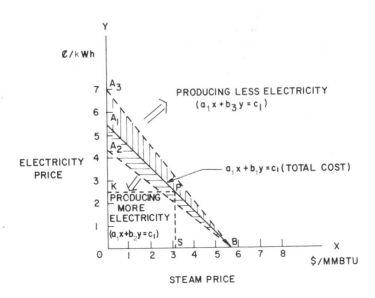

Fig. 13–21. Impact of electricity rates on cogeneration.

any of the following three categories if it is based on the utility's avoided cost instead of cogeneration costs for electricity:

- Equal to the required electricity price for recovering cogeneration cost, *OK*
- Greater than *OK*
- Smaller than *OK*

Note that the utility is a regulated industry, and the rate set for purchasing cogenerated electricity will not be responsive to cost changes in cogeneration.

If the utility's electricity rate is equal to the electricity price required to recover cogeneration costs for electricity, then cogenerators will be neither subsidized nor "taxed" by the utilities, and the production ratio between electricity and steam would be unchanged, as originally designed in Equation 13-1. This price is sufficient to encourage cogeneration economically without misallocating resources.

If the utility's purchasing rate is greater than the required electricity price, then the cogenerators will be subsidized and need to produce less electricity to recover their costs for producing electricity ($b_3 < b_1$ in Fig. 13-21). If they decide to produce the same amount of electricity as in case 1 because the production schedule is fixed, then they will get A_1BA_3 amount of subsidy from the rate payers (see vertically shaded area in the figure).

On the other hand, if the utility's purchasing rate is less than the required electricity price, then congenerators need to produce more electricity in order to recover allocated electricity cogeneration costs ($b_2 > b_1 > b_3$ in the figure). If they decide to produce the same amount of electricity as in case 1 for technical reasons (the electricity is a by-product of steam generation) and the production is fixed, the cogenerators will be "taxed" at the amount of A_2BA_1 (see horizontally shaded area in the figure).

The important point to infer here is how high the utility rate for purchasing cogenerated electricity should be so that cogenerators can recover distributed electricity generation costs. Clearly, the magnitude of the utility rate depends on how cogenerators treat the incremental cogeneration cost: as by-production or joint-production cost. The treatment of cogenerators in allocation of the incremental cogeneration costs affects the cogeneration potential and benefits discussed.

COGENERATION LEGISLATION AND ITS IMPACTS

There are three major pieces of legislation that affect cogeneration: the National Energy Act (NEA), the Clean Air Act (CAA), and the Windfall Profits Tax. The NEA is a blanket term covering the following five separate pieces of legislation:

- National Energy Conservation Policy Act of 1978 (NECPA)
- Public Utility Regulatory Policies Act of 1978 (PURPA)
- Natural Gas Policy Act of 1978 (NGPA)
- Energy Tax Act of 1978 (ETA)
- Powerplant and Industrial Fuel Use Act (FUA)

Of the five NEA acts, PURPA, NGPA, and FUA significantly affect cogeneration. The Clean Air Act includes prevention of significant deterioration (PSD) provisions and defines the degree of emission control that must be incorporated into any new fossil-fired plant facility. The Windfall Profits Tax legislation provided a new 10% tax credit before 1982 for cogeneration equipment.

Public Utility Regulatory Policies Act of 1978 (PURPA)*

Sections 201 and 210 of PURPA are the key legislation guiding the implementation of cogeneration. Section 201 contains definitions of cogeneration; Section 210 provides the authority for requiring that electric utilities offer to buy power from, and sell power to, qualifying cogenerators at fair rates. It also requires FERC to prescribe and, from time to time, to revise rules in order to encourage cogeneration. In establishing such rules, FERC is required to ensure that the rates are just and reasonable, in the public interest, and not discriminatory against qualifying cogenerators. However, the act prohibits any rule requiring the utility to purchase power from a qualifying cogenerator at a rate that exceeds the cost of the utility's purchasing the electricity from another utility or generating it itself.

FERC issued final rules (Federal Register/Vol. 45, No. 38), effective March 20, 1980, for utilities to sell and buy power from qualifying cogenerators. Under these regulations, utilities must buy this power at rates equal to their avoided costs (i.e., what it would cost them to generate the electricity themselves or to buy it elsewhere). In addition, utilities must provide these facilities with supplemental or back-up power "on a non-discriminator basis" and at "just and reasonable" rates.

Because of difficulties in the translation of the principle of avoided-capacity costs from theory into practice, and because of the diversity within state utility commissions, FERC left many of the details of these regulations to the states and therefore granted them "flexibility for experimentation and accommodation of special circumstances" to establish precise rules within a year. In establishing rules, two factors must be considered: Who would

---

* PURPA was recently challenged in a Mississippi federal district court and the judge ruled parts of the act unconstitutional. However, he did not issue an injunction. The Mississippi regulatory commission is working on the rates for cogenerators.

be qualified as cogenerators, and what would be included in the avoided costs?

Qualification of Cogenerators. In March 1980, FERC allowed small power producers to certify themselves as "cogenerators," unless it should enter a specific denial (see Federal Register/Vol. 45, No. 56). The self-certification would stand unless an intervenor entered an objection to it and FERC determined there were sufficient grounds for denial of the certification.

FERC forbids utilities to own "more than 50% of the equity interest in the cogeneration facility" either by themselves or their holding companies and allows them to refuse to purchase power "from a facility with a design capacity of 500 kW" or more until 90 days after such facility has either notified a utility of its qualification as a cogenerator or applied to FERC for such status.

In addition, if oil or gas is used in cogeneration, the following efficiency standards must be met for qualifying as cogenerators:

- For oil- and gas-fired topping-cycle cogeneration facilities, "the useful power output plus one-half of the useful thermal energy output must be, during any calendar year, no less than 42.5% of the energy input" of natural gas and oil.
- For bottoming-cycle cogeneration facilities, generally no efficiency standard is required. However, facilities that use oil or gas as supplemental fuel must meet an annual standard under which useful power output must be no less than 45% of the energy input of oil or gas during any calendar year.

Components of Avoided Costs. Utilities must take into account both fixed and operating costs by buying power from cogenerators. If power from a photovoltaic installation is purchased during a period of peak demand, the price must be based on the utility's costs of running its peaking units, not on its average system cost. However, the cogenerators will have to pay the cost of interconnecting with the utility grid. Also, the price paid by utilities must include the capital costs for new construction that will be prevented because of cogeneration, if the following condition is met:

If a qualifying facility offers energy of sufficient reliability and with sufficient legally enforceable guarantees of deliverability to permit the purchasing electric utility to avoid the need to construct a generating unit [or] to enable it to build a smaller, less expensive plant, or to purchase less firm power from another utility, then the rates for such a purchase will be based on the net avoided capacity and energy costs. (FERC)

Powerplant and Industrial Fuel Use Act (FUA)

The Powerplant and Industrial Fuel Use Act* restricts the use of oil and natural gas according to user classes. It revises some of the original provisions of the Energy Supply and Environmental Coordination Act (ESECA).

ESECA originally banned the use of oil and gas in industrial boilers larger than 1000 MMBtu/hr. DOE, however, is required under FUA to grant an exemption when the cost of using coal or some other alternative fuel "substantially exceeds" the cost of using imported oil. Congress was silent on the precise quantitative meaning of the phrase "substantially exceeds," presumably leaving it up to DOE to make the decision based on an economic analysis of alternative definitions. In applying the law, DOE has great latitude in defining when oil and natural gas use exemptions are permissible for cogeneration. One provision in the act prohibits new major fuel-burning boiler installations from using petroleum and natural gas as a primary energy source. However, as an incentive to cogeneration, the act permits exemption of cogeneration facilities from this prohibition provided they demonstrate that the economic and other benefits of cogeneration can only be obtained by using oil and natural gas.

Because of their size, most new industrial cogenerators will have to demonstrate their eligibility for an exemption in order to use oil or natural gas. The act defines major fuel-burning installations as capable of using fuel at an input rate of at least 100 million Btu/hr. Industrial cogeneration facilities generally need a steam demand of at least 100 million Btu to be economically attractive. Therefore, most new industrial cogenerators will be classified as major fuel-burning installations and will be required to comply with the act.

Interim rules to carry out the exemption provisions of the act have been issued by ERA. These interim rules, effective May 8, 1979, permit oil and/or natural gas use for cogeneration if a petitioner demonstrates to the satisfaction of ERA that it meets the following criteria: (1) the oil or gas to be consumed will be less than that which would otherwise be consumed in the absence of the cogeneration facility; or (2) it would be in the public interest because of specific circumstances such as technical innovations or maintaining industry in urban areas. ERA officials have indicated, however, that the predominant requirement is oil and/or gas savings.

If a petitioner who plans to operate a cogeneration facility cannot qualify to use oil or natural gas under the cogeneration exemption, he, like any other industrial facility, can still petition for exemption status under some other category—such as lack of adequate capital or environmental requirements.

---

* For more details, see Chapter 9.

ERA has attempted to encourage cogeneration by generally requiring less eligibility documentation for the cogeneration exemption than what is required from noncogenerators seeking permanent exceptions. However, ERA's interim rules do not provide any preferential treatment for cogenerators seeking an exemption under some other category. For example, ERA can grant an exemption from the prohibition on petroleum and natural gas use due to the lack of an alternate fuel supply at a reasonable cost, if the cost of using the alternate fuel substantially exceeds the cost of using imported oil. All facilities petitioning exemption status under these conditions must prepare cost comparison data in accordance with ERA specifications. Accordingly, a potential cogenerator would also have to prepare and submit for approval the necessary documentation to obtain an economic exemption to use oil and natural gas.

Thus, petitioners with small facilities that can economically use only oil or natural gas, but cannot show oil or gas savings or otherwise qualify for the cogeneration exemption, are not encouraged to cogenerate. Small facilities usually require disproportionately high investments before being able to burn coal. A cogenerator in this situation who seeks a cogeneration exemption but cannot prove oil or natural gas savings, must then seek an exemption under some other category, as explained above.

Another drawback of ERA's interim rules is that large potential cogenerators who can prove oil or natural gas savings are not required to disprove that they could have used alternative fuels in their facility. It is believed the rules should not encourage oil or gas cogeneration in a large facility if that facility could use coal or alternate fuels to economic and environmental advantage, and that the rules should encourage and not burden potential small cogenerators.[9]

Natural Gas Policy Act of 1978 (NGPA)

Provisions of the Natural Gas Policy Act of 1978 will incrementally increase gas prices for certain industrial users to the Btu equivalent of substitute fuel oil. FERC is required to develop implementing rules applicable to industrial boiler fuel facilities and other industrial users designated as subject to incremental pricing. The act provides, however, that a qualifying cogeneration facility is exempt from these incremental pricing provisions to the extent allowed by FERC rules.

The exemption of a cogeneration facility from these price increases can contribute to improving its economic attractiveness. FERC has thus granted natural-gas-fired cogenerators a near-total exemption from the incremental pricing regulations authorized by Title II of NGPA.

Clean Air Act of 1970*

The Clean Air Act of 1970, as amended (42 U.S.C. 7401, *et seq.*), requires any new fossil-fired plant to meet the emission control and PSD provision of the act. Emission control requirements include the BACT for clean air or attainment areas, and the Lowest Achievable Emission Rate (LAER) for nonattainment areas.** A strict interpretation of the Clean Air Act would prevent the siting of all new air-polluting facilities in nonattainment areas. Once existing nonattainment areas came into compliance, new facilities could be sited as long as the new pollutants did not interefere with maintenance of the standards or prevention of significant deterioriation of air quality requirements.

However, in December 1976, EPA announced an offset policy setting forth conditions under which new facilities could be sited in nonattainment areas while conforming to the requirements of the Clean Air Act. The policy allows new sources to be located in nonattainment areas as long as, among other things, the new pollutants are more than offset by a reduction in emissions of the same pollutants from existing facilities in the same area. In addition, individual states which have the responsibility to implement Clear Air Act requirements can set stricter new source regulations than those of the federal government.

The effect of the environmental-protection requirement is substantial. For example, the California Energy Resources Conservation and Development Commission states, in a report on issues affecting cogeneration in California, that cogeneration in California is predicated primarily on air-pollution regulations. Under the New Source Review rules in force in California, in areas where pollution levels currently exceed state or national ambient air quality standards, new pollution is allowed only when an existing pollution source can be decreased by an amount that exceeds the additional new pollutants involved. In areas where air pollution regulations dictate stringent controls, such pollution tradeoffs can be difficult to obtain. According to the report, even where such emission reductions are possible through in-plant changes, industry might choose to use the tradeoff to implement higher-priority projects such as increasing production facilities rather than adding cogeneration capability.[9]

---

* See Chapter 9 for the status of the act.
** EPA established primary and secondary standards for six classes of pollutants—sulfur dioxide, particulate matter, carbon monoxide, hydrocarbons, nitrogen oxides, and photochemical oxidants. In many areas of the country, neither the primary nor the secondary standards have been attained. These areas are called nonattainment areas.

Windfall Profits Tax Legislation

The primary purpose of windfall profits tax legislation is to tax oil companies $227 billion from their windfall profits. However, nearly buried in this legislation is an array of tax credits designed to encourage investments in unconventional energy sources and conservation. A new 10% tax credit was provided for investments in qualified cogeneration equipment, effective until 1982; U.S. tax revenue loss in 1980–1990 due to this credit is estimated at $356 million. (See Chapter 9 for details.)

APPLICATIONS AND R&D EFFORTS

This section describes industrial applications and research and development efforts on cogeneration in DOE and in the Electric Power Research Institute (EPRI).

Industrial Applications of Cogeneration

Table 13–12 shows a list of cogenerators selected by Synergic Resources Corporation for EPRI[11] according to the following criteria:*

- Currently operational
- At least one year old
- Its electrical capacity greater than 5 MW (preferably greater than 10 MW)
- Operational for at least eight hours per day
- Currently providing both electrical and thermal energy
- In the industrial sector, preferably in the energy-intensive manufacturing industries
- Willing to cooperate and provide data

Figure 13–22 indicates the location of these cogenerators, and Table 13–13 shows their characteristics. The capacity of the cogeneration systems studied varied from 7.5 MW to 90 MW, with an average load factor of 63%.

R&D Efforts in DOE

DOE's research and development efforts include: cogeneration fuel cell applications, cogeneration case history studies, new cogeneration technologies,

---

* The candidates were also selected to represent a diversity of prime movers, utility interfaces, primary fuels, ownership, industrial groups (SIC), and geographic locations.

Table 13–12.   List of Cogeneration Systems Studied.

1. American Enka Company, Lowland, Tennessee
2. Anheuser Busch, Inc., St. Louis, Missouri
3. Bowater Southern Paper Corporation, Calhoun, Tennessee
4. (Name withheld),* Pennsylvania
5. General Foods Corporation, Woburn, Massachusetts
6. Gulf States Utility Company, Baton Rouge, Louisiana
7. Holly Sugar Corporation, Brawley, California
8. Pacific Gas and Electric Company, Avon, California
9. Potlatch Corporation, Lewiston, Idaho
10. Riverside Cement Company, Oro Grande, California
11. St. Regis Paper, Houston, Texas
12. Shell Oil Company, Deerpark, Texas
13. Celanese Chemical Company/Southwestern Public Service Co., Pampa, Texas
14. Phillips Petroleum Company/Southwestern Public Service Co., Borger, Texas
15. Southern California Edison, Pomona, California
16. Stauffer Chemical Company, Henderson, Nevada
17. Union Carbide Corporation, Texas City, Texas

* Because of confidentiality restrictions, the name of this company cannot be disclosed.
Source: Synergic Resources Corp.[11]

an industrial cogeneration optimization program, and a cogeneration demonstration program.* They are summarized below.

Feasibility Study for Industrial Cogeneration Fuel Cell Applications.

The objective of this program was to identify and analyze specific industrial sites for the possible application of a 4.8-MW phosphoric acid fuel cell power plant in a cogenerating mode, to generate electrical power and hot water in plants where the demand for these energy forms is balanced with the capability of the fuel cell. Four specific industrial sites were selected. The study evaluated the potential for conservation and the economic feasibility of the system. A demonstration program plan was prepared as one product of this effort. This project was primarily funded by Energy Technology and supported by the Office of Industrial R&D Programs of DOE.

Cogeneration Case History Studies.

The objective of this program is to conduct a series of industrial cogeneration case history studies to develop a picture of financial, regulatory, institutional, and environmental factors

---

* The existence of these programs is in doubt because of the Reagan Administration's intention to abolish DOE.

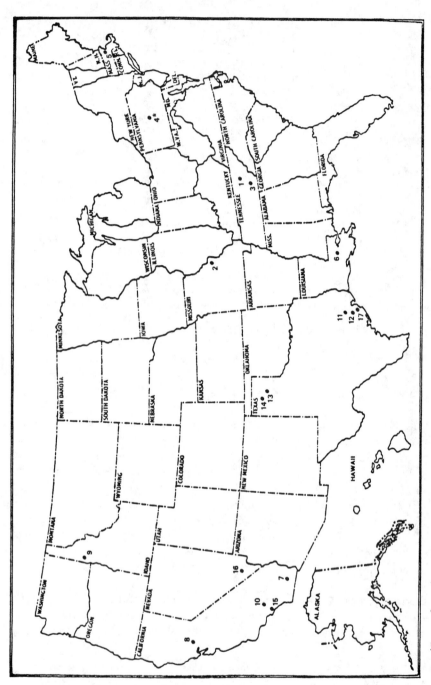

*Exact location disguised

that influence the original decision to invest, and a summary of the industrial experience with the systems to date. Five cases representing the following industrial sectors were selected: food, textiles, pulp and paper, chemical, and petroleum refining.

Brayton/Glass-Alpha. The exhaust gas stream of a glass furnace is at a very high temperature and flow rate, so considerable energy is wasted, even with the use of conventional recuperators. A proven technique for increasing the production rate of a glass furnace, with no additional local consumption of fuel, is to electrically "boost" it by passing current through electrodes submerged in the molten-material tank. The Brayton-cycle glass project was to develop a turbomachinery device powered by the energy now wasted in the exhaust gas stream which would generate, on the spot, the electric current for "boosting" the tank.

Table 13-13. Characteristics of Cogeneration Systems Studied.

NAME OF FIRM	1 YEAR COGENERATION BEGAN	2 YEAR LAST EXPANSION	3 ELECTRIC CAPACITY (MW)	4 ANNUAL ELECTRIC GENERATION (Million KWH)	5 UTILITY RELATIONS[a]	6 SYSTEM TYPE[b]	7 OPERATIONS MODE[i]	8 FUELS USED	9 1979 FUEL COST ($/10 Btu)	10 1979 ELEC. PURCHASE COST (¢/KWH)
AMERICAN ENKA	1947	1960	20	88	PO	STT	Thermal, Electric	Coal	1.44	2.4
ANHEUSER BUSCH	1929	1951	27.6	72	PO	STT	Thermal	Coal, NG	2.25	NA
BOWATER SOUTHERN	1954	--	45	378	GI	STT	Thermal	Wood wastes, Oil, NG	1.00	2.5
(NAME WITHHELD)	1934	1948	11	65	PO	STT	Thermal	Oil	2.40	NA
GENERAL FOODS	1978	--	7.5	26	GI	STT	Thermal	NG	2.50	3.8
GULF STATES UTILITY COMPANY	1930	1954	190	1,078	GI	STT	Thermal	NG, Refinery Gas	2.60	--
HOLLY SUGAR	1949	1978	7.5	16	I	STT	Electric	NG	2.40	NA
PACIFIC GAS & ELECTRIC	1939	--	50	209	GI	STT	Thermal	NG, Refinery Gas	Propri- etary	--
POTLATCH	1951	1977	20	102	PO	STT	Thermal	NG, Black Liquor	1.20	0.8
RIVERSIDE CEMENT	1954	--	15	126	GI	STB	Electric	Waste heat, Oil	(d)	3.75
ST. REGIS	1967	--	95	557	GI	Combined Cycle	Thermal	NG, Wood waste	1.15	2.3
SHELL OIL	1941	1979	60	342	PO	STT, GT	Thermal, Electric	Refinery Gas, NG	2.40	2.2
SOUTHWEST PUBLIC/ CELANESE[c]	1979	--	30	189	GI	STT	Electric	Coal	1.20	3.0
SOUTHWEST PUBLIC/ PHILLIPS	1966	--	33	NA	GI	GT	Electric	NG	2.40	3.0
SOUTHERN CALIFORNIA EDISON	1960	--	14.5	78	GI	GT	Electric, Thermal	NG, Oil	Propri- etary	--
STAUFFER CHEMICAL	1968	--	27	180	PO	Combined Cycle	Thermal	NG	2.70	2.9
UNION CARBIDE	1941	1967	70	386	PO	STT, GT	Thermal	NG	2.50	2.8

See Footnotes on the following page.

Table 13-13.   (Continued).

NAME OF FIRM	11 CAPACITY FACTOR	12 SYSTEM EFFICIENCY	13 ALPHA VALUE Btu(e)/ Btu(t)[e]	14 NET HEAT RATE Btu/KWH	15 DEMAND MET BY COGENERATION (%) THERMAL	16 ELECTRICAL	17 AVAILABILITY (%)	18 UNSCHEDULED OUTAGES (hrs/yr)	19 FUEL COST THERMAL ($/10^6 Btu)[f]	20 FUEL COST ELECTRIC (¢/KWH)[g]
AMERICAN ENKA	0.50	0.62	0.14	6,430	100	23	99.9	0	2.45	0.9
ANHEUSER BUSCH	0.30	0.77	0.11	5,600	100	41	99.9	0	2.86	1.3
BOWATER SOUTHERN	0.96	0.55	0.27	11,190[i]	82	39	99+	0	1.43	1.1
(NAME WITHHELD)	0.67	0.76	0.12	5,750	100	71	99.9	0	3.06	1.4
GENERAL FOODS	0.40	0.73	0.15	7,650	62	100	85	0	3.16	1.8
GULF STATES UTILITY COMPANY	0.65	0.73	0.17	5,830	--	--	99.9	1	3.43	1.5
HOLLY SUGAR	0.24	0.70	0.10	7,550	73	100	99.9	0	3.32	1.9
PACIFIC GAS & ELECTRIC	0.47	0.60	0.46	9,100	--	--	99.9	NA	--	--
POTLATCH	0.58	0.64	0.03	5,700	82	23	95+	NA	1.86	0.7
RIVERSIDE CEMENT	0.96	(h)	(h)	11,320[h]	0	87	99.9	NA	(d)	(d)
ST. REGIS	0.67	0.67	0.21	5,200	100	NA	99+	48	1.72	0.6
SHELL OIL	0.65	0.76	0.07	4,500	NA	21	99.7	0	3.15	1.1
SOUTHWEST PUBLIC/ CELANESE[c]	NA	0.67	0.07	4,800	NA	NA	NA	NA	1.80	0.6
SOUTHWEST PUBLIC/ PHILLIPS	NA	0.70	0.48	7,030	NA	NA	99.9	0	--	1.7
SOUTHERN CALIFORNIA EDISON	0.61	0.65	0.34	6,630	--	--	94	NA	--	--
STAUFFER CHEMICAL	0.76	0.74	0.30	5,050	100	50	98	NA	3.51	1.4
UNION CARBIDE	0.63	0.68	0.08	6,600	100	100	99.9	NA	3.60	1.7

[a] Utility relations: I—isolated; PO—parallel operation; GI—grid interconnected.

[b] STT—steam turbine topping cycle; STB—steam turbine bottoming cycle; GT—gas turbine topping cycle; STT, GT—steam turbine and gas turbine operating in parallel; CC—combined cycle, gas turbine exhaust used to produce steam for steam turbine.

[c] Celanese numbers have been estimated from numbers given by utility.

[d] Riverside Cement utilizes waste heat; no price could be assigned.

[e] Alpha value is the ratio of the Btu value of electricity generated to the Btu content of steam produced for process.

[f] Fuel Cost Thermal is the cost of the fuel needed per million Btu of process steam.

[g] Fuel Cost Electric is the cost of the fuel needed to generate 1 kWh of electricity.

[h] Riverside is bottoming cycle; steam efficiency and alpha value not applicable.

Additional abbreviations: NA—not available; NG—natural gas.

At the time of this writing, engineering and materials studies had been under way, component tests had been made, and several small turbomachinery units had been built and tested in a scale-model glass furnace and on a commercial container glass furnace. A major decision was required on the configuration of the full-scale turbomachinery. A full-scale demonstration unit was prepared and installed in an existing on-line commercial glass furnace and tested to determine performance and economics, and to obtain further design data. No further progress was reported.

Industrial Cogeneration Optimization Program (ICOP). ICOP was oriented toward determining optimum conditions for cogeneration systems on a site-specific basis. Two contractor teams, headed by TRW[14] and A. D. Little,[16] were working on identical efforts, which were completed late in FY 1979, after approximately six man-years of work.

Each contractor was to collect energy consumption data in five industries— food processing, pulp and paper, textiles, chemicals, and petroleum. They were to characterize all available and near-term future cogeneration systems considering both topping and bottoming cycles. The use of abundant fuels and the resultant ramifications for these system types and the resulting economics were to be explored under various energy-pricing scenarios.

The generic information was used together with institutional, regulatory, and environmental constraints to produce preliminary cogeneration system designs on a site-specific basis. The schedule for implementing a system, including design, manufacture, and installation, was formulated, the system economics determined, and the energy savings estimated.

Cogeneration Demonstration Program. The primary objectives of this program were to provide near-term, highly visible, and credible hardware demonstrations to increase industry's interest in cogeneration systems and to expedite technology transfer. Emphasis was placed on cogeneration concepts that represent a significant advance in applied conversion system technology, process interface, or industry/utility relationship, and can be shown to be not permanently dependent on natural gas or distillate fuel forms. Another objective was to provide industry and DOE with first-hand experience in dealing with institutional cogeneration. These evaluation projects would also contribute to future development of better and more advanced cogeneration systems. The program considered large and small applications for a wide range of industrial applications. Industries included wood products, and pulp and paper. Conversion system technologies included externally fired gas turbines, steam turbines, diesels, and fluidized-bed combustion. Coal, wood waste, biomass waste, and residual oil were considered. No further results were reported due to organizational difficulties in DOE.

Other Cogeneration R&D Efforts in DOE. Other R&D efforts, not mentioned above, are summarized as follows:

- Integrated Industry Cogeneration Program (IICP) consisting of five basic elements: research, component development, program integration and system analysis, demonstration, and technology transfer.

- Integrated Community Energy System (ICES) Program to develop community-scale cogeneration systems in residential/commercial/institutional sectors.
- Integrated Coal Conversion and Utilization Systems (ICCUS) to increase the use of coal in baseload electrical generation power plants through more efficient integrated gas turbines such as fluidized-bed combustion and coal gasification.
- Advanced Cogeneration Systems (ACS) to improve fuel utilization and efficiency through advanced engines such as Stirling engines and direct-fired gas turbines.
- Cogeneration Technology Alternatives Study (CTAS) to identify opportunities for cogeneration in six energy-intensive industries.[15]
- Total Energy Technology Assessment Study (TETAS) to define the most promising advanced cogeneration systems for commercialization in the residential/commercial/industrial sectors.

R&D Efforts at the Electric Power Research Institute

There are four research projects in congeneration at EPRI, all in software areas. They are summarized as follows:

- Assessment of Dual Energy Use System (DEUS) fuel cell application completed in February 1979, report no. EM381.
- Evaluation of alternative technologies for DEUS, to identify the technical, economic, and institutional factors determining the feasibility of DEUS; to develop a methodology with which the utility industry can evaluate DEUS; to identify the preferred generic DEUS; and to identify the R&D needed to further characterize and implement beneficial applications of DEUS.
- Inclusion of cogeneration in electric utility models, to provide methods for assessing the economic impact of cogeneration (dual energy use systems) and to include cogeneration in electricity forecasting models.
- Forecasting in-plant cogeneration in the industrial sector, to develop behavioral models that can be used to produce conditional forecasts of the share of electrical energy produced in-plant for industry.

SUMMARY

This chapter presents the major issues of cogeneration: concepts, technologies, potential impacts, pricing and costing, legislation, applications, and R&D

efforts. Rates for cogeneration electricity, fuel availability, electricity demand profile, industrial steam need, and environmental constraints will determine the future of cogeneration development.

Evidently more cogeneration efforts will be encouraged by public utility commissions and other state energy planners in the near future. In advocating cogeneration, attention must be paid to the following:

- The adverse impact of increasing oil and gas use due to cogeneration. Will this be in the interest of national energy and security?
- The benefit to the rate payer. Will the rate payer benefit from cogeneration? Or will the cogenerators get all the benefit of cogeneration?
- The balance between energy development and environmental protection. Undoubtedly, some environmental regulation will be relaxed in order to encourage cogeneration development. It is important to maintain a balance between these two national goals.

REFERENCES

1. Department of Energy. "Cogeneration: Technical Concepts, Trends, Prospects," DOE-FFU-1703, Sept. 1978.
2. Dow Chemical Company. "Energy Industrial Center Study," prepared for the National Science Foundation, PB-243 823, June 1975.
3. Edison Electric Institute (EEI). *Statistical Yearbook: Historical Series.* Washington, D.C.: EEI.
4. International Federation of Industrial Cogenerators. 1972 report.
5. Streb, A. and Harvey, D. "Industrial Cogeneration: An Assessment of System Alternatives," in *Workshop Proceedings: Dual Energy Use Systems,* EPRI EM-718-W, May 1978.
6. Resource Planning Associates, Inc. (RPA). "The Potential for Cogeneration Development in Six Major Industries by CR–04–60172–00, December 1977.
7. Electric Power Research Institute. "Open-Cycle MHD Systems Analysis," EPRI, AF-1230, November 1979.
8. Davis, H. et al. "Potential for Cogeneration of Heat and Electricity in California Industry—Phase II," final report, prepared for U.S. Department of Energy by Jet Propulsion Laboratory, January 1979.
9. U.S. General Accounting Office. "Industrial Cogeneration—What It Is, How It Works, Its Potential," EMD–80–7, April 1980.
10. Icerman, L. and Staples, D. "Industrial Cogeneration Problems and Promise," *Energy,* Vol. 4, 101–117, 1979.
11. Synergic Resources Corp. "Industrial Cogeneration Case Studies," final report to Electric Power Research Institute, EPRI EM-1531, April 1980.
12. Electric Power Research Institute. "Fuel Cell Power Plants for Dispersed Generation," EPRI TS-1/54321, May 1979.

13. Electric Power Research Institute. "Technical Assessment Guide," EPRI PS-1201-SR, July 1979.
14. Department of Energy. "Industrial Cogeneration Optimization Program," prepared by TRW, Inc., DOE/CS/4300–1, January 1980.
15. General Electric Company. "Cogeneration Technology Alternatives Study," 6 vols., prepared for NASA and DOE, DOE/NASA/0031–80–1 to 6, January 1980.
16. United Technologies Corporation. "Cogeneration Technology Alternatives Study," Vol. 1 (Summary Report), prepared for NASA and DOE, DOE/NASA/0030–80/1, January 1980.

14
Cogeneration: Power Plant Reject Heat Utilization

INTRODUCTION

Reject heat utilization, a form of cogeneration, is an application of waste heat recovery and utilization. Various applications of waste heat recovery and utilization were discussed in Chapter 12, and in-plant cogeneration was detailed in Chapter 13. The technical, economic, environmental, and institutional aspects of reject heat utilization are examined in this chapter.

Reject heat utilization is a concept of combining power generation and heat generation at the power plant site, in contrast to in-plant power generation. We may call it "in-plant heat production." Thus, principles of cost allocation between electricity and steam discussed in Chapter 13 for in-plant power generation can be applied to in-plant heat production.

This chapter considers the source of reject heat and methods of utilization, including district heating, agricultural use of reject heat, aquacultural use of reject heat, and integrated energy systems; legislative, environmental, and institutional constraints for reject heat utilization; and current research and development reject heat utilization.

SOURCE OF REJECT HEAT AND METHODS OF UTILIZATION

The electric power industry consumes about 25 quadrillion Btu (quad) in 1980, approximately 30% of the total U.S. energy consumption. Of the total

25 quad, 77% is in fossil fuel, and 11% is in nuclear energy. Thermal efficiencies of power generation plants range from about 33% to 40%. Older fossil-fueled plants are typically 33% efficient, with recent systems improving to about 40% (Fig. 14–1, upper chart). Light water reactor (nuclear) systems are generally 33 to 35% efficient (Fig. 14–1, lower chart). The remaining 60 to 67% of the energy consumed is dissipated to the environment as reject heat. In other words, U.S. thermal power plants release about 15 quad heat, which is about 20% of the total U.S. energy consumption. However, this energy is low-grade and is diluted in large volumes of water.

Energy Cycle—Coal-Fired Steam-Electric Plant

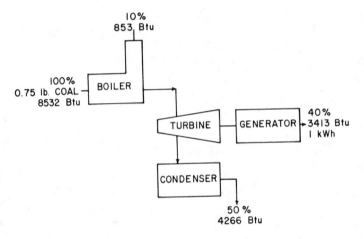

Energy Cycle—Nuclear Steam-Electric Plant

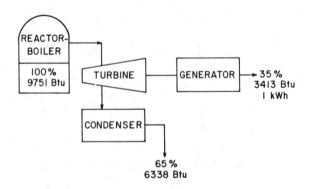

Fig. 14–1.   Thermal efficiencies of coal and nuclear power plants. Source: EPRI.[1]

A fossil-fueled plant rejects about 50% of the energy input through cooling condensers, and a nuclear-powered plant rejects about 65%. Waste heat is herein defined as the heat energy contained in power plant condenser cooling water.

This low-grade thermal energy is dissipated to the environment through two basic cooling systems, the "once-through" and the "closed-cycle" condenser cooling systems (Fig. 14–2). In a once-through system, water is pumped from a large water source (i.e., a river, lake, or ocean), transported through the condensers to condense the turbine exhaust steam, and returned to the source. In closed-cycle systems, the water is cooled onsite and reused through the condensers. Blowdown is used to maintain acceptable water quality. Cooling canals or ponds, or cooling towers can be used for this purpose.

This low-grade thermal energy can be effectively utilized through dual energy use systems, which offer the advantage of combining power generation and heat production simultaneously in one facility. The main problems in using reject heat from large power stations for dual energy use systems arise from the reduction in availability of generation capacity and from the need to transmit large amounts of heat over relatively long distances.

Currently, in the majority of research and commercial reject heat projects, condenser cooling water from fossil-fueled open-mode power plants has been used directly in aquacultural systems to produce organisms. The four most significant applications, including aquaculture, are as follows:

- District heating
- Agricultural use of reject heat
  - Greenhouses

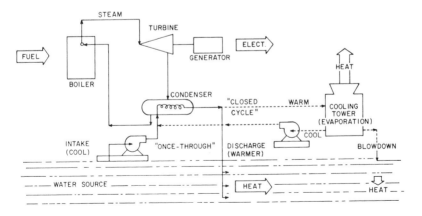

Fig. 14–2. Energy cycle including condensing water loop. Source: EPRI.[1]

- Soil warming, frost protection, and irrigation
- Reclamation of livestock waste
- Crop drying, food processing, and livestock housing
- Aquacultural use of reject heat
- Integrated energy systems

District Heating

District heating (DH) started in 1877 in Lockport, New York, when Birdsill Holly laid a continuous 700-foot pipe underground and improvised a small boiler in his basement for steam. By 1881 it was a growing industry.

DH systems in the United States and Canada are generally located in the large cities; were started by, and subsidized in their greatest growth period by combined electric and steam operations; and have drifted from what is currently termed "cogeneration" into separate energy supply facilities. In most areas, district steam systems and electric supply systems are in competition.

The growth of district heating has always been modest in comparison to that of electric or natural gas heating. Although in-plant cogeneration has been applied successfully in U.S. industrial plants, district heating is negligible and continues to decline. In contrast, combined heat and power generation (CHP) by public utilities is a large, successful, expanding enterprise in many European countries. It supplies DH service by means of circulating hot water, along with electric service.

Some of the factors that cause Europe in particular to surpass the United States in the application of district heating are the trend in Europe toward clusters of factories and residences, the absence of air conditioning and summer peaking of electric plants, government ownership or control of the utilities which often are not profit-oriented, and the absence for many years of low-cost gas and oil.

Benefits of District Heating. Development of feasible CHP/DH technologies for the United States is highly worthwhile. Fuel utilization efficiencies of over 80% are achievable, compared to 38% for electric generation alone; and thermal discharges into waterways are reduced or eliminated. Other benefits of district heating are as follows:

- It increases fossil fuel economy and reduces air pollution, because of substitution of central stations for many small inefficient heating plants with low chimneys.
- Conversion to coal is feasible, and fuel transportation by truck is reduced.
- Space is saved in buildings, maintenance and operating costs of buildings and fire hazards are reduced, and service reliability is improved.

Industrial Statistics. The International District Heating Association (IDHA) reports statistics on the operation and utilization of distribution systems of hot and chilled water. About 40 district heating utilities report to the IDHA. Some important points are summarized as follows:

- The total steam sold in 1976 by district heating systems was about 3×10^{12} pounds. Of this total system, 39.4% was supplied from topping turbines. Primary fuels used for generation were as follows: 62.9% from oil, 19% from gas, 15.2% from coal, and 1% from solid waste.
- District heating is highly concentrated in a few utilities; one company accounts for more than 40% of all steam sales (Consolidated Edison), about 15 others comprise another 40%, and the remainder is spread among 25 or more small companies.
- Universities were the major users of district heat. Including estimates for heat use not reported to the IDHA, district heating would barely approach 1% of the U.S. and Canadian energy use.

A mail survey of 59 district heating electric utilities conducted for EPRI[5] revealed almost the same results. Some of these results are shown in Table 14–1.

Economics of District Heating. The above EPRI survey indicated that average steam price ranged from $0.63 per thousand pounds (San Diego) to $9.25 per thousand pounds (Pittsburgh), and the mean steam price for all respondents was $5.07 per thousand pounds. Of the utilities that sold steam, 64% sold it at prices ranging from $4 to $7 per thousand pounds.

Estimates of the cost to extend distribution system piping, as provided by the utilities, range from $17 per foot ($56 per meter) for a 6-inch (15-centimeter) pipe to $1280 per foot ($4200 per meter) for a 24-inch (61-centimeter) pipe.

Two Methods For District Heating. Depending on power plants, there are two methods for district heating: district heating using reject heat from closed cycle turbines and block heating power plants.

District Heating Using Reject Heat from Closed-cycle Turbines. The closed-cycle turbine coupled with either a nuclear or coal heat supply is one of several alternative systems for district heating:

- Fluidized-bed system for coal-fired power plants (see Chapter 13)
- Gas turbine high temperature gas reactor (GT-HTGR) for nuclear power plants (see below)

Table 14–1.   Summary of Mail Survey Results.

Number of utilities receiving questionnaire	59
Number of utilities responding	50
Number of utilities providing services other than heat and electricity	
Gas	36%
Other	28%
Utility type (percent of utilities responding)	
Investor-owner	62%
Municipal	38%
Percent of utilities with cogeneration	50%
Percent of steam produced through cogeneration	43%
Distribution system type	
Condensate return (steam)	60%
No condensate return (steam)	36%
Other	4%
Range of steam sendout pressures	5–400 psig
Average steam sendout per utility	1.841 billion lb/yr
Average load factor	33%
Average gross revenue per utility from steam	$10.6 billion
Average number of customers per utility	344
Total fuels used (billion Btu/yr)	
Oil	95,251 (59%)
Coal	37,945 (24%)
Gas	25,810 (16%)
Other	1,466 (1%)
Total	160,472 (100%)
Average selling price increase (1973–1978)	125%
Potential for expansion (percent of utilities responding)	
Utilities with excess distribution capacity	68%
Utilities with excess generation capacity	68%
Utilities with both excess distribution and excess generation capacity	60%

Source: EPRI.[5]

Figure 14–3 shows the cycle diagram for a 3000-MW(t) GT-HTGR with a dry cooling tower and heat rejection adapted to district heating. As indicated in the figure, the predicted electricity production is 1200 MW(e) for a 3000-MW(t) reactor. The thermal efficiency of electricity conversion is 40% with present HTGRs. Higher efficiencies are possible with higher helium temperatures. The incorporation of district heating with a water supply temperature of 350°F decreases the heat rejection to the sink by a factor of 4.7, or from 1800 MW(t) to 384 MW(t). The utilization of the reactor heat is then 87% when the electricity and district heat are both used to their maximum capability, 1200 MW(e) of electrical power and 1416 MW(t) of heat.

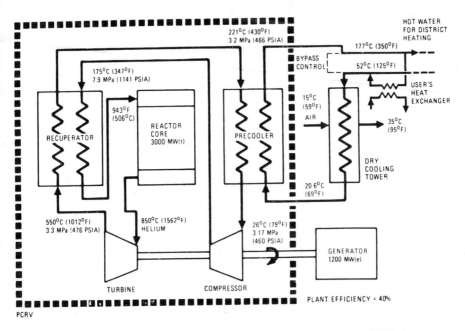

Fig. 14–3. GT-HTGR cycle diagram with district heating. Source: EPRI.[1]

A GT-HTGR with dry cooling has a 14% power generating cost advantage over an LWR with wet cooling towers and a 24% advantage when both are functioning with dry towers. There is no penalty in capacity or fuel cost in providing waste heat for district heating, and it is an economical means for providing electrical power. The only disadvantage is the possibility of radiation contamination of the water. This environmental problem must be handled very carefully.

The most current benefit and cost data for district heating using reject heat from closed-cycle gas turbines are from a study done by S. Bros. for the Swiss Federal Bureau of Energy in 1970.[2] He compared seven alternative systems: central heating station, extraction–condensing steam turbine station, back pressure turbine station, nuclear power station, open-cycle gas turbine power station, closed-cycle gas turbine power station, and large central thermal power station. The first is a heating-only system; the remaining six are combined power and heating systems.

By using 1970 cost data and a thermal output of 116 MW(t), S. Bros. compared the costs of the seven systems as shown in Figs. 14–4 and 14–5. The figures indicate that the large thermal power station plant was the only favorable system compared with the same size heating-only station in 1970.

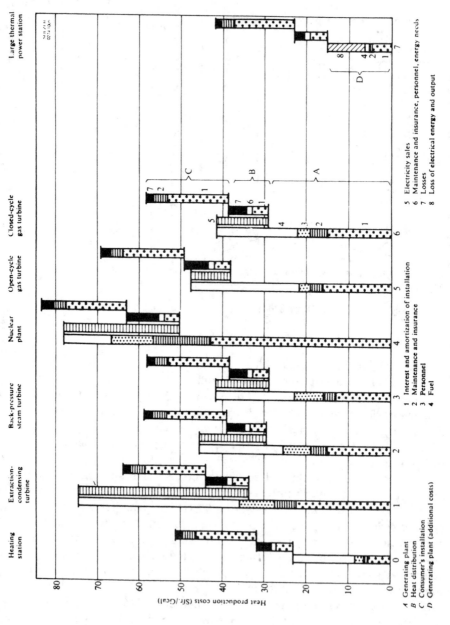

Fig. 14-4. Breakdown of overall production costs for district heating systems. Source: EPRI.[1]

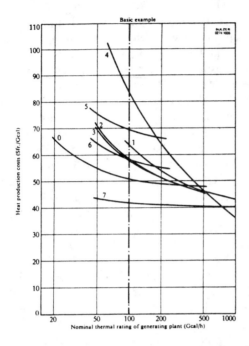

Fig. 14-5. Overall heat production costs for plant configurations 0 to 7 (as shown in Fig. 14-4). Variation of the nominal thermal load for the basic example. Source: EPRI.[1]

The cost advantage of a coal-fired large power cycle system would be greater than others indicated in this study.[3]

Block Heating Power Plants. The block heating power plant represents a new concept, the intent of which is to use power plant reject heat and to avoid the difficulty of district heating transportation and transmission.

An experimental block heating power plant, the first of its kind, was constructed to replace an existing oil-fired heat production plant to heat 283 apartment units in an area approximately 350 × 250 meters, adjacent to a municipal electric utility.[1]

The design capacity of the block heating power plant was based on the average annual load curve for heat consumption. The plant was designed to cover approximately 50% of the maximum heat demand of the consumer, and supplies about 90% of the annual heat energy production. The rest of the heat production is provided by a peaking boiler and a heat storage device, which can also be used to level the daily capacity fluctuations. Natural gas was used as primary energy. Heat and electricity in the block heating power plant can be decoupled temporarily (i.e., there is a possibility of a temporary supply of electric power independent of the required heat demand).

Noise insulation for the motor noise level (between 105 and 112 db(A)) was emphasized to ensure that the noise level is kept at 35 db(A) at a distance of 1 meter from the immediately adjacent houses.

Installation of the block heating power plant is shown in Fig. 14–6. The installation was compact. The cooling water of the gas otto motor (943 hp, mechanical power) is led over two cooling cylinders (connected in series) and a heat exchanger. In the first cooling cylinder, the exhausts of the motor are cooled from approximately 540°C to 250°C. In the second cylinder the exhausts are cooled from 250°C to 105°C; the cooling water is thus heated up to 70–90°C, according to the selected water flow. Two heat storage devices are connected to the same circuit. The secondary side of the heat exchanger is connected to the "heat bus," which supplies heat to, and receives returned hot water from the consumer.

The peaking boiler also supplies this "heat bus." Figure 14–7 shows the ground plan for the combined heat and power plant. The motor generator block is installed in a noise-insulated case that is accessible from all sides by doors. The motor is a gas otto motor with 16 cylinders in a V-shape and 900–1300 rpm.

The cost of first-generation block heating is much higher than that of closed-cycle turbine systems; however, it can be reduced to the levels of turbine and steam heating power plants with the use of intermediate reheater and fresh-water cooling if standardized components are used.

Problems in District Heating. District heating requires a large transportation and transmission system from the power generating stations, and it is generally agreed that the system must be almost universally accepted if it is to be economical. District heating has found application in northern Europe where the heating load is high and there is an absence of high-quality low-cost fossil fuels such as natural gas and oil.

Since the transmission and distribution piping systems require a large capital investment, their cost optimization is important. Experiments conducted for the city of Philadelphia indicated that the percentage cost of the transmission portion of the system is reduced from 34.3% to 18.9% as the temperature difference between heat supply and heat return is increased by a factor of 3.[4]

Agricultural Use of Reject Heat

Warm water discharged from power plants is available throughout the United States, as shown in Fig. 14–8. However, past opportunities for utilizing reject

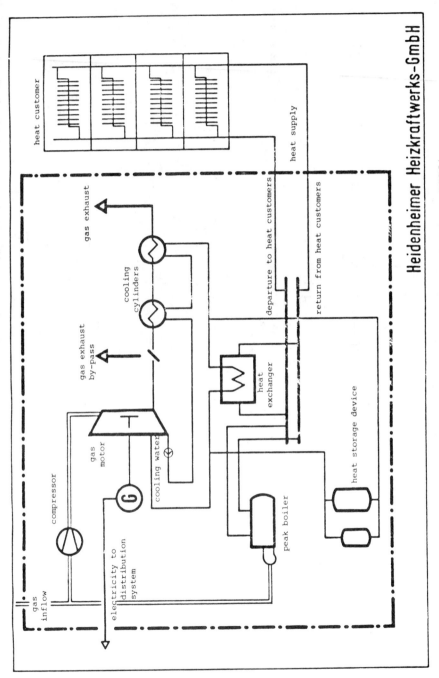

Fig. 14–6. Schematic illustration of block heating power plant. Source: EPRI.[1]

Heidenheimer Heizkraftwerks-GmbH

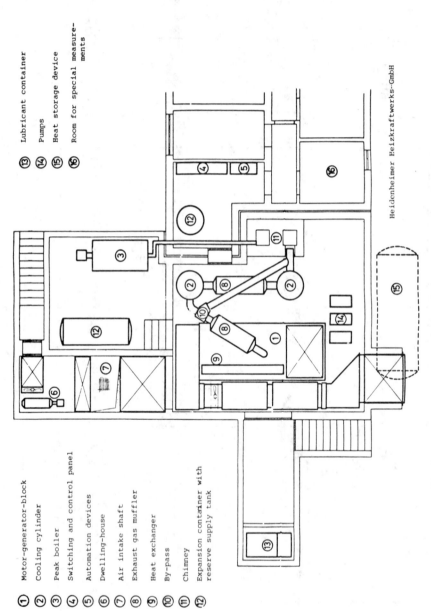

① Motor-generator-block
② Cooling cylinder
③ Peak boiler
④ Switching and control panel
⑤ Automation devices
⑥ Dwelling-house
⑦ Air intake shaft
⑧ Exhaust gas muffler
⑨ Heat exchanger
⑩ By-pass
⑪ Chimney
⑫ Expansion container with
    reserve supply tank

⑬ Lubricant container
⑭ Pumps
⑮ Heat storage device
⑯ Room for special measure-
    ments

Heidenheimer Heizkraftwerks-GmbH

Fig. 14–7.   Ground plan of the block heating power plant. Source: EPRI.[1]

Fig. 14—8.   Electrical generating capacity in megawatts for 1975 by county (based on FPC data). Source: EPRI.[1]

heat were not significantly realized, the reason being that, in the past, fuel was relatively inexpensive, power station sites were remote, utilization costs such as steam pipeline costs were excessive, markets for reject heat were small, and there were no convincing demonstrations of reject heat utilization.

As shown in Tables 14–2 and 14–3, reject heat can be used for heating enclosed animal and plant growing areas and for aquaculture if its benefits significantly exceed transmission costs for low-temperature hot water (Fig. 14–9).

Major agricultural applications of reject heat are in: greenhouses; soil warming, frost protection, and irrigation; biological recycling; and crop drying, food processing, and livestock housing. These application are described in the following pages.

Table 14–2. Minimum (January) Temperatures of Condenser Outlet Water at Several Locations.

Location	Utility	Condenser Outlet Temperature (°C) Once through C_1 Tower	
Minnesota Sherco Station	Northern States Power		30
Tennessee Watts Bar	Tenn. Valley Authority		40
North Carolina Keowee Station	Duke Power	18	
Alabama Browns Ferry Station	Tenn. Valley Authority	21	43

Source: EPRI.[1]

Table 14–3. Optimum Growth Temperatures for Several Plants and Animals.

Plant or Animal	Optimum Temperature (°C)
Lettuce	13–17
Tomatoes	18–22
Cucumbers	26–30
Chickens (broilers)	13–18
Swine	18–24
Trout	12–15
Shrimp	25–32
Catfish	28–32

Source: EPRI.[1]

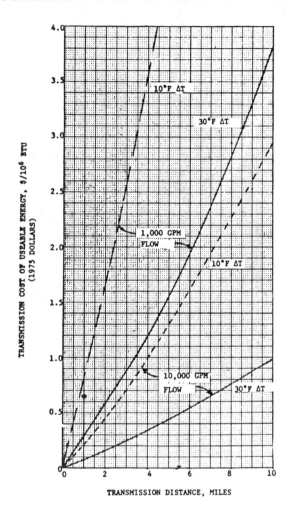

Fig. 14–9.   Estimated costs for low-temperature hot water transmission versus distance (single-pipe system, 1975 dollars, maximum $\Delta P = 250$ psi). Source: EPRI.[5]

Greenhouses.   The experimental use of power plant condenser cooling water in environmental control systems for greenhouses has been attempted by a number of utilities. There are four advantages to this application: (1) Greenhouses can be used to supplement or reduce cooling towers or cooling ponds at some power plants; (2) optimum greenhouse temperature (50–65°F) are in a range that can be theoretically maintained with condenser effluent, even from open-cycle plants with condenser discharges 11.1°C (20°F) or more above natural water temperatures; (3) an alternate low-cost energy source is needed for greenhouses owing to the large amount of energy they consume; and (4) use of a significant amount of reject energy in greenhouses will improve power plant energy efficiencies.

Because of the large land requirement and the cost of constructing and operating such greenhouse complexes, it is doubtful that green houses can be justified on the basis of cooling condenser discharge. The major cost of greenhouse construction would have to be justified on the market potential for greenhouse crops. Unless vegetable production in large waste-heat greenhouse complexes can become more competitive with field-grown crops, there will probably be insufficient demand in the near future to justify construction of complexes large enough to dissipate reject heat from power plants.

Technologies for Using Reject Heat in Greenhouses. There are three practical heat-exchanging systems for transferring reject heat from power plants to greenhouses: (1) the direct contact system, (2) the forced air system, and (3) the rock floor/porous concrete system.

The direct-contact system performed satisfactorily in Oak Ridge National Laboratory (ORNL) and Tennessee Valley Authority (TVA) experiments, maintaining average greenhouse air temperatures of 14.4°C (58°F) with 21°C (70°F) water when ambient temperatures were as low as −19.5°C (−3°F). High humidity associated with this method aggravated production problems and necessitated close attention to disease prevention and control programs. However, with proper management, yields comparable to those produced in conventional greenhouses were achieved. The initial system cost and electrical requirements for fans were higher than those for conventional systems, but potential fuel savings indicate that the system tested by TVA would have an overall cost advantage.

The forced air system demonstrated in the Northern States Power Company's Sherco greenhouse in Minnesota was shown to be capable of maintaining a temperature of 14.4°C (58°F) when outside air temperature was −41.4°C (−42.6°F), and 32.2°C (90°F) water was used. The primary operational problem was fouling of the heat exchangers caused by deposits of silt and organic material on the pipe and heat exchanger walls. Chlorination and acid cleaning were used to alleviate this problem. Although this system required a higher initial investment and more fan power than conventional heating systems require to obtain adequate heat transfer, it offers significant potential for reducing fuel costs.

A rock floor/porous concrete system, developed by Rutgers University, to be tested by the Public Service Electric and Gas Company, may have an advantage over the above systems because it does not require expensive high-volume air-moving equipment. This system, as do the others, needs to be more thoroughly evaluated in order to establish capabilities and operational limitations. A potential problem that must be addressed in this system is contamination of condenser cooling water by pesticides and fertilizer nutrients in irrigation drainage.

Eight major R&D projects in using reject heat in greenhouses are summarized in the following paragraphs.

Some of the earliest work in the United States to indicate that reject heat from power plants might be used for environmental control of greenhouses was begun in the 1960s when the University of Arizona Environmental Research Laboratory initiated research to develop an integrated system for providing power, water, and food for coastal desert areas.[6] The system was devised to generate electricity, evaporate and condense pure water from seawater, and produce food crops in controlled-environment facilities. This system were subsequently demonstrated in research facilities in Tucson, Arizona; Penasco, Mexico; and Abu Dhabi.

The purpose of an ORNL project in 1968 was to study possible urban applications of reject heat from power plants.[7] One application was to heat greenhouses.

TVA began to evaluate the potential for large-scale greenhouses in the TVA region in 1968, started construction in 1972, and completed the pilot greenhouse at the National Fertilizer Development Center in 1973.[8] Costs and benefits of this project will be discussed in the next section.

The Eugene, Oregon thermal water utilization project was initiated in 1968 to take warm water from the cooling condensers of the electrical generating plant at the Weyerhaeuser Company's Springfield Mill.[9] A plastic greenhouse was constructed in 1972. Crop production was not good when outside weather was extremely cold.

As shown in Fig. 14–10, the Northern State Power Company (NSPC)/ University of Minnesota (UOM) Greenhouse Project is the nation's largest experimental greenhouse using reject heat from a power plant. In this project, NSPC and UOM attempted to investigate ways to use reject heat in condensers, and to demonstrate that the reject heat can be used to grow top-quality vegetables and flowers during the harsh Minnesota winters. If the demonstration proves economically feasible, private commercial growers could begin operating their own facilities at the Sherco plant site. NSP would sell heat from its warm water and lease land to those growers. The study showed that water temperatures from "once-through cooling plants" were too low in Minnesota for most potential beneficial uses, but that closed-cycle plants at temperatures of 29.4°C (85°F) or above did have potential for development.[10,11]

The Public Service Electric and Gas Company (PSE&G)/Rutgers University project (1977) involved construction of a small greenhouse at PSE&G's Mercer generating station near Trenton, New Jersey to evaluate the use of

Fig. 14–10. Northern States Power Company's new Sherburn County power plant (Sherco) is producing more than electricity; it is also demonstrating that the waste heat produced during the power generation process can be used to grow top-quality vegetables and flowers during the winter. This is the nation's largest experimental greenhouse using reject heat from a power plant. Source: DOE photo.

reject heat from the plant.[12] Since a solar greenhouse heating system and a porous-capped concrete floor were used, the greenhouse used less electricity for the fans in the heat exchangers than did TVA and NSP greenhouses.

Two polyethylene-glazed greenhouses were constructed in 1976 at the Vermilion Power Plant for a University of Illinois project. One was heated by spraying power plant cooling water over the outside surface, and the other was unheated. Warm water for the heating experiment was obtained from a cooling tower supply line. Data collected were used to develop equations relating greenhouse, heated water, and outside ambient temperatures.[13]

Two plastic greenhouses were constructed in 1977 for the Fort Valley State Project in Georgia. The purpose was to compare the performance of a waste heat system with that of a conventional heating system.[14] The project is still in progress.

Economics of Using Reject Heat in Greenhouses. The cost-effectiveness of using reject heat in greenhouses is location- and power-plant-specific. The TVA reject heat greenhouse project is analyzed below in order to yield some insights because this project has the most complete data available.

The structure of the TVA reject heat greenhouse is shown in Fig. 14–11. The greenhouse was a 24 × 100-foot glass-glazed structure, equipped with a combination evaporative pad and finned-tube heat exchange system, a fiberglass attic and motorized shutters for air recirculation, attic vent fans,

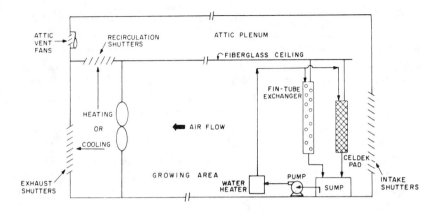

Fig. 14–11. Schematic drawing of reject heat research greenhouse, Muscle Shoals, Alabama.
Source: EPRI.[5]

and a water boiler to provide simulated power plant cooling water. The
evaporative pad system was used to accomplish both heating and cooling.
The heating and cooling system was designed to use 21.1°C (70°F) minimum
water to maintain a minimum greenhouse temperature of 12.8°C (55°F) at
an outside minimum of −8.9°C (16°F).

Cost Comparisons. Initial capital investment costs are compared for the
reject heat and conventional greenhouses in Table 14–4. The table shows

Table 14–4. Initial Capital Investment Costs Comparison of Components that are Different for the Muscle Shoals Waste Heat Greenhouse (223 m²) and a Conventional Greenhouse, 1975.

Item*	Waste Heat	Conventional
CELdek system	$ 1,573	$ 975
Conventional heating system	—	1,353
Fin-tube heater and piping	650	—
Fiberglass attic and recirculation chamber	3,250	—
Exhaust fans and shutters	3,627	1,779
Extra space required	1,200	—
Extra doors	195	—
Total	$10,495	$4,107

* Installation cost was assumed to be about 30% of materials cost. Projecting the savings to a
1-ha (2.5-ac.) house shows an annual cost advantage of about $44,000 and conservation of over
823 m³ (29,071 ft³) of fuel even after adjustment is made for the extra electricity used by the
waste heat system.

Source: EPRI.[5]

that the initial investment cost for the reject heat greenhouse is about $6,388 more than that for the conventional one. However, as shown in Table 14–5, the annual costs were $980 less for the waste heat greenhouse as a result of fuel savings. When fuel prices become higher, the difference will become greater. Note that much of the extra capital cost for the waste heat system was associated with the recirculation attic, exhaust fans, and shutter system. No charge was made for the warm water in these calculations.

Soil Warming, Frost Protection, and Irrigation.   Soil heating, frost protection, and irrigation are among the beneficial uses of power plant condenser cooling water. Ideally, these practices would result in economic dissipation of power plant waste heat so that the undesirable environmental effects of cooling towers and ponds could be reduced.

Soil Warming.   Thermal characteristics of soils, including specific heat, conductivity, and effects of soil moisture on heat transfer, have been found to be similar. Most differences among soils are due to variations in organic matter content, moisture levels, bulk density, and porosity.[14] Heat conductiv-

Table 14–5.   Estimated Annual Cost Comparisons of Items that are Different in the Muscle Shoals Waste Heat Greenhouse (223 m²) and a Conventional Greenhouse of Similar Size, 1975.

Item	Cost Per Year ($)		
	Waste Heat System	Conventional System	Compared with Conventional
Initial capital (amortized at 9% for estimated life of item)*	1,418	618	+800
Operating costs: Fuel (LP gas at $100/m³)	—	2,000	−2,000
Electricity (2.5¢/kWh)	466	316	+150
Maintenance (1 or 5%)**	220	150	+ 70
Total cost	2,104	3,084	−980

* Ten years was used for estimated life of CELdek, pumps, and motors; 5 years was used for shutter motors; 15 years was used for fiberglass. Other major structure items were assumed to last 20 years. The salvage values were assumed to be zero.

** One percent of the item cost including installation was used for maintenance on structural items, and 5% was used for pumps, motors, etc.

Source: EPRI.[5]

ity seems to be the most important soil characteristic, since it determines the rate of lateral movement of heat away from pipes, to deep layers, and to the surface where it can be transferred to the air. It increases with bulk density and decreases with porosity. Moisture content is the primary factor causing variable conductivity among soils.

Effective heat dissipation or soil warming requires that the reject heat be brought into contact with the soil. Direct applications of hot water from power plants would result in a saturated soil before any meaningful amount of heat had been added. Pipe grids should be used to conduct heated water through the soil; polyethylene pipes would be most desirable for the lateral lines in a soil heating grid because of their low cost and flexibility.

The temperature profile in soils is the basic factor considered in pipe grid designs for heated water distribution. Computer models can be used to determine either the maximum acreage that could be heated from a given size power plant or the minimum acreage needed for successful heat dissipation.

Heat is radiated rather uniformly around a heat source in soil, and temperature drops as the distance from the heat source increases. Soil temperature decreases rather rapidly above the 90-cm (35.4-in) deep heat source, resulting in surface temperatures below the optimum for good response, particularly from young shallow-rooted plants.[15]

Fairly extensive research to measure crop responses to heated soil has been conducted in Oregon, Pennsylvania, Alabama, North Carolina, and Maryland. With the exception of the Maryland work, these experiments have been oriented specifically toward using waste heat from power plants as a crop production resource. Also, Germany and France recently started some extensive soil heating studies. The German study is a large-scale experiment of 25 hectares heated with power plant waste heat. No data are available yet. Results in Oregon, Pennsylvania, and Alabama are briefly discussed to exemplify crop responses on warmed soil.

Experiments to evaluate waste heat as a production were begun in 1969 at Corvallis, Oregon[16] and at Springfield, Oregon.[17] At Corvallis, heating was done with electrical cables with a capacity of 65.6 W per linear meter (68.2 Btu/ft) buried 92 cm (36 in) deep and 183 cm (72 in) apart. At Springfield, waste heat was distributed through a 6.25-cm (2.5-in)-diameter PVC pipe buried 60 cm (24 in) deep with 150-cm (60-in) spacing.

Tables 14-6 and 14-7 indicate various effects of soil heating on grain and vegetable yields. Examination of these results shows that the benefit of soil heating is not stable, and, in the case of cantaloupe, the soil heating reduced yields.

Research was begun in 1972 at The Pennsylvania State University to determine the feasibility of soil warming as an alternate power plant heat dissipation system. A computer model and an experimental 15- by 60-m (50- by 200-

Table 14–6. Effect of Soil Heating on Corn Grain and Silage Yields in Oregon.

Year	Plant Component*	Yield, Metric t/ha Unheated	Yield, Metric t/ha Heated	Percent Increase
1969	Total	12.3	17.9	45
	Grain	7.2	9.6	34
1970	Total	22.7	23.8	5
	Grain	11.7	13.7	17
1971	Total	14.3	19.3	35
	Grain	4.8	7.6	57
1972	Total	13.2	15.8	20
	Grain	7.6	9.2	22

* Total yields were on dry basis, while grain yields were calculated at 15.5% moisture.
Source: EPRI.[5]

ft) heated field plot were used to calculate that a 1500-MW(e) power plant in Pennsylvania would require 1820 ha (4500 ac.) of land for year-round heat dissipation using 5-cm (2-in) pipe buried 30 cm (12 in) deep and 60 cm (24 in) apart.

Experiments showed that soil warming induced earlier spring growth and hastened maturity. However, in some instances, the very early new growth promoted by soil warming was killed by subsequent frost. Soil warming delayed development of cold tolerance in alfalfa, wheat, and barley in the fall and hastened loss of cold tolerance in the spring.

As shown in Table 14-8, soil warming reduced crop yield for alfalfa, orchard grass, and tall fescue by 12 to 45%, and increased the yield for winter wheat straw and sudangrass by 21 to144%.

In 1970, soil heating investigations were begun at Muscle Shoals, Alabama.[18] Electric cables were used as the heat source, buried about 30

Table 14–7. Effect of Soil Heating on Vegetable Yields at Springfield, Oregon.

Crop	Yield, Metric t/ha Unheated	Yield, Metric t/ha Heated	Percent Increase
Tomato (avg. of 2 var.)	136.2	136.2	0
Sweet corn (ears)	13.0	16.6	28
Cantaloupe	10.5	5.3	−49
Acorn squash	55.7	62.9	13

Source: EPRI.[5]

Table 14–8.   Soil Temperature Effects on Crop Yield.

Crop	Yield, Metric t/ha		Percent Increase
	Heated	Unheated	
Alfalfa	9.6	10.9	−11.9
Orchard grass	7.3	11.7	−37.6
Tall fescue	5.7	10.4	−45.2
Winter wheat grain	2.5	1.4	78.6
Straw	4.6	3.8	21.1
Sudangrass	3.9	1.6	143.8

Source: EPRI.[5]

cm (12 in) apart and 25 to 30 cm (10 to 12 in) deep and providing 110 W of heating capacity/m^2 (10 W/ft^2). In some experiments, an irrigation variable was included. Attempts were made to grow vegetables from early April to late November.

In 1971 and 1972, sweet corn, green beans, and summer squash were planted in early April when ambient soil temperatures were below 15°C (59°F). Table 14–9 shows that the effects of heat or irrigation alone were similar in magnitude, and that the two inputs when combined resulted in total yield increases of 131% with squash, 178% with beans, and 268% with sweet corn. Responses to heat were more significant with corn and beans than with squash, because the squash continued to produce fruit throughout the midsummer period when ambient soil temperatures were sufficient for maximum growth.

The three most common Tennessee Valley field crops—cotton, soybeans, and corn—were grown for two years on soil warmed to 29°C (84°F). Of

Table 14–9.   Effect of Soil Heating on Vegetable Yields at Muscle Shoals, Alabama, 1971–72.

Treatment	2-Year Average Yield in Metric t/ha and % Increase Due to Treatment					
	Sweet Corn		Green Beans		Summer Squash	
	Yield	Increase %	Yield	Increase %	Yield	Increase %
Heat plus irrigation	19.5	268	18.6	178	68.6	131
Irrigation only	11.3	113	12.5	86	49.5	66
Heat only	11.6	118	12.7	90	41.2	39
No heat or irrigation	5.3	0	6.7	0	29.7	0

Source: EPRI.[5]

these, only cotton appeared to show a level of response that might be economically significant (Table 14–10).

Frost Protection. Sprinkler irrigation is commonly used to protect early-blooming fruit crops from late spring frosts. Since continuous sprinkling over the entire area is necessary during subfreezing periods, portable and center pivot systems are unsuitable. Berry and Miller[17] conducted a five-year experiment using thermal water for frost protection and irrigation at Springfield, Oregon. No frost damage occurred in the experiments. However, sprinkling for frost protection is not a beneficial use of water heat for power plants. Because the water cannot be collected and returned to the plant, water used must be replaced by makeup water from another source in closed-cycle plants.

Irrigation. Thermal water from power plants or other industrial sources has been suggested for irrigation. Both sprinkler and surface irrigation methods have been studied.[17-19] Although thermal water irrigation is technically feasible, it is not beneficial to power plant operation because it consumes water. Irrigation can be accomplished more economically by using conventional surface and underground water sources.

Concluding Comments. The prospect of using reject heat for soil warming, frost protection, and irrigation in the United States is not promising in the next 10 to 20 years. It is unlikely that yield increases in response to warm water will be sufficient to amortize the investment costs for water distribution systems and to pay for the loss of power plant water.

Biological Recycling (Reclamation of Livestock Waste). Biological recycling is defined here as an economic utilization of reject heat from power

Table 14–10. Effect of Soil Heating on Field Crop Yields at Muscle Shoals, Alabama, 1973.

| | Yield, kg/Plot | | |
| | --- | --- | --- |
Crop	Unheated	Heated	% Increase
Corn stalks	58.1	67.5	16
Corn grain	71.7	74.6	4
Soybean stalks	42.7	46.2	8
Soybean grain	29.7	29.9	1
Seed cotton (Seed + lint)	8.6	12.3	43

Source: EPRI.[5]

plant condenser discharge water to enhance nutrient recycling and use livestock manures to grow aquatic and terrestrial plants. It can also include municipal wastewater treatments.

R&D Projects. Most knowledge of reject heat biological recycling was gained from experiments conducted on municipal wastewater. Four of the major reject heat biological recycling studies are discussed in the following paragraphs.

The objective of the Oregon State University swine waste management project was to determine the value of waste heat used in a swine waste management system for nutrient recycling.[20] The system relied on aerobic digestion and nutrient recovery to produce bacteria and algae in shallow fiberglass basins by using simulated waste heat. The simulated waste heat was provided to the digester and algal basins by electrical resistance heaters and transferred to the cultures with stainless steel heat exchangers. Chlorella Vulgaries 211/8K, a high-temperature tolerant strain of algae, was selected for growth in this facility.

This system demonstrated potential for pollution abatement, high-protein production, methane generation, and water conservation. However, mechanical harvesting of microbial biomass is energy-intensive and requires a significant capital investment. Additional operating experience is needed on a commercial scale, possibly as large as a 500-head swine facility.

The objective of the original Richmond poultry waste recycling project at the University of California was to develop a poultry waste recycling system; the chicken excreta were flushed down manure troughs into a sedimentation tank.[21] An integrated waste heat/waste treatment system was then proposed to include a total ponding area of 4046 ha (10,000 ac.)/1000 MW(e) to accomplish the cooling needs of the power plant and sewage treatment for a city of one million people. A closed-loop heat exchanger system would be required to transfer heat.

The supernatant from the primary sedimentation was pumped to an algal pond, and the solids were transferred to a heated anaerobic digester; the supernatant from the digester was discharged into the algal pond, and digester sludge was washed periodically to the environment; and pond effluent or supernatant from the algal separation process was pumped to the manure troughs for use as flushing water to complete the cycle. Dried algae were fed back as a supplement to the chickens' diet. Yields of algae (*Scenedesmus*) were projected to be equivalent to 68 to 90 metric t/yr/ha (30 to 40 t/yr/ ac.) on a dry-weight basis.

TVA's Division of Agriculture Development, located at the National Fertilizer Development Center at Muscle Shoals, Alabama, initiated a biological recycling project to:

- Develop and evaluate methods that use waste heat from power plant condensers and manures from confined livestock facilities to grow aquatic and terrestrial plants.
- Provide technical assistance to commercial users of reject heat.[22]

Three state-of-the-art systems of aquatic bio-recycling initially proposed were: (1) *Lemna minor* L. (or some other duckweed) consumed by grass carp; (2) algae consumed by a fresh-water shellfish, such as Asiatic clams; and (3) algae consumed by a phytoplanktivorous fish, such as the silver amur. The end product of aquatic culture (grass carp, Asiatic clam, or silver amur) would be used as livestock feed or protein feed supplement. Initial attempts to produce freshwater Asiatic clams with swine waste fertilization were not successful. Other species were tested with mixed results.

TVA plans to continue testing more plant and animal species for potential application in the waste recycling system. A heat exchanger system will separate the nuclear power plant cooling water from the culture water. Heat exchanger technology, power-plant-interfacing-land constraints, pumping cost, and product marketing are viewed as major hurdles to be overcome prior to commercial application.

The Kaplan Industries animal waste recycling and methane production project was cofunded by Kaplan Industries of Barton, Florida, and the U.S. Department of Energy. It was aimed at using animal (beef) wastes. Waste heat was used to improve the waste recovery efficiency.[23] Animal wastes fall through slotted floors onto a concrete floor and are pumped to an open holding tank. Vibrating screens separate undigested solids, waste feed, and hay. The solids are washed, treated with formalin, and stored in compost piles. These solids are used to make up 25 to 35% of the feed ration and take the place of cottonseed hulls or other roughage. The clarified liquid effluent from the wastewater stream is pumped to a series of holding ponds. Shallow ponds with mechanical aerators are operated as aerobic lagoons, and some aerobic ponds are stocked with tilapia, which are harvested periodically, taken to a fish processing plant, and sold as fillets. Fish processing waste is returned to Kaplan's rendering plant for feed processing into fish meal, which is recycled to the cattle ration.

A methane generator was constructed by Hamilton Standard, a division of United Technology of Connecticut; its methane production is projected to be capable of replacing 88% of Kaplan's 6250 gallons of fuel oil consumption per week. The undigested solids from the methane gas generator will be processed into animal feed. This is expected to yield $158,000 worth of methane gas and $35,000 worth of 23% protein animal feed annually. It will probably take 3½ years to recover investment costs.

Biological Recycling Potential. Biological recycling potential depends on three factors: (1) quality of plant nutrients in animal manures, (2) product (e.g., algae) potential, and (3) product market value.

Quality of Plant Nutrients in Animal Manures. Table 14–11 shows the major and minor plant nutrients in animal manures. Successful dilution of manures to crop-tolerant levels for plant or animal production generally results in nontoxic levels of heavy metals. Livestock waste treatment technology has developed sufficiently to permit on-farm management of nitrogen for optimum destruction or conservation.[24] For in-depth discussions of livestock waste management, the reader is referred to references 25 and 26.

Product Potential. The most common products of biological recycling are algae, aquatic macrophytes, duckweeds, methane, fish, and invertebrates.

The types of algae suitable for biological recycling in reject heat systems are *Scenedesmus obliquus, Chlorella pyrenoidosa, Anacystis nidulans,* and *Plectonema boyanum.* Temperature plays an important role in the respiration rates of bacteria and algae, as respiration quotients of waste-grown biomass are elevated by warm water. This temperature effect plays a major role in the production of CO_2 from organic substrates by the microbial population. Optimum temperatures for most algae are between 20°C and 25°C (68°F and 77°F), but thermophilic strains of *Chlorella* and *Anacystis* grow best at

Table 14–11. Kilograms of Organic and Major Fertilizer Elements in the Complete Animal Excrement Per 1000 kg Live Weight.

Manure Composition	Poultry	Swine	Beef
Wet manure	32,200	22,400	20,600
Nitrogen (N)	333	185	138
Phosphorus (P_2O_5)	253	110	41
Potassium (K_2O)	118	172	112

Grams of Secondary and Minor Elements in 1000 l of Fresh Animal Manures.

Livestock	Ca	Mg	S	Fe	Zn	B	Cu
Poultry	33,549	2,683	2,907	436	83.9	55.9	13.4
Swine	5,256	738	1,341	257	55.9	39.1	14.5
Beef	1,901	972	648	36.9	13.4	13.4	4.47

Source: EPRI.[5]

about 40°C (104°F). Temperature optima for biomass production can vary with light intensity and concentration of certain culture nutrients, and adaptation to different temperatures occurs in algal cultures. Photosynthesis of the algae cultures is affected by water depth and the nitrogen-to-phosphorus ratio.

Techniques for harvesting algae from wastewater facilities have been extensively researched. The following common processes are suggested by Middlebrooks et al.:[27]

- Centrifugation
- Microstraining
- Coagulation—flocculation
- In-pond removal of particulate matter
- Complete containment
- Biological disks, baffles, and raceways
- In-pond chemical precipitation of suspended matter
- Autoflocculation
- Biological harvesting
- Oxidation ditches
- Soil mantle disposal
- Dissolved air flotation
- Granular media filtration
- Intermittent sand filtration

If used as a food or as feed for livestock, the algae must be cleanly harvested, and must be relatively free of harmful material and foreign matter, such as sand and filtration aids that have been used for algae collection.

Vascular aquatic plants have been used in treating wastewater when land and climate are suitable.[28] Therefore, much information on nutrient uptake rates, productivity stands, and waste-grown aquatic plants is available; little information on thermal responses of various aquatic vascular plants to waste fertilization is available.

With regard to wastewater treatment methods, aquatic macrophytes are divided into four groups: water hyacinth, submerged plant, floating plant (duckweed), and emergent plant.[28] Composted aquatic plants are used for fertilizer and to improve soil texture.

Harvesting large volumes of aquatic waste weeds for forage processing or compositing presents a problem, and engineering designs for increasing mechanical harvesting rates are still evolving. Water weeds are bulky and heavy owing to their high water content, and any system for harvesting and processing must be mobile and waterborne. Standing crop, productivity, and yield data for four common species in natural ecosystems of the Southeast are shown in Table 14–12.

Table 14–12. Standing Crop, Productivity, and Yield Data for Four Species of Aquatic Vascular Plants.

Species	Standing Crop (t dry wt./ha)	Maximum Productivity (g dry wt./m²/day)	Maximum Yield (t/ha/yr)
Eichhornia crassipes (water hyacinth)	12.8	14.6	54.7
Justicia americana (water willow)	24.6	31.1	113.5
Alternanthera philoxeroides (alligator weed)	8.0	17.0	62.0
Typha latifolia (cattail)	15.3	52.6	192.0

Source: EPRI.[5]

Production of methane from wastes has been uneconomical because of the high capital cost of equipment and the availability of cheap energy sources. A combination of concerns over energy shortages and pollution control has refocused attention on this process of producing a clean-burning and transportable fuel—methane. Methane generation largely depends on operating temperature. The temperature optimum for a given waste should be established only by experimental methods. Table 14–13 shows same samples of methane production from various wastes.

Table 14–13. Bio-gas Production (60% CH_4 and 40% CO_2).

Animal	Volatile Solids kg/Animal/ Day	Probable Volatile Solids* Destruction %	Gas m³/ Day	Btu/Day**
Beef	2.7	45	0.85	18,000
Dairy	3.9	48	1.24	26,000
Poultry, layers	4.3	60	2.04	43,000
Poultry, broilers	5.4	60	2.60	55,000
Swine (growing— finishing)	2.2	50	0.82	17,400

* Percent destruction of volatile solids will vary, depending primarily upon detention time and digester temperature.

** Calculated at 17 Btu/m³ (heat content will vary depending upon quality of gas). For comparison, some other heating values are: gasoline, 454 Btu/m³; diesel fuel, 503 Btu/m³; natural gas, 24–28 Btu/m³; propane, 348 Btu/m³.

Source: EPRI.[5]

Jewell et al.[29] estimated the cost of generating methane on a 40-cow dairy as \$22 to \$80/10^6 kcal (\$5.54 to \$20.16/10^6 Btu, 1975 dollars assumed).

Because of seasonal variation of reject heat effluents, heat exchangers are needed in using reject heat for methane production.

Fish and other aquatic organisms show potential for biological recycling. The concept of using reject heat and livestock manures to produce aquatic products has been applied in only a limited sense in foreign countries.

Product Market Value. An important constraint on the economic viability of biological recycling is the acceptability of its products to the public. This problem has not been fully addressed and researched, even though various economic studies on the potential use of reject heat in commerical aquaculture have been completed.[30] The future of biological recycling depends on the progress of research in the area of product acceptance.

Crop Drying, Food Processing, and Livestock Housing. No commercial or research projects have been conducted on using actual or simulated power plant cooling water in crop drying and livestock housing. However, this application has been discussed with great interest.[31,36] More research is needed on reject heat exchange and economic feasibility before it can be commercialized.

Crop Drying. Fossil fuel is used for heated-air drying of agricultural products. Drying grains uses the largest portion of fossil fuel; drying tobacco, hay, and other nongrain crops also requires substantial amounts of fuel. Two grain-drying system are traditionally used: a high-temperature, high-speed process when grain is harvested; and a low-temperature, long-period process using a large volume of air to dry grain when it is held in storage bins.

A low-temperature grain-drying system can use either solar energy or waste heat. Temperatures that would be available with efficient heat exchange systems using waste heat would be comparable to those available with some solar drying systems. However, the use of waste heat would require drying facilities located near the waste heat source.

The amount of waste heat required for crop drying is insignificant; that is, the annual energy use for crop drying is about 100 trillion Btu, less than 1% of the waste heat now produced.[32] Also, use of waste heat for grain drying would be seasonal, and heat exchange systems would be idle for most of the year. Therefore, it would be more efficient to integrate crop drying with greenhouse heating. Since crop drying does not have a critical need for constant heat, a back-up heating system would not normally be required as long as air flows could be maintained through drying containers.

Grain and tobacco are transported from the harvest location to central drying stations. Transportation costs and energy consumption en route to a waste heat utilization center must be included if cost–benefit analyses are conducted for any drying applications. Furthermore, the possibility of grain deterioration is much greater in low-temperature drying systems than in conventional systems, and further expansion of the low-temperature systems is hampered by insufficient knowledge of their effects on grain quality.

Food Processing. The food system of production, processing, distribution, and home preparation consumes about 17% of total U.S. energy, with food processing the largest component, accounting for approximately 30% of the food system's energy use. Process heat requirements for the U.S. food processing industries vary significantly, in the range of 38°C to 100°C.[33]

Table 14–14 shows the food processing industries that appear to be feasible candidates for potential applications of power plant waste heat with some augmentation. These industries require less than 60°C process temperatures, and the process heat is used mostly for hot water for scalding, product washing, and cleanup.[33]

In general, there is little possibility for industrial application of power plant waste heat without augmentation. The temperature of power plant condenser cooling water, typically in the range of 38°C to 49°C, must be augmented to temperatures near 60°C for the needs of food processing industries.

A combination of various agricultural reject heat applications (greenhouse, food process plant, and hot and cool water ponds) was studied at Cornell University[34] by using computer simulation. The studies assumed a 1000-MW(e) steam–electric generating plant, a hot water pond of 40 ha (98.8 ac.), a cool water pond of 400 ha (988 ac.), a greenhouse complex of 120 ha (296.5 ac.), and a food processing plant capable of processing 98 metric t (108 t) of vegetables per day. These facilities would be constructed to use the waste heat flow of 36.7 m³ sec of 49°C water (581,240 gpm at 120°F) from the power plant. The net present value approach was used to compare alternatives of natural draft (wet) tower, mechanical draft (wet) tower, and a greenhouse–hot and cool water ponds–food processing complex.

The comparison period was 30 years, using an 11% after-tax cost of capital and a corporate tax bracket of 48%. The sum of the years-digits method was used to calculate depreciation. The vegetable processing plant was assumed to operate 16 hours a day, six days a week, and use 18.9 m³ (5000 gal) of warm water per hour. The results of this study indicated that the waste heat utilization system would be economically justifiable.

Livestock Housing. The growth rates and feed efficiencies of cattle production are not greatly affected by fluctuations in the comfort zone of −7°C to 16°C

Table 14–14. Potential Food Processing Applications of Power Plant Waste Heat at Process Temperatures Less Than 60°C (140°F) in Tennessee, Alabama, and Kentucky.

SIC Code	Industry	Process Media	Usage (1974), 10^6Btu	Percent of Total Process Heat Requirement	Potential for Power Plant Waste Heat Application
2011	Meat packing plants	Water	.62	54	High
2013	Sausages and other prepared meats	Water	1.63	94	High
2016	Poultry dressing plants	Water	.29	100	High

Source: EPRI.[5]

for most European breeds of cattle. Direct heating is normally not used, even in cold climates, for environmental control of beef production facilities.[35] There is little potential for the use of power plant reject heat in these production systems.

Egg production requires low direct heating, but poultry requires a large amount of heat energy, ranging from 4270 to 8350 Btu per bird in typical naturally ventilated houses and from 2750 to 4930 Btu per bird in well-insulated, ventilated houses.

Environmentally controlled facilities are also used extensively for swine production. Estimates for direct heating energy required to maintain a desirable air temperature for swine farrowing and brooding have ranged from about 71 to 97 thousand Btu per pig produced. However, direct heating is not required in many areas of the country for growing–finishing swine.

The only areas of livestock housing capable of using power plant reject heat are poultry brooding and swine farrowing and brooding. Since low supplemental heat is required in animal production, its energy cost is relatively low. Therefore, it is not economically attractive to invest in large reject heat utilization projects.

Utilization of all the condenser cooling water from a modern power plant would require extremely large animal production facilities. Olszewski[31] has suggested a 320-ha (791-ac.) broiler or swine production facility to use the wintertime waste heat from a 1000-MW(e) power plant. Even if all supplemental heat used in poultry brooding and swine production were obtained from power plant waste heat, less than 1% of the waste heat being rejected would be required. The rate of heat use would vary greatly with climatic factors, animal type and age, and other animal production management criteria. This fluctuating heat demand would not lead to efficient power production if this use were intended to perform reliable cooling for power plants.

Aquacultural Use of Reject Heat

The objective of using power plant reject heat in aquaculture is to control water temperature in facilities in ranges so that maximum growth and food conversion efficiency are attained. The reason for this temperature control is that oxygen consumption, food consumption, conversion efficiency of food to flesh, growth rate, and a variety of other factors that determine production efficiency are greatly affected by water temperature.

Optimum temperature differs substantially among culture species, and changes of a few degrees above or below optimum can significantly alter food production, food conversion, and growth rate. Application of heated water to aquaculture in temperature regions could yield several benefits:

- A lengthened or year-round growing season
- Optimization of the aquaculture facility with resultant reduction in production costs
- Production of commercial species near marketing sites
- Production of tropical organisms in temperate climates
- Enhanced rates of maturation and metamorphosis so that increased survival is experienced in early life stages

As shown in Tables 14–15 and 14–16, heated water has been used in the United States to culture a wide range of aquatic organisms in both marine and fresh-water systems. Thermal effluents from power companies have been used commercially for a long time. The oldest user in the United States is probably Long Island Oyster Farms, which has produced American oyster seed using effluent from Long Island Lighting Company's Northport plant since the mid-1960s (see Table 14–15). Since the applications are numerous and each application is a specific case associated with a particular power plant, it serves no purpose to describe these projects. Only the constraints and research needs in furthering aquacultural applications of power plant reject heat are presented in the following sections.

Constraints in Aquacultural Use of Reject Heat. Because waste heat aquaculture uses thermal effluents from power plants, constraints are related to power plants. They are as follows:

- Power plant site-specific factors—geographic location, plant size, fuel source, cooling mode, etc.
- Background water quality—ambient temperatures; suspended and dissolved solids; chemical and biological pollutants from municipalities, industry, and agriculture; availability of nutrients for supporting productivity; salinity; and endemic diseases and parasites.
- Cooling water quality—saturation of various gases, solids, chemical oxygen, heavy metals, biocides and toxic chemicals, radionuclides, and other chemical inputs.
- Environmental constraints—U.S. Environmental Protection Agency's water quality standard and effluent limitations applicable to the power plant as shown in Table 14–17 (level II limitations for new aquaculture facilities). (For details, see the later section on environmental constraints.)

The above factors affect the ability to culture various organisms, the types of organisms, and the cost–benefit of waste heat aquaculture.

Table 14–15. Heated Water Aquaculture Projects Identified in the United States with Marine Organisms.

Organization	Location	Activity
Oysters, Mussels, or Clams		
Long Island Oyster Farms	Northport, NY	Commercial
International Shellfish Enterprises	Moss Landing, CA	Commercial
University of Maine	Orono, ME	Research
University of California	Livermore, CA	Research
Northeast Utilities	New London, CT	Research
Texas A&M University	Baytown, TX	Research
University of Connecticut	Norwalk, CT	Research
Oregon State University	Corvallis, OR	Research
Maine Dept. Marine Resources	Wiscasset, ME	Research
Massachusetts Institute of Technology	Cambridge, MA	Feasibility
University of Massachusetts	Amherst, MA	Feasibility
Virginia Institute of Marine Science	Gloucester Point, VA	Feasibility
Lobsters		
San Diego State University	San Diego, CA	Research
Boston Edison Company	Boston, MA	Feasibility
Penaeid Shrimp		
Ralston Purina Company	Crystal River, FL	Research
Marifarms, Inc.	Panama City, FL	Commercial
Texas A&M University	Corpus Christi, TX	Research
Texas A&M University	Baytown, TX	Research
University of Miami	Miami, FL	Research
Salmon		
Maine Salmon Farm	Wiscasset, ME	Commercial
Oregon State University	Corvallis, OR	Research
University of Washington	Seattle, WA	Research
Weyerhaeuser Company	Springfield, OR	Commercial
Boston Edison Company	Boston, MA	Feasibility
Puget Sound Power and Light	Seattle, WA	Feasibility
Alaska Department of Fish and Game	Juneau, AK	Research
Flatfish		
University of New Hampshire	Durham, NH	Research
Striped Bass		
Public Service Electric and Gas Co.	Trenton, NJ	Research
San Diego State University	San Diego, CA	Research
Warm Water Marine Finfishes		
Texas A&M University	Baytown, TX	Research
University of Miami	Miami, FL	Research
Tampa Electric Company	Tampa, FL	Research

Source: EPRI.[5]

Table 14–16. Heated Water Aquaculture Projects Identified in the United States with Fresh-water Organisms.

Organization	Location	Activity
Fresh-water Prawns		
Public Service Electric and Gas Co.	Trenton, NJ	Research
Farm Fresh Shrimp Company	Miami, FL	Research
Texas Electric Company	Monahans, TX	Research
Sierra Pacific Power Company	Xerington, NE	Research
Catfish		
Tennessee Valley Authority	Gallatin, TN	Research
Kansas Gas and Electric Co.	Colwich, KS	Research
Kansas Power and Light Co.	Hutchinson, KS	Research
Aquarium Farms, Inc.	Freemont, NE	Commercial
Kraft, Inc.	Harrisburg, PA	Research
Fish Breeders of Idaho	Buhl, ID	Commercial
Cultured Catfish, Inc.	Colorado City, TX	Commercial
Texas A&M University	College Station, TX	Research
Mississippi Power and Light Company	Jackson, MS	Research
Clemson University	Clemson, SC	Research
Washington Agriculture Exp. Sta.	Pullman, WA	Feasibility
Eels		
Public Service Electric and Gas Co.	Trenton, NJ	Research
Tilapia		
Aquarium Farms, Inc.	Freemont, NE	Commercial
Fish Breeders of Idaho	Buhl, ID	Commercial
Weisbart and Weisbart, Inc.	Alamosa, CO	Commercial
Idaho State University	Pocatello, ID	Research
Yellow Perch		
Idaho State University	Pocatello, ID	Research
Vermont Yankee Nuclear Power Corp.	Rutland, VT	Feasibility
Trout		
Public Service Electric and Gas Co.	Trenton, NJ	Research
Maine Salmon Farms	Wiscasset, ME	Commercial

Source: EPRI.[5]

Research Needs in Aquacultural Applications of Reject Heat.

Research is needed in water treatment, indirect use of waste heat, and optimum aquaculture systems designs. Research needs in the area of water treatment can be summarized as follows:

- Information on waste characteristics of cultured species is limited.
- Effluent treatment technology for heated water aquaculture facilities

Table 14-17.   Proposed EPA Effluent Limitations for Aquaculture Facilities.

| Culture System | Parameter | Level I Effluent Limitations (July 1, 1977) | | Maximum Instantaneous (mg/l) |
| | | kg/100 kg Fish on Hand/ Day | | |
		Max. Daily	Avg. Daily	
Native fish				
a. Flow through	Suspended solids	2.9	2.2	—
	Settleable solids[1]	—	—	0.2
	NH$_3$–N	0.12	0.09	—
	Fecal coliforn[2]	—	—	200 organisms/100 ml
b. Ponds	Settleable solids[1]	—	—	3.3
	Fecal coliform[2]	—	—	200 organisms/100 ml
Nonnative fish		No discharge		
Level II Effluent Limitations (July 1, 1984) and Level III Effluent Limitations (New Sources)				
Native fish				
a. Flow through	Suspended solids	1.7	1.3	—
	Settleable solids[1]	—	<0.1	0.2
	NH$_3$–N	0.12	0.09	—
	Fecal coliforn[2]	—	—	200 organisms/100 ml
b. Ponds	Settleable solids[1]	—	—	3.3
	Fecal coliforn[1]	—	—	200 organisms/100 ml
Nonnative fish		No discharge		

[1] Reported as ml/l. Limitation applies to cleaning culturing units containing fish or cleaning after fish have been removed.

[2] Salmonid operations are excluded from this effluent limitation.

Source: EPRI.[5]

must be developed that allows discharge standards to be met without incurring excessively high production costs.

- The possibility of recycling aquaculture effluents to the power plant and subsequent effects upon condensers, cooling tower operations, and overall efficiency have not been investigated.

- Information is needed on treatment process optimization and costs prior to commercial effluent recycling efforts.

Problems with background water quality, the influence of plant operations on cooling water quality, or other restrictions may require indirect use of waste heat in providing temperature control. Substantial problems exist relative to the economic transfer of heat from a power plant cooling system to an aquaculture system. Heat exchangers of high efficiency and low cost should be developed.

Optimum designs for commercial waste heat aquaculture systems require innovation and blending of several technology disciplines. Specific engineering problems that need further research and development include:[5]

- Evaluation of the effects of biological and chemical fouling on the performance of pumps, piping, heat exchangers, etc.
- Definition of the influences of minor elements leached to cooling waters from cooling systems and other water-conveying mechanisms.
- Development of economical back-up systems for water supply, heating, and aeration.
- Development of technologies for economically moving water against high and low heads.
- Evaluation of mechanisms of temperature control in aquaculture systems that are not subject to variable generating loads, plant shutdowns, and ambient temperature changes.
- Definition of the needs and availability of land relative to aquaculture systems requirements.
- Design of containment structures (raceways, tanks, ponds, etc.) that facilitate culture operations, are efficient in metabolite removal, and are cost-effective.

Integrated Energy Systems

An integrated energy system is defined as an energy system that optimally integrates electricity generation and reject heat utilization into community energy need for various activities (e.g., cooling, heating, and lighting). There are two methods for this integration: the select energy system and the total energy system. The difference between the two methods is in the degree of on-site electricity generation. A select energy system generates on-site a portion of the electricity required by the institution that owns it, whereas a total energy system generates on-site all of the electricity. A total energy system has adequate back-up capacity, but a select energy system operates in parallel with the local power utility.

These two methods of integration are shown in Fig. 14–12. Recovery of low-quality heat from the power generation cycle and its subsequent usage in other subsystems are the key to energy savings and attendant economic feasibility.

Integrated energy systems can help reduce both utility resource requirements of universities and hospitals, and the cost of utility operations. The magnitude of energy and cost savings is site- and system-specific. Some have claimed that these systems can reduce overall energy requirements as much as 30% and costs of purchased energy by 40%.[37] However, Was and Golay indicated that, with many economic uncertainties involved and the dependence on oil, a total energy system is not economically attractive for the Massachusetts Institute of Technology* owing to low demand-density and low capacity factors.[38]

The costs and benefits of integrated energy systems depend on the characteristics of individual communities' energy need and the composition of energy

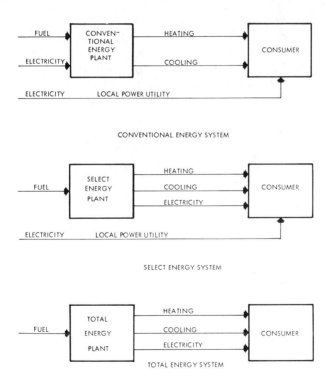

Fig. 14–12. Comparison of conventional, select, and total energy systems. Source: EPRI.[1]

---

* There is another example of unfavorable total energy systems that will be discussed below.

sources. Three examples of these systems are the Integrated Community Energy System at Trenton, New Jersey; the Total Energy System at Walt Disney World; and the Integrated Energy Utility System at the University of Florida.[5] The Integrated Community Energy System is analyzed to illustrate possible costs and benefits of such energy systems.

Integrated Community Energy System (ICES) at Trenton, New Jersey.
This project was developed by the Public Service Electric and Gas Company, Department of Planning and Development of the City of Trenton, New Jersey. The U.S. Department of Energy provided the funding for an evaluation of the feasibility of applying the grid-connected (ICES) concept as a means of supplying electric and thermal energy to a group of office and housing buildings in downtown Trenton. The Trenton ICES plant is thermal-controlled (i.e., the demand for thermal energy controls the output of the plant and, in turn, the quantity of electricity produced).

Community Served. The community intended to be served by the Trenton (ICES) plant is comprised of 1,884,025 ft² of floor area in existing buildings, and 1,251,500 ft² floor area of new construction projected for completion by 1982. This results in a demonstration community of 3,135,525 ft² floor area. The total distribution distance to all existing buildings is approximately 3400 feet.

Plant Design. The plant is a combination of turbine-waste heat boilers. Four combustion turbine generators, each with a 2.5-MW capacity, were selected; the waste heat of each turbine supplies a heat-recovery boiler with a steam capacity of 14,200 lb/hr at 150 psig. In addition, an auxiliary back-up boiler was provided, with a capacity of 32,000 lb/hr. The low-temperature recovery stages of the heat-recovery boilers will also generate a total of 480 gpm of 320°F hot water as an additional heating supply. Figure 14–13 shows a schematic representation of the ICES plant.

In determining the steam and electric capcity needs of the ICES plant, appropriate thermal storage was included at six steam distribution receiving terminals in the community.

Economic Evaluations. The economic evaluations were based on minimum revenue requirements.* Major annual cost components for such economic evaluations were:

---

* Minimum revenue requirement is the minimum amount of money that must be obtained to cover all expenses associated with, and including, the minimum acceptable return on any capital required by a plan.

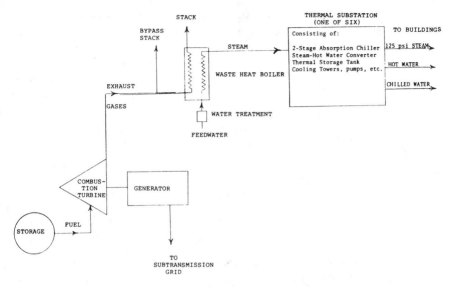

Fig. 14–13. Schematic diagram of ICES plant showing heat recovery and storage. Source EPRI.[1]

- Those costs associated with the production of energy from a cogeneration facility, including capital, fuel, operating, and maintenance.
- Net utility savings resulting from operation of the cogeneration facility.
- The costs associated with the conventional heating (and cooling, as applicable) facilities to provide thermal output equivalent to that of the cogeneration facility, including capital, fuel, operation, and maintenance.

The economic screening of the ICES shows a marginal penalty of about $100,000/year, when compared to the conventional oil-heated system. (See Table 14–18.)

Concluding Comments. A comparative analysis of the fuel requirements of the ICES versus those of a conventional system is summarized in Fig. 14–14. ICES in design was expected to require 73,364 barrels of oil to supply the same final energy demand for which the conventional system required 68,754 barrels of oil plus 6,440 tons of coal (equivalent to 27,903 barrels of oil) or a total of 96,657 barrels of oil equivalent. The conventional system consumes less oil by about 4,610 barrels per year (about 6%).

If, in the future, the coal savings of ICES can compensate the cost increase due to oil price increases, ICES can be economically attractive without subsidies. However, the possibility is insignificant because of fast oil price increases.

Natural gas has then been used to replace oil when ICES is implemented.

Table 14-18. Cost Comparison of Alternate Energy Supplies to Trenton Downtom Community.

	Levelized Annual Costs *ICES Installation 1982*
A. *ICES (PSE&G Ownership) in Design*	
1. ICES energy production	
(a) Capital	$ 1,900,000
(b) Fuel	2,380,000
(c) Operating and maintenance	1,410,000
2. Reduced electrical energy	(300,000)
(ICES vs. conventional)	
3. Replacement energy and capacity credit	(1,550,000)
4. Transmission and distribution credit	
(a) Energy losses	(70,000)
(b) Long-range investment savings	(110,000)
SUBTOTAL	$ 3,660,000
B. *CONVENTIONAL SYSTEM* (New Buildings Oil Heated)	
1. Steam production	
(a) Capital	$ 560,000
(b) Fuel	1,440,000
(c) Operating and maintenance	1,560,000
SUBTOTAL	$ 3,560,000
C. *ICES PENALTY*	$ 100,000

Source: EPRI.[1]

The economics of this substitution will depend on the price of natural gas, in particular, after deregulation.

LEGISLATIVE, ENVIRONMENTAL, AND INSTITUTIONAL CONSTRAINTS FOR REJECT HEAT UTILIZATION

The outlook for power plant reject heat utilization depends on its economics and ability to overcome legislative, environmental, and institutional constraints. These constraints have been seriously blockading widespread applications of power plant reject heat.

Legislative Constraints

Legislative constraints are examined from three aspects: congressional legislation, DOE's budget, and the Administration's energy policy.

Congressional Legislation. Reject heat utilization was mostly ignored in the National Energy Act (NEA). Although some earlier congressional

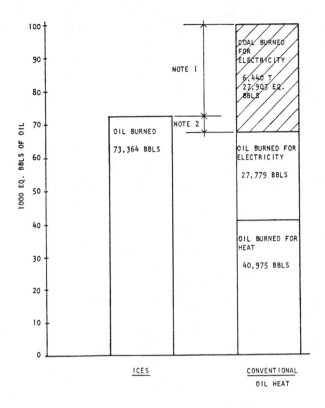

Fig. 14–14. Fuel requirements: ICES versus oil heating. Source: EPRI.[1]

initiatives* were taken in the area of reject heat utilization, relatively little attention has been paid to the subject. Most legislative references to reject heat utilization are for in-plant (industrial) cogeneration. The only other relevant portion of the NEA for reject heat utilization is the provision of an additional 10% investment tax credit for reject heat utilization equipment.

---

* The bill introduced by Mr. Hart and Mr. Sasser in the first session of the 95th Congress was titled "Cogeneration and Waste Heat Utilization Act," but the only part of that legislation that survived was the portion that became Sections 201 and 210, i.e., cogeneration provisions of the Public Utility Regulatory Policy Act (PURPA) in the National Energy Act.

DOE's Budget. Another evidence of lack of government interest in reject heat utilization is the insignificance of DOE's budget in these areas. Even though there are several DOE programs dealing with various projects in reject heat utilization, the level of funding has been low, and they are relatively unemphasized. In the view of both the Administration's and Congress's energy policy-makers, reject heat utilization plays only a small role of energy conservation.

The Administration's Energy Policy. Reject heat utilization was almost unmentioned in National Energy Plan (NEP) I, which was the cornerstone of the Carter Administration's energy policy.

Three solutions for energy problems were prescribed in NEP I:

- Use conservation to reduce dependence on foreign oil.
- Substitute abundant fuels for oil and gas.
- Develop renewable energy sources, primarily solar energy.

Because electricity was considered to be inefficient (two-thirds of the primary energy sources are wasted), uses oil and gas in peak electricity generation, and competes with solar energy, it was not favored in NEP I. Power plant reject heat utilization was not emphasized owing to its connection with electricity.

The Reagan Administration emphasizes a free market policy in NEP III, which may not help the development of reject heat utilization either: The initial development of reject heat utilization requires assistance from the government because of its capital-intensiveness.

Other Constraints. The definition of power plant reject heat utilization is so broad that it can hardly be fully comprehended. Its technologies and applications range from massive, centralized district heating or total energy systems to small, individual agricultural or aquacultural applications such as greenhouses and methane production. This broad and vague definition thus has made clear communication and effective legislation difficult.

Environmental Constraints

Prior to 1955, quantities of power plant reject heat were relatively small because the average size of U.S. generation units was about 35 MW. However, in recent years most of the units constructed have been 500 MW or more, and the volumes of condenser cooling water required at such large facilities have created both the potential for severe environmental problems related to thermal discharges and an awareness of the vast reservoirs of reject heat available for utilization. Environmental constraints are divided into two categories: environmental legislation and effluent limitations for aquaculture facilities.

Environmental Legislation. Congress enacted P.L. 92–5000 in 1972 to provide far-reaching amendments for the Federal Water Pollution Control Act (FWPCA). Congress further amended FWPCA with the Clean Water Act of 1977 (P.L. 95–217). This legislation also initiated the National Pollutant Discharge Elimination System (NPDES), which is used for the issuance of discharge permits to various categories of point source dischargers. Discharge limitations for these permits were based on phase and levels of technology, and then attainment dates were specified as follows:[39]

- Effluent limits providing for best practicable control technology (BPT) were to be met by July 1, 1977.
- Effluent limits providing for best available technology (BAT) were to be met between July 1, 1984, and July1, 1987, depending on the category of pollutant.
- The 1977 amendments (P.L. 95–217) provided for an additional category termed best conventional pollutant control technology (BCT), which was to be implemented by July 1, 1984.

EPA was thus required by law to publish effluent limitations for the classes of treatment technologies specified above and to publish performance standards for "new sources." A new source was defined as any source for which construction was begun after the publication of the new source performance standards.

Table 14–17 (see above) shows the proposed EPA effluent limitations for aquaculture facilities. In the table, level I effluent limitations are for existing dischargers, and level III (the same as level II) limitations are for new sources.

The effluent limitations for BPT and BAT primarily affect the issuance of thermal discharge permits and the use of cooling lakes. The only limitations that affect reject heat utilization are those applied to aquaculture facilities.

Effluent Limitations for Aquaculture Facilities. In Table 14–17, one example is given of proposed EPA effluent limitations in the discharge of pollutants associated with aquaculture projects. In addition, according to the published EPA regulation for the discharge of pollutants, permits issued to aquaculture projects must meet at least the following primary requirements:[39]

- Migration of pollutants outside the project area will not violate water quality standards or effluent limitations applicable to the supplier of the pollutant.

- Any modifications caused by construction of a containment structure shall not alter the tidal regimen of an estuary or interfere with migrations of unconfined species.
- The applicant must demonstrate that the pollutant will result in increased harvest.
- The crop will not have potential for human health hazard resulting from its consumption.

Institutional Constraints

Institutional constraints include regulatory issues, and social and political factors. The regulatory issues come mainly from the Federal Energy Regulatory Commission (FERC) and state regulatory agencies. Social and political factors are related to the problem of public acceptance of power plant reject heat utilization ventures.

Regulatory Issues. Electric utilities are engaged in a "business affected with public interest," and they should be regulated by a system "designed to ensure an adequate flow of service to the consumer, at prices which correspond to costs of production. . . ."[40] Therefore, the independent regulatory commission was created to monitor the investment, construction, financing, distribution, and pricing of public utilities. This monitoring by the regulatory commission is to ensure the implementation of legislative mandates.

Regulatory agencies that have the authority to regulate the use of cogeneration facilities and to set rates for electricity and steam production by such facilities are state utility commissions and FERC. Although federal regulation was limited to the use of the facility itself and to setting the rate for interstate electric sales, PURPA provides for federal review of some state regulations regarding in-plant cogeneration facilities. However, this review may not affect the use of power plant reject heat because it was not covered in PURPA.

Federal Jurisdiction. FERC has jurisdiction over the interstate transmission of electric power and interstate sales for resale of electric power. It sets rates for electricity but not for steam, even though the jurisdiction appears to extend to all facilities, including the electricity and steam production components. A public utility may not sell, lease, merge, consolidate, or otherwise dispose of facilities subject to FERC jurisdiction prior to FERC's approval. Under the Federal Power Act, a public utility is a person who owns or operates facilities subject to federal jurisdiction.

FERC's jurisdiction over a facility that produces both electricity and steam for sale depends on the nature of the ownership of the facility and the type of sale of electric energy. For example, if the facility supplies power only to a small number of intrastate industrial users (not for resale), FERC would not have jurisdiction over the facility or the rates for such sales regardless of the ownership arrangement. However, if the facility provides wholesale power to a public utility system subject to FERC jurisdiction, it is subject to FERC regulations.

FERC sets rates for electric energy only and not for steam. If the electricity generation component is owned by a public utility and the steam production component is owned by industrial companies, then the steam owners are not subject to the Federal Power Act.

State Jurisdiction. The definition of what constitutes a public utility subject to state regulations may vary from state to state. The usual characteristic of a public utility subject to state regulations is that its services are available to the public generally and indiscriminately. Thus, companies supplying electricity, natural gas, heating and cooling, telephone, water, and sewer services are included. However, municipally owned and cooperative organizations may be exempted from state regulations.

Depending on the state, the authority of regulation may be exercised by the Public Utility Commission, the state energy agency, an independent power siting board, or other agency. Determination of which agency has which, if any, authority over electricity- and steam-supplying public utilities in a given state must be made for each state individually. Normally, the state Public Utility Commission has authority to regulate rates for both electric and steam service when either service is offered to the general public. If the service is not offered to the general public, the commission may regulate the service, depending on the nature of the cogeneration facility ownership and on the type of sale. For example, the commission sets rates for power plant reject heat sales to a retail customer, but rates for large primary industrial customers may be set by negotiation between the utility and the customer, subject to commission approval.

As discussed under "Federal jurisdiction," for a cogeneration facility where the electricity generation components are owned by a public utility but the steam production components are owned by an industrial firm, the industrial owner would not be subject to state regulation except when he sells his steam to the public generally and indiscriminately.

Utility Obligation to Serve. One of the problems in reject heat utilization is utility obligation to serve. This issue is of particular importance in district heat. An economically viable district heating system must allow maximum utilization of plant capacity and a rough balance between thermal and electrical production. Thus, a district heating utility may have to restrict service to the facilities within a well-defined area exhibiting the requisite characteristics. This economic consideration clashes with most state utility codes containing a statutory mandate to service. It also clashes with a U.S. Supreme Court decision stating:

> Corporations which devote their property to a public use may not pick and choose, servicing only the portions of the territory covered by their franchises which it is presently profitable for them to serve, and restricting the development of the remaining portions by leaving their inhabitants in discomfort without the service which they alone can render.[41]

This conflict between economic and institutional factors can be resolved, and potential obstacles to reject heat utilization (in particular, adoption of district heating) can be overcome, but solutions differ from jurisdiction to jurisdiction.

Social and Political Factors. Besides the consideration of technical and economic feasibility of reject heat utilization, social and political factors can determine its final outcome. These factors are in the area of public acceptance of reject heat utilization, and they can be divided into three areas: market acceptance, community acceptance, and regulatory acceptance.

Market Acceptance. Market acceptance is a twofold issue:

- Ventures in reject heat utilization are accepted in capital markets.
- Products of reject heat utilization are accepted in consumer markets.

Economic viability is the key to capital formation for reject heat utilization projects. Properly conceived joint-venture organizations are needed to share risks over participants, projects, and time. On the other hand, the economic viability of reject heat utilization projects may depend on consumer acceptance of their products. A core organization should be established and empowered to research markets for the products of reject heat utilization projects, to identify qualified entrepreneurs who can use reject heat, and to assist them in getting established.

Community Acceptance. Obtaining a community's acceptance of environmental and siting effects of the reject heat utilization projects requires careful planning and early continued involvement of local groups. This can be achieved by educating community people about what has been done, the benefits and costs, and how they can get involved to influence the future of reject heat utilization projects (e.g., in particular, district heating systems).

Regulatory Acceptance. Many public utility commissions may not be ready to approve waste heat ventures of a utility. Regulations regarding joint-venture agreement and classification of reject heat users may need revision, and rate schedules may have to be set for reject heat sales. Regulatory agencies need to be accustomed to reject heat utilization ventures, and appropriate regulations need to be established.

CURRENT R&D IN REJECT HEAT UTILIZATION

Research and development projects in reject heat utilization are usually included in building and community systems and industrial energy conservation programs. These projects may involve DOE, the Department of Health and Human Services, and other relevant organizations (e.g., the Electric Power Research Institute).

Currently ongoing projects include district heating and cooling,* grid-connected integrated community energy systems (grid-ICES),* energy-integrated industrial parks,* and evaluation of alternative technologies for dual energy use systems (DEUS).

District Heating and Cooling

The projected progress of district heating and cooling in fiscal years 1981–1986 is shown in Table 14–19. This project, which was sponsored by DOE's Buildings and Community System Programs, emphasizes the cost-effective utilization of reject heat from existing power plants near urban areas. Assessments of its costs/benefits were made in two states (Minnesota and Wisconsin), one region (Northern New Jersey), three metropolitan areas (Detroit, Philadelp..ia, and Minneapolis/St. Paul), and one small town (Piqua, Ohio).

These assessments show positive benefits for energy conservation, and environmental and socioeconomic effects. Current efforts by local community/energy teams concentrate on the type of arrangements (ownership, financing, regulatory, and operation). A National District Heating Plan has been developed to enhance these efforts.

Grid-Connected Integrated Community Energy Systems (Grid-ICES)

The projected progress of the grid-ICES project is shown in Table 14–19. The University of Minnesota (Minneapolis campus), in close cooperation

---

* Subject to change because of the Reagan Administration's energy policy.

Table 14–19. Key Projects in Buildings and Community Systems Programs of DOE (Basic Level).

Buildings and Community Systems	Fiscal Year**					
	1981	1982	1983	1984	1985	1986
District Heating and Cooling	Go/no go for small systems. Initiation of a national DH plan	Go/no go for large systems. Go/no go for a national DH implementation	Start construction on small DH. Start operation on DH plan	First phase construction for large DH. Massive number of DH projects started	Finish construction of small DH. Go/no go decision on DH projects	Finish construction of 1st phase of large DH
Grid-Connected Integrated Community Energy Systems (Grid-ICES)	U. of Minn. finishes construction	Trenton ICES starts const. 1st year of operating results from U. of Minn.	Trenton ICES finishes construction. 2nd year of operating results from U. of Minn.	Go/no go decision on nat'l plan for Grid ICES implementation	Start of nat'l implemented plan	Massive number of grid-ICES projects under way
Other Demos (ICES and district heating/cooling)	Heat pump–ICES finishes first phase. Solar ICES concepts developed	Go/no go decision on HP* ICES implementation. Solar ICES concepts evaluated and assessed	Solar ICES RFP issued. HP ICES detailed design	Solar ICES finishes first phase. HP ICES construction	Go/no go decision on solar ICES and HP ICES	Go/no go decision HP ICES nat'l implementation plan

* Heat pump.
** Subject to change because of the Reagan Administration's new energy policy.
Source: DOE.

with the local power company (Northern State Power Company—NSP), is recycling a small retired power plant (previously owned by NSP) to a coal-firing grid-ICES plant to provide the large campus with heating, cooling, and hot water. The construction was expected to be finished by the beginning of FY 1982, whereupon a sophisticated monitoring/evaluation system will provide data on energy, environmental, and operational economics. At the time of this writing the City of Trenton was pursuing a grid-ICES based on a combination of gas turbines and diesel engines as discussed before. A monitoring/evaluation system was installed in order to gather data similar to that of the University of Minnesota project. The monitoring/evaluation systems average $300K to $400K each.

Energy-Integrated Industrial Parks

This project was sponsored by DOE's Office of Industrial R&D Programs. In this project, an extensive study of the advantages, favorable conditions, and limitations of energy-integrated industrial parks was completed. An additional study effort was being conducted to examine and qualify several new factors uncovered during the initial effort.

Evaluation of Alternative Technologies for DEUS

This project has been sponsored by the Electric Power Research Institute since 1978. Dual energy use systems (DEUS), the combined generation of heat and power, show promise for applications like district heating. The objectives of this project are to identify the technical, economic, and institutional factors determining the feasibility of DEUS; to develop a methodology with which the utility industry can evaluate DEUS; to identify the preferred generic DEUS; and to identify the research and development needed to further characterize and implement beneficial applications of DEUS.

CONCLUSIONS

We have evaluated various methods of using reject heat from power plants. These applications include district heating, agricultural use, aquacultural use, and integrated energy systems. The analysis indicates that most of those applications are not economically attractive under the current energy price structure. Although district heating has been practiced successfully in Europe, it is not economically feasible in the United States because of the latter's comparatively inexpensive fuel, high distribution cost, and complex utility regulations. Agricultural and aquacultural uses of reject heat do not indicate significant potential because they only use a small portion of the reject heat, and most of these uses are not substantial. Under the current energy price structure, experiments for integrated energy systems do not indicate any significant economic advantage over conventional energy systems.

In addition to the economic consideration, advocates of reject heat utilization must work to gain public acceptance of reject heat applications. This public acceptance includes acceptance in capital markets, markets for reject heat products, communities, and regulatory agencies. The economic feasibility of reject heat applications can be significantly strengthened if the public can accept the reject heat applications.

REFERENCES

1. Electric Power Research Institute (EPRI). "Workshop Proceedings: Dual Energy Use Systems," EPRI EM-718-W, May 1978.
2. Ecabert, R., Helbling, W., and Leiner, H. J. "Technical and Economic Possibilities of District Heating in Switzerland," study for the Federal Bureau of Energy, Sulzer Technical Review No. 2, 1974.
3. Karkheck, J., Powell, J., and Beardsworth, E. "Prospects for District Heating in the United States," *Science,* Vol. 195, pp. 948–955, 1977.
4. Ibid.
5. EPRI. "State-of-the-Art Waste Heat Utilization for Agriculture and Aquaculture," EPRI EA-922, October 1978.
6. Hodges, C. N. and Hodge, C. O. "An Integrated System for Providing Power, Water and Food for Desert Coasts," *HortScience,* Vol. 6, No. 1, 30, February 1971.
7. Miller, A. et al. "Use of Steam-Electric Power Plants to Provide Thermal Energy to Urban Areas," Oak Ridge National Laboratory, ORNL-HUD-14, January 1971.
8. Burns, E., Pile, R., and Madewell, C. "Waste Heat Use in a Controlled Environment Greenhouse." in *Proceedings of the Conference on Waste Management and Utilization,* Vol. 2, Miami, Florida, May 9–11, 1977.
9. Berry, J. and Miller, H., Jr. "A Demonstration of Thermal Water Utilization in Agriculture," U.S. Environmental Protection Agency, EPA/660/2–74–011, April 1974.
10. Ashley, G. C. and Hietala, J. "The Sherco Greenhouse: A Demonstration of the Beneficial Use of Waste Heat," p. IX-A-3 in *Waste Heat Management and Utilization,* Vol. 3, Miami, Florida, May 1977.
11. Boyd, L. et al. "Use of Waste Heat from Electric Generating Plants for Greenhouse Heating," University of Minnesota, ASAE Paper No. 77–4531, 1977.
12. Godfriaux, Bruce. Personal Communication by E. Burns, R. Pile, and C. Madewell.
13. Walker, Paul N. Personal Communication by E. Burns, R. Pile, and C. Madewell.
14. Crumbly, I. J. Personal Communication by E. Burns, R. Pile, and C. Madewell.
15. DeVries, D. "Thermal Properties of Soils," p. 210 in *Physics of the Plant Environment.* Amsterdam: North Holland Publishing Company, 1963.

16. Rykbost, K. et al. "Crop Research to Warming Soil Above Their Natural Temperatures," Oregon Agricultural Experiment Station Special Report 384, 1974.
17. Berry, J. and Miller, H. "A Demonstration of Thermal Water Utilization in Agriculture," Environmental Protection Agency Technology Series, EPA 660/24–74–011, 1974.
18. Mays, D. "Use of Waste Heat for Soil Warming," in *Soils for Management of Organic Waste and Waste Waters.* Soil Science Society of America, Madison 1977.
19. Wierenga, P., Hagan, R., and Nielsen, D. "Soil Temperature Profiles During Infiltration and Redistribution of Cool and Warm Irrigation Waters." *Water Resources Research,* Vol. 6, 230, 1970.
20. Borsma, L. et al. "Animal Waste Conversion Systems Based on the Use of Thermal Discharges with By-Product Recovery and Recycling," Oregon State University, October 1973.
21. Dugan, G., Golueke, C., and Oswald, W. "Recycling System for Poultry Wastes," *Journal of Water Pollution Control Federation,* Vol. 44, No. 3, 432, 1972.
22. Madewell, C., Martin, J., and Isom, B. "Environmental and Economic Aspects of Recycling Livestock Wastes—Algae Production Using Waste Products," Tennessee Valley Authority, Bulletin Z-36, 1971.
23. Wilcox, J. "Waste Heat Utilization in Aquaculture: Some Futuristic and Plausible Schemes," p. X-A-35 in *Waste Heat Management and Utilization,* Vol. 3, May 1977.
24. Hill, D. and Barth, C. "Nutrient Conservation in Animal Waste Management," Clemson University, Agricultural Engineering Department, ASAE Paper No. 75–4031, 1975.
25. American Society of Agricultural Engineers. "Managing Livestock Wastes," in *Proceedings of the 3rd International Symposium on Livestock Wastes,* University of Illinois, April 1975.
26. Loehr, R. "Food, Fertilizer, and Agricultural Residues," in *Proceedings of the 1977 Cornell Agricultural Waste Management Conference.* Ann Arbor: Ann Arbor Science Publishers, Inc., 1977.
27. Middlebrooks E. et al. "Techniques for Algae Removal from Wastewater Stabilization Ponds." *Journal of Water Pollution Control Federation,* Vol. 46, No. 12, 2676, 1974.
28. National Academy of Sciences. *N.A.S. Making Aquatic Weeds Useful: Some Perspectives for Developing Countries.* Washington, D.C., 1976.
29. Jewell, W. et. al. "Bioconversion of Agricultural Wastes for Pollution Control and Energy Conservation," National Technical Information Service, U.S. Department of Commerce, September 1976.
30. Olszewski, M. "An Assessment of the Potential for Wide Scale Implementation of Power Plant Reject Heat Utilization" (draft), Oak Ridge National laboratory, ORNL/TM-5841, September 1977.
31. Olszewski, M. "Power Plant Reject Heat Utilization: An Assessment of the Potential for Wide-Scale Implementation," Oak Ridge National Laboratory, ORNL/TM-5841, December 1977.
32. Olszewski, M. "An Assessment of the Potential for Wide Scale Implementation

of Power Plant Reject Heat Utilization" (draft). Oak Ridge National Laboratory, ORNL/TM-5841, September 1977.

33. Kim, B., Wilkinson, W., and Rosenberg, H. "Technical Assessment of Industrial and Residential Uses of Power Plant Waste Heat," Battelle Columbus Laboratories, April 24, 1978 (draft).

34. U.S. Government Printing Office. *A Study of Energy Conservation Potential in the Meat Packing Industry.* Washington, D.C., 1976.

35. Witz, R. et al. "Livestock Ventilation with Heat Exchange," North Dakota State University, ASAE Paper No. 74–4525, 1974.

36. Oak Ridge National Laboratory. *Waste Heat Utilization Proceedings of the National Conference,* edited by Marvin M. Yarosh, CONF-711031, 1971.

37. Kirmse, D. and Coxe, E. "State of the Art for Integrated Energy/Utility System," prepared for HEW, PB-269-684, February 1977.

38. Was, G. and Golay, M. "Cogeneration—An Energy Alternative for the U.S.?," *Energy,* Vol. 5, 1023–1031, 1979.

39. U.S. Environmental Protection Agency. "Development Document for Effluent Limitations, Guidelines and New Source Standards for the Steam Electric Power Generating Point Source Category," EPA 440/1–74 029a, 1974.

40. Gies, T. "The Need For New Concepts in Public Utility Regulation," p. 90 in (W. G. Shepard and Thomas G. Gies, eds.) *Utility Regulation: New Directions in Theory and Policy.* New York: Random House, 1966.

41. *New York and Q. Gas Co. v. McCall,* 245 U.S. 345, 351, 1918.

PART IV

PROSPECTS OF ENERGY CONSERVATION

This section presents a discussion of the future of energy conservation.

15

The Future of Industrial Energy Conservation

INTRODUCTION

Basically, the progress of industrial energy conservation depends on the relative price and availability of energy (in particular, petroleum products). As discussed in Chapter 3, the future price and availability of oil are substantially undertain and can hardly be controlled by the United States. Therefore, industrial energy conservation will become more and more important if stable economic growth is to be maintained. This chapter considers six aspects of the future of conservation: redefinition of industrial energy conservation efforts; importance of industrial energy conservation in the economy; the time to make decisions for conservation; direction of industrial conservation policies; agreement instead of argument in implementing conservation; and the roles of government and industry in conservation.

REDEFINITION OF INDUSTRIAL ENERGY CONSERVATION EFFORTS

In the past, both industry and government have focused on retrofitting and bookkeeping, auditing, and data collection in conservation. The success of these efforts has been limited because resulting energy savings have not been substantial. Replacement of old facilities in the energy-intensive industries with new energy-efficient processes should be the new direction of conservation. Furthermore, new efficient processes that can also conserve petroleum should be the first priority for investment.

For government, this new definition of industrial energy conservation efforts includes: (1) letting relative energy prices increase enough that new conservation investments are economically attractive; (2) providing sufficient fiscal incentives for the new investments; and (3) assuring the balance between a safe environment and efficient energy use. Moreover, the definition of industrial energy conservation should also be expanded to include investments in facilities that produce more energy-efficient equipment and appliances because these improved products will, in turn, be used by other firms to produce goods and services.

IMPORTANCE OF INDUSTRIAL ENERGY CONSERVATION IN THE ECONOMY

Because labor cost in the United States is high, especially when compared with Japanese and other producers, investing in energy-efficient processes is urgently needed. Conservation will aid competition because most of these processes also make more energy-efficient and higher-quality products, save labor costs, and reduce pollution. The growth of the U.S. economy depends on the competitiveness of U.S. products, and the U.S. competitive position will largely depend on the efficiency of production processes, and goods and services. However, with the present uncertainty in world oil supply and prices, today's investments in energy conservation measures will also assure better control of tomorrow's energy supply.

THE TIME TO MAKE DECISIONS FOR CONSERVATION

Replacement of old facilities with new and more efficient processes requires three to ten years for completion, even after the investment decisions have been made. The long lead time demands that industry make decisions for conservation early enough that the efficiency of new processes will quickly increase the ability of U.S. producers to compete with imports, in particular, the Japanese products.

On the other hand, industry should make decisions on facility investments in order to improve the energy efficiency of products before the consumer decides to purchase imported energy-efficient equipment and appliances. The consumer will become aware of the energy efficiency of equipment and appliances when the energy cost is sufficiently high. This point is exemplified by consumer attitudes toward the purchase of an automobile. The U.S. automobile industry did not expect the consumer to give up the inefficient big cars so quickly and therefore was not prepared for the change; it has lost large shares of the market to Japanese imports.

DIRECTION OF POLICY FOR ENERGY CONSERVATION

Industry's Direction

Industry should carefully evaluate investments for energy conservation and those for nonconservation on the basis of overall cost–benefit comparisons. A comprehensive method discussed in the previous chapters can be used for this evaluation. Although oil and energy prices have been rising significantly, industry should not rush into conservation. As discussed in Chapter 10, for example, if energy prices do not increase 100% over those of 1974 (other nonenergy prices assumed to be constant), capital-intensive conservation technologies may not be economically attractive: Any large investments in new processes may not be economically rational if cost is the only consideration.

In considering investments in new energy-efficient processes, attention must be paid to the processes that can significantly improve the energy efficiency of goods and services to the produced. Old facilities must be replaced as soon as possible if either of the following two conditions exists:

- Near-term energy prices are expected to increase so significantly that conservation technologies will become economical.
- The consumer's awareness of energy efficiency of goods and services is so strong that there is a danger that products and services produced by existing facilities will be forced out of the market.

Government's Direction

The government's energy conservation policy must enhance the improvement of the competitiveness of U.S. producers because competition is the key to U.S. economic growth. Any contrary policy will be considered undesirable. A monitoring system to measure the impacts of governmental energy policy on U.S. competitiveness should be established.

The following three governmental actions may improve the country's competitive position:

- Providing fiscal incentives to replace old facilities with more energy-efficient processes that also produce more energy-efficient goods and services
- Maintaining appropriate energy prices so that price, rather than governmental regulations, will determine the economics of conservation investments
- Focusing conservation efforts on industries and businesses with the greatest economic motivation and technical capabilities for conservation

AGREEMENT INSTEAD OF ARGUMENT IN IMPLEMENTING CONSERVATION

It is possible that, when an energy conservation decision is made, conflict may arise between environmentalists and energy conservationists, between industry and government, and between industry and the consumer. Agreement and coordination rather than argument and confrontation are needed to obtain a compromise among interest groups. In resolving conflicts, common interests and goals should be identified and pursued so that the efforts of various interest groups are constructive rather than destructive. One such common interest is to provide sufficient energy supply at competitive cost in a safe and clean environment so that U.S. products are competitive internationally. It is important to emphasize competitiveness repeatedly.

As discussed in previous chapters, a conflict between energy conservation and environmental protection exists in many conservation technologies (e.g., cogeneration and reject heat utilization). A balance between better use of energy and appropriate protection of the environment must be worked out in the very near future if conservation technologies are to be advocated. Also, better understanding of the impacts of various types of pollution on human life is needed for determination of the balance. Commonly agreeable methods of research on these impacts are thus urgent.

There has been a constant battle between government agencies and U.S. industry regarding conservation and other regulations. Industry and government have grown accustomed to communicating with one another through a formalized legal process that only serves to exaggerate differences and discourage compromise solutions. However, industry and government are not inevitable adversaries over energy conservation. The government must guard the public welfare to ensure efficient energy use, clear air, and a safe environment; and to maintain a dependable and adequate energy supply. Yet, if American industry is to survive in an increasingly competitive world economy with high-cost energy, these are precisely the goals that it must strive to attain. Industry and government must take joint responsibility for improving energy efficiency and competitiveness.

The conflict between industry and the consumer can be illustrated by the automobile case. The U.S. automobile industry has not been able to produce the efficient and high-quality cars that American customers have demanded. Consequently, the market share for U.S. car makers has shrunk significantly. This is only the beginning if U.S. industry does not change its production facilities to produce energy-efficient and high-quality goods and services. U.S. industry is losing the battle because many American products are less efficient, less reliable, and less durable than imports, either in the consumer's perception or in reality. Many imports are purchased because they are only as good as the domestic in quality but cheaper in cost. On the other hand, while many Japanese basic metal industries use advanced technologies that consume less than half of the energy used by conventional U.S. mills to produce higher-quality metals, the U.S. mills still use more-than-20-year-old facilities. With their high energy costs, high wages, and old facilities, many U.S. industries cannot compete with Japanese exporters. Industry must pay more attention to the consumer's need for more energy-efficient and reliable products if it wants to survive in international markets.

THE ROLES OF INDUSTRY AND GOVERNMENT IN CONSERVATION

Industry will play the key role in energy conservation, and government will be a catalyst that basically supports industry in replacing old facilities with new energy-efficient processes. These efficient processes will produce energy-efficient products and services at lower costs.

Conservation investment is one of many ways in which industry can choose to spend its capital. Industry will carefully evaluate all alternative investments for conservation and nonconservation technologies and choose the investments that will provide efficient processes to produce more efficient and better-quality goods and services. However, industry should not rush into conservation just through fear of energy shortage.

Government can foster conservation by providing fiscal incentives to replace energy-inefficient facilities that cannot make energy-efficient and high-quality products, by balancing the needs for efficient energy use and a safe environment, and by monitoring and collecting data for efficient technologies. Government should not become a regulator in conservation. Its primary function is to help industry use energy more cost-effectively and thus increase its competitive position in the world market.

CONCLUSIONS

Industrial energy conservation includes using energy more efficiently and producing more efficient goods and services. Thus U.S. industry can increase its competitive position in the world market. It requires agreement and coordination among industry, government, environmentalists, and consumers. The major thrust for industry is to evaluate conservation investment carefully and choose the most cost-effective, so that old facilities can be replaced with new, energy-efficient processes that produce more energy-efficient and higher-quality goods and services. Government should not become a regulator in industrial energy conservation, but should be a catalyst that provides fiscal incentives to foster conservation investments and, at the same time, maintain a safe environment.

Appendix A
Capital Recovery Factor

INTRODUCTION

The capital recovery factor is a term frequently used among engineers to determine the annual capital cost of an investment. It is defined as a uniform payment schedule over a certain period of time to recover a sum of initial investment. The mathematical expression is as follows:

$$A = P\left[\frac{i(1+i)^n}{(1+i)^n - 1}\right] = P(A/P_{i,n}) \qquad \text{(A-1)}$$

where A is a uniform payment over n periods at interest rate i to recover initial investment P. The capital recovery factor is commonly abbreviated as $(A/P_{i,n})$; it is used in both personal financing and in comparing investment alternatives.

DERIVATION OF CAPITAL RECOVERY FACTOR

A uniform payment schedule over n periods at interest rate i will result in the following present value (P):

$$P = \sum_{k=1}^{n} A(1+i)^{-k} \qquad \text{(A-2)}$$

where A is the amount of payment in each period. Equation A-2 can be rewritten as the following form:

$$P = \frac{A}{(1+i)}\left[\sum_{k=0}^{n-1} (1+i)^{-k}\right] \qquad \text{(A-3)}$$

449

Therefore $\left[\sum\limits_{k=0}^{n-1} (1 + i)^{-k}\right]$ is a geometric series with an incremental of $(1 + i)^{-1}$ and can be transformed as:

$$\sum_{k=0}^{n-1} (1 + i)^{-k} = \frac{(1 + i)^n - 1}{(1 + i) - 1} = \frac{(1 + i)^n - 1}{i} \tag{A-4}$$

Substitution of Equation A-4 into Equation A-3 yields:

$$P = A\left[\frac{(1 + i)^n - 1}{i(1 + i)}\right] \tag{A-5}$$

By reversing Equation A-5, we obtain Equation A-1.

Appendix B
An Industrial Energy Conservation Model

This appendix presents a partial example of exercising the copper energy conservation model for Company A. The example is the final loop of the linear programming corresponding to the final choice of facilities' utilization rates for the model as shown in Step 5 of Fig. 8–10. The appendix discusses the following three computer programs:

1. The main program to calculate the final set of unit production cost (c_{ijk}) and energy requirement (e_{ijk}) for each production path and valid upper and lower boundaries for each facility. These data correspond to the final conjecture of the optimal utilization rates for company A's facilities. This main program also calls the next two subroutines.
2. A subroutine for preparing an IBM MPS/360 coefficient matrix for the LP problem. In using this subroutine, the left-hand side of every equation must be smaller than or equal to the right hand side, i.e., $ax \leqslant b$.
   If the left-hand side is larger than or equal to the right-hand side,

$$ax \geqslant b \qquad \text{(B-1)}$$

then Equation B-1 must be converted into

$$-ax \leqslant -b \qquad \text{(B-2)}$$

If a constraint equation is in the following form:

$$ax = b \qquad \text{(B-3)}$$

then it should be expressed by the following two equations (B-4 and B-5):

$$ax \leqslant b \qquad \text{(B-4)}$$
$$ax \geqslant b \text{ or} \qquad \text{(B-5)}$$
$$-ax \leqslant -b \qquad \text{(B-6)}$$

Equations B-4 and B-6 must be used to replace Equation B-3.

3. A subroutine for calling the IBM MPS/360 package. The manuals for this package should be followed carefully.

Appendix C
Overview of Energy Conservation Technologies

This appendix summarizes major industrial energy conservation technologies reported in public in alphabetical order. The majority of these technologies were supported by the Department of Energy.* All the costs in this appendix are in 1977 dollars except as otherwise noted. The list of the technologies is not exhaustive.

ABSORPTION CHILLERS

Closed-cycle absorption chillers use either lithium bromide–water or water–ammonia absorbent–refrigerant fluid combinations. They are commercially available in sizes ranging from approximately 3 tons to 1700 tons. In this type of system, relatively high-temperature heat in the form of steam or hot water at 180°F to 400°F is applied in the generator to boil refrigerant vapor out of a strong refrigerant–absorbent solution (at constant pressure). This type of system needs redesigning and reoptimizing in order to be operable with source temperatures below 170°F to 190°F so that it can be better used for waste heat and solar applications. Capital and O&M costs vary, depending on system configuration. No cost estimate is available.

ALUMINUM SMELTER MODIFICATION

The aluminum smelter modification uses an improved cathode for the basic Hall cell. It can be retrofitted onto existing facilities and applied to new facilities (not

---

* Except as otherwise mentioned, more detailed discussion of the technologies can be found in the DOE report.[4]

including direct aluminum reduction facilities). The energy saving rate is estimated to be about 25% over the conventional Hall cell process. DOE estimated a capital cost of $165,000, including a $40,000 installation charge, for an ideal application of this technology. No incremental O&M is expected. Its market fraction is not certain.

BASIC OXYGEN FURNACE

The basic oxygen furnace (BOF) represents the latest process improvement in steel making, primarily replacing the traditional open hearth. Except for electricity (about 61 kWh/ton) needed for auxiliaries and for oxygen production, the BOF consumes no other fuel; that is, heat produced in the furnace by the oxidation of carbon, silicon, and other trace materials is sufficient to bring the metal and slag to 2900°F. Efficient waste heat boilers can be installed over BOFs to produce steam. Because the BOF has a limited capacity for scrap, electric furnaces are needed to process remaining scrap. The effectiveness of the basic oxygen furnace is about 92%, the highest of the three steel-making processes (open hearth, basic oxygen, and electric furnaces).[5] No cost estimate is available.

BLAST FURNACE GASIFIER

This technology retrofits existing blast furnaces not currently in operation with an oxygen plant to produce medium Btu gas for industrial applications. In the United States, there are currently 40 such retired units capable of being modified to consume coal, coke breeze, slag, and scrap materials and to produce fuel gas, molten iron, and slag. Based on a DOE estimate for a 20-foot furnace using 630 tons/day of molten metal and 1575 tons/day of coal, the average blast furnace gasifier unit will produce 10 trillion Btu of medium Btu gas per year. It will consume about 20% more energy in the coal input than what is produced by the process.

The capital cost would be less than $75 million for installing a coal feeder, and oxygen and modifying furnaces. O&M costs are uncertain. In addition to medium Btu gas, this technology will receive a credit for producing iron. The credit is about $4.2 million/year. Competing in the steam and steel reheat service sectors, it could have a maximum market fraction of about 30%.

BLENDED CEMENT

According to Portland Cement Association data,[3] approximately 7% of the current U.S. cement production is blended with pozzolanic material. The blending of pozzolanic material, such as blast furnace slag or fly ash, into Portland cement can reduce the overall energy consumption in cement production. Based upon a 250,000 ton/year blended cement plant using a 30% substitution of pozzolanic material, its annual energy savings are about 148.4 billion Btu/year. Based on an DOE estimate of $5.00/ton of capacity for a blended cement facility, its minimum capital cost is $1.25 million. O&M costs were estmated at 5% of capital costs. Maximum market fraction for this technology would be about 8% of cement production.

BOILER AND HEATER AIR/FUEL CONTROL

The boiler and heater air/fuel control technology uses microprocessor controls supported by a stack gas analyzer and possible spectral flame analyzers at each burner. By minimizing excess air and operating major combustion equipment at near-stoichiometric levels, fuel savings of 1 to 2% are possible.

Based upon a four-burner 200,000-lb/hour boiler operating at 7920 hours/year, its energy savings would be between 13 and 26 billion Btu/year. The total installed cost for a 200,000 lb/hour air/fuel control system was estimated to be between $115,000 and $180,000. O&M costs were estimated to be about 4% of total installed costs. Maximum market fraction is uncertain for this technology.

BRAYTON CYCLES

See Chapter 13 for details.

CEMENT BLOCK DRYING TECHNOLOGY

The cement block drying technology consists of adding pozzolanic material to the cement mix, insulating the curing unit, and using the exothermic reaction in the block to drive the curing process. This technology will minimize energy consumption. DOE estimated that a typical cement block kiln (operating at 2000 hours/year) using this process could save 1.3×10^9 Btu/year. The capital cost for modifying existing cement block curing chambers is between $1,750 and $3,000. O&M costs were estimated at 5% of capital costs. The maximum market fraction for cement block curing in the brick firing service sector could be about 3%.

COAL IN ALUMINUM REMELT

Conventional remelt furnaces in the aluminum industry burn oil or natural gas. This technology uses coal as fuel, competing in the clean direct heat service sector for the aluminum industry (SIC 3334). Based upon a 300×10^6 lb/year aluminum remelt plant, this technology can replace 320×10^9 Btu/year oil or natural gas and requires a capital investment in the range of $10 to $14 million. O&M costs were estimated to be about 2% of total capital costs. The maximum market fraction for this technology was estimated at more than 50% of the remelt furnaces in the aluminum industry.[4] These figures are not very reliable owing to the uncertainty of the technology.

COGENERATION TECHNOLOGIES

See Chapters 13 and 14 for details of various technologies.

CONTINUOUS CASTING OF STEEL COILS

This technology provides continuous casting of thin sheet stock omitting the soaking pits and slab preheaters to save fuel. Based on a 700,000 ton/year plant operating at 6000 hours/year, there will be a saving of $10 or $15 million installed capital costs with no increase in O&M charges, compared with conventional processes.[9] However, this is a new process, and its market penetration is unknown.

CUPOLA FURNACE MODIFICATION

This technology is used in the iron-making service sector. It allows a conventional cupola furnace to operate without using natural gas in incinerating large quantities of carbon monoxide. DOE estimated that a typical cupola furnace modification could save 35 million cubic feet/year of natural gas. Estimated capital cost for this technology is between $60,000 and $90,000 with no incremental O&M costs. The maximum market fraction could be about 15% of the iron-making service demand.

DIRECT REDUCTION OF ALUMINUM

The direct reduction of aluminum is a radical change from the convention Hall cell aluminum-making process. This new method would save 50,000 Btu/lb over the electrolytic process. Based on a 300,000 ton/year direct heat process, which could potentially displace 38×10^{12} Btu/year of electric energy, the direct reduction process will use 4676 MMBtu/hr, compared with 7384 MMBtu/hr, a savings of 37%. The capital cost was estimated to be between $150 and $250 million. O&M costs were estimated at 12% of capital cost. The maximum market fraction could be about 15% of 1976 total aluminum production with great uncertainty.

DRY MILK HEAT PUMPS

This technology is a high-temperature industrial heat pump that uses low-level waste heat and delivers process heat for milk drying. The proposed heat pump system would use two heat exchangers, the first from a 140°F waste water source and the second from 192°F air flow off a baghouse. With this system, ambient air will be elevated from 300°F to 350°F for use in the milk-drying process. This heat pump has a potential fuel saving of 88×10^9 Btu/year. The installed capital cost was estimated to be between $225,000 and $500,000. O&M charges were estimated to be 5% of total capital costs. The maximum market fraction was estimated to be about 5% of the clean direct heat service demand. (See Chapter 12 for more details about heat pumps.)

EXPANDERS

An expander is used to extract power directly from high-pressure gas or liquid streams. The three commonly used basic expander designs are reciprocating piston, rotary, and turbine. Expanders have flow rate capacities ranging from 200 to 8000 cfm with a capability of operating at efficiencies between 60% and 90%. Possible efficiency and design improvements are constrained by high development cost and limited markets. The installed capital cost varies from $50 to $1,000/hp, depending on the type of expander used.[10] (See Chapter 13 for more details.)

FLASH CALCINATION OF ALUMINUM HYDROXIDE

The flash calciner replaces rotary kilns for calcining aluminum hydroxide to form alumina used in smelting. It reduces energy alumina by 30% (from 2000 to 1400 Btu/lb) and also reduces the total energy needed to produce fabricated aluminum by 1%. The process was developed by the Aluminum Company of America (Alcoa), and 24 flash calciners have been installed at seven Alcoa and affiliate company bauxite refining plants throughout the world.[6]

FLASH SMELTERS

See Chapter 7 for details.

FLAT GLASS ENERGY REDUCTION TECHNOLOGY

This technology uses optical sensors and microprocessor controls to monitor and control glass output, and thus reduces energy consumption in the flat glass industry. Assuming a 0.4 million Btu/ton energy savings with a 128.7 thousand ton/year capacity, annual energy savings would be about 15.4 billion Btu/year for this technology. The capital cost was estimated to be between $70,000 and $140,000, with an O&M cost of 10% of capital cost. Its maximum market fraction would be about 1% of the glass melting service demand.

FOAM FIBER PROCESS

A foam fiber process would be used to replace the traditional pad bath in the textile industry to decrease the demand for steam. The steam is used to heat water required in fabric finishing. An A. D. Little study estimated that 674 technically comparable applications exist in the industry. Therefore, the maximum market fraction could be about 15%. Energy savings potential was estimated at 25×10^9 Btu/year. The capital cost was estimated to be between $65,000 and $600,000. No difference is anticipated between O&M costs of the conventional and new systems.

FUMES-AS-FUEL IN CURING COIL COATED MATERIALS

This technology requires a retrofit system, in which paint fumes are incinerated within the curing ovens and thereby used as fuel for curing paint. The total installed capital cost to retrofit an 80 million Btu/hour plant operating at 6000 hours/year was estimated to be $850,000 with no increase in O&M charges.[10] Its market share is uncertain.

GLASS CONGLOMERATES

The glass conglomerate technology is a method for preheating the pelletized glass container batch (with potential applications in other glass processes, e.g., flat glass production) to save melting energy, increase furnace life expectancy, and reduce air pollution. In addition, a furnace operated with this technology would require less frequent furnace wall rebuilds, and, under some circumstances, this technology could also significantly increase the capacity of an existing furnace.

Based on a 150 ton/day furnace operating at 7920 hours/year, the pelletized glass preheat furnace consumes 307×10^9 Btu/year, compared with 328×10^9 Btu/year for the conventional furnace. Therefore, the energy savings are 21×10^9 Btu/year. Its capital cost was estimated to be between $1 and $1.8 million. It was estimated to have a net O&M savings of 12% with a long wall liner life and less air pollution. Its maximum market fraction was estimated to be 6% of the glass melting market.

HEADBOX MODIFICATION FOR PAPER

This technology modifies the headbox in the paper-making process to increase the solid content of the paper slurry from .5% to 2.0%. The process modification will thus decrease the steam service need in the drying cycle. Based upon paper-making units of over 200 inches, the headbox modification can save $.1 \times 10^{12}$ Btu/year and requires a capital investment in the range of $.7 to $1.2 million. O&M costs are expected to be equal to those of the conventional system. The maximum market fraction for this technology was estimated to be more than 4% of paper-making processes.

HEAT EXCHANGERS

See Chapter 12 for details.

HEAT RECYCLE UREA PROCESS

The heat recycle urea process transfers heat directly between process streams and requires less energy transfer across the process boundary. Consequently, it uses about .85 ton of 150 psig steam to produce a ton of urea, compared with about 1 ton of 350 psig steam for stripping processes and even more for other conventional processes.[7] The process was developed by Ivo Mavrovic and licensed to Technip Inc.[8] It recovers more heat through heat exchanges than conventional processes can recover. No cost estimate is available.

HIGH-TEMPERATURE RECUPERATORS

High-temperature recuperators are used to recover heat in gases exiting high-tempera-ture industrial processes, for preheating incoming combustion air. According to the American Iron and Steel Institute (AISI), on a soaking pit recuperator that uses an 1800°F waste gas system to preheat combustion air from 70°F to 800°F, these recuper-ators would save 24.03×10^9 Btu/year of fuel, assuming a 30% savings over a cold air condition.[1] Based on ideal application and involving no unusual installation costs, the total installed cost was estimated at $62,200; this includes $41,500 for a radiation recuperator, $6,500 for piping, $2,000 for installation crane rental, $1,700 for engineering, and $10,500 for labor. Depending on facilities, the total cost can reach $167,000. O&M charges were estimated at $4,000/year or 6.4% of installed cost. Based on the survey of waste heat recovery in 73 industry groups by Garrett Airesearch, its maximum market fraction was estimated at 5% for calcining, 13% for glass melting, and 17.5% for iron-making and steel preheating.[2]

HYPERFILTRATION

This technology was supported jointly by DOE and EPA because it saves energy and reduces pollution control costs. The technology is used in the closed-cycle operation of textile plants to recycle hot process water, thereby reducing energy need and water consumption, and recovering chemicals and dyes. Based on a 30-gallon/ minute filtration unit recycling 90% of 180°F water, this technology saves 13×10^9 Btu/year, and requires a capital investment between $130,000 and $180,000. With a typical O&M charge of 13% of the capital cost, the hyperfiltration unit provides an average nonfuel savings of about 11% yearly. Maximum market fraction was estimated at 20% of textile plants.

LOW-LEVEL HEAT PUMPS

The low-level heat pump, a reverse Rankine cycle, is designed to reclaim waste heat from industrial processes. It has a potential fuel savings of $.98 \times 10^9$ Btu/year, based on the fraction of steam service demand for which an adequate waste heat source is available. A cost breakdown for this technology was estimated as: power recovery system, $303,000; piping, steam evaporator, $448,000; and installation, $195,000. The actual capital cost for this technology can be between $750,000 and $1.3 million, depending on facilities retrofitted. Annual O&M charges were estimated at about 7% of the total capital cost. (See Chapter 12 for more details about heat pumps.)

LUBE OIL RECOVERY

DOE sponsored the development of an improved process for recovering lube oil. The lube oil recovery technology competes for liquid feedstock service demand in SIC-29, petroleum refining, and related industries. Based on a 10-million gallon/ year re-refining plant at a 70% efficiency, this technology will save about .15 trillion Btu/year because the re-refined oil would consume 1.28×10^6 Btu/bbl compared with 2.2×10^6/bbl for virgin lube oil. The capital cost was estimated to be between $3 and $5 million for a re-refining facility. O&M cost is expected to be 12% of the total installed capital cost. Assuming 25% of the annual consumption of 1.35 billion gallons of lube oil in the United States could be collected from various recycling centers, the maximum market fraction is only about 1% of the total liquid feedstock service need.

MEAT-PACKING MODIFICATION

This technology is retrofitted to process meats before cooling in order to save refrigeration and transportation of nonedible portions of meat. Based on a 200 head/day plant operating at 2080 hours/year, the installed capital cost was estimated to be

$360,000 with an incremental O&M cost of 3% of the installed cost. The maximum market fraction is unknown because it might be difficult to maintain meat tenderness and flavor with this technology.

MOVING BEAM FURNACES

The moving beam technology reduces the vibration in a steel reheating furnace, thereby maintaining the integrity of the insulation, a major problem with conventional walking beam or pusher furnaces. The AISI data[1] estimated that the moving beam furnace uses 2.25×10^6 Btu/ton, compared with 2.4×10^6 Btu/ton for a conventional walking beam or pusher furnace. The installed capital cost was estimated to be between $2.0 and $3.5 million with no incremental O&M costs. Maximum market fraction was estimated to be about 3% of steel reheating service demand.

OXYGEN-ENRICHED SMELTERS

See Chapter 7 for details.

PAPER PULP SLUDGE DRYING

The paper pulp sludge drying consists of three steps to dispose of pulp sludges: (1) sludge thickening, (2) solvent drying, and (3) steam recovery combustion. This technology replaces the traditional two-stage disposal method (i.e., mechanical dewatering and fuel-oil-assisted incineration in hog fuel boilers). The energy saving potential was estimated at 115.8×10^9 Btu/year, based on a 50-ton/day dry weight plant. The installed cost was estimated to be between $3 and $5 million dollars. No incremental O&M costs were expected for the new system. The potential market fraction is less than 1% of paper drying service demand.

POLYPROPYLENE TO FUEL

The function of this technology is the thermal cracking of waste atactic polypropylene into a low-sulfur No. 6 fuel oil. Based on a 200 million lb/year plant, 3.46 trillion Btu/year of No. 6 can be produced, but 2.4 million kWh/year of electricity will also be consumed. Assuming the plant is operated at 70% efficiency, the annual energy savings would be about 2.4 trillion Btu. The capital cost was estimated to be between $2 and $3 million, and O&M is about 1.5% of the total capital cost. Maximum market fraction was estimated at 17% of atactic polypropylene production.

POULTRY PROCESS MODIFICATION

This technology includes modification of the scald tank, installation of heat recuperators on hot process water overflows, and adaptation of a heat exchanger to the refrigeration units to preheat water. The technology was supported by DOE and is mainly applied in the food industry (SIC 20). Equipment capital costs and energy savings for a double-shift poultry plant are broken down in Table C-1.

Table C-1. Equipment Capital Costs and Energy Savings for Double-Shift Poultry Plant.

Component	Cost*	Energy Savings 10^9 Btu//year*
Initial housekeeping	$10,000	19.2
Modified scalder tank	35,000	7.5
Heat recuperators	35,000	8.9
Modified cleaning system	7,000	3.6
Total	$87,000	39.2

* DOE estimation.

O&M costs are expected to be about 5% of the installed capital cost. The maximum fraction for this technology was estimated at more than 3% of the total food industry.

PULP PAPER CHARACTERIZATION

The National Bureau of Standards developed an image analysis device to characterize paper in order to: (1) increase recycling of poorly utilized mixed pulp paper, (2) optimize existing recycling processes by accurately measuring pulp quality prior to paper-making, and (3) increase control and use of materials in the virgin pulp industry. Maximum energy savings was estimated at 70×10^9 Btu/year or a 10% to 12% energy reduction, based on an average consumption of 20 to 30 million Btu/ton for paper. The equipment capital cost could range from $110,000 to $160,000, and O&M costs are estimated to be 5% of capital cost. Maximum market fraction for this technology was expected to be larger than 2%. These figures are extremely preliminary.

RANKINE CYCLE

See Chapter 13 for details.

RANKINE-CYCLE DRIVEN CHILLERS

In this type of system, waste heat is used in a Rankine-cycle engine to drive a vapor compression chiller. For large systems where high-temperature waste heat sources are available, generated steam is used to drive a turbine that in turn drives a centrifugal-type chiller employing a refrigerant. Where waste heat sources are low, a double-loop system can be used. Although most of these systems are in the experimental or prototype state of development, some have been fabricated for commercial use in limited sizes. For example, systems providing 200, 800, and 2000 tons of cooling have been operating in Japan for several years.[10] The largest systems are also equipped with a generator for producing electrical power. No cost data were available.

REFUSE IN CEMENT KILN

The purpose of this technology is to substitute refuse-derived fuel extracted from municipal solid waste for conventional fuels (mainly coal) in the calcining sector for SIC 32, the cement industry. Annual energy savings for a 305,000-ton/year kiln were estimated at 181 billion Btu. The capital cost was estimated at $2 million for a modification of a 305,000-ton/day cement kiln; O&M cost would be about 5% of the capital cost. Maximum market fraction would be about 8% of conventional fuel use in the cement industry.

STEAM HEAT PUMPS

A steam heat pump is used to generate steam from low-temperature waste heat sources. DOE estimated that this heat pump would produce 8000 lb/hr of steam. Assuming 1200 Btu/lb for steam and 7000 hours/year of operation, the heat pump would displace 67.2×10^9 Btu/year of steam service demand. The capital cost was estimated to be between $350,000 and $600,000 with an O&M cost of 10% of installed capital costs. The maximum market fraction is uncertain. (See Chapter 12 for more details about heat pumps.)

STIRLING ENGINES

See Chapter 13 for details.

STEAM JET SYSTEM

A steam jet system uses waste heat to generate steam; the steam is then used to drive an ejector pump that compresses low-temperature water vapor from the evaporator. Steam jet system performance is often quoted in terms of demand (e.g., pounds of steam/hour/ton). Steam at 5 to 150 psig can be used to drive the ejector. Steam demand generally decreases with increasing steam pressure (and temperature) but remains relatively constant above 100 psig. Steam jet systems are not manufactured as packaged units but are constructed on-site for commercially available components. Most of the systems in use are employed for air conditioning purposes and have capacities greater than 100 tons.[10] Their market appears to be limited. No cost estimate is available.

THERMALLY DRIVEN SYSTEMS

Heat engines are thermally driven systems used to generate power from low-pressure waste streams. They have been designed to operate on both open- and closed-cycle configurations. Three principal types of heat engines are Stirling engines, Rankine cycles, and Brayton cycles. Their conversion efficiencies range from 10% to 60%, depending on the type of heat engines and systems operating temperature. (For more details, see Chapter 13.)

WASTE HEAT RECOVERY TECHNOLOGIES

See Chapter 12 for details of various technologies.

REFERENCES

1. AISI. "Handbook on Energy Conservation in the Steel Industry," American Iron and Steel Institute, May 1976.
2. Garrett Airesearch. "Survey of Potential Energy Savings Using High Effectiveness Recuperators from Waste Heat Recovery from Industrial Flue Gases," Garrett Airesearch Manufacturing Company, October 1977.
3. FEA. "Energy Conservation Potential in the Cement Industry," FEA Conservation Paper 26, June 1975.
4. Energy and Environmental Analysis, Inc. "Industrial Sector Technology Use Model (ISTUM), Industrial Energy Use in the United States, 1974–2000," Vol. 4, Technology Appendix, final report, U.S. Department of Energy, October 1979.
5. Hyftopoulos, E. et al. *Potential Fuel Effectiveness in Industry.* Cambridge, Mass.: Ballinger Publishing Company, 1974.
6. Sheldon, A. "Energy Use and Conservation in Aluminum Production," in (Y. Chang et al., eds.) *Energy Use and Conservation in the Metals Industry.* New York: The Metallurgical Society of AIME, 1975.
7. Fidler, R. and Hoffman, E. "Heat Recycle Urea Process Offers Energy Savings, Greater On-Stream Reliability," *The Oil and Gas Journal,* March 14, 1977.
8. Mavrovic, I. "Improved Urea Process is Developed," *Chemical Engineering Progress,* February 1974.
9. U.S. Energy Research and Development Administration (ERDA). "Market Oriented Program Planning Study (MOPPS): Technology Characterizations," review draft, Industrial Working Group, June 13, 1977.
10. U.S. ERDA. "Industrial Application Study," CONS/2862, prepared by Drexel University, United Technologies Research Center and Mathematic Inc., December 1976.

Appendix D

An Example of a Power Sales Agreement of Cogeneration (by Pacific Gas and Electric Company)

This appendix contains an example of a power sales agreement of cogeneration, prepared by Pacific Gas and Electric Company. Included in this example are a sample blank power sales agreement form, and definitions and terms (including prices) for power sales of cogeneration and small power production. This agreement form was originally issued on February 4, 1980 and some of the terms, e.g. prices, have been revised subsequently.

POWER SALES AGREEMENT
BETWEEN

_____

AND
PACIFIC GAS AND ELECTRIC COMPANY

THIS AGREEMENT is entered into as of the _____ day of _____, 19____, by and between _____ ("Seller"), a _____, and PACIFIC GAS AND ELECTRIC COMPANY ("PGandE"), a corporation organized and existing under the laws of the State of California, hereinafter sometimes referred to collectively as "Parties" and individually as "Party."

RECITALS

WHEREAS:
A. Seller owns or will own and operate a Co-generation* Facility or Alternative Fuel Generation Facility, and
B. Seller wishes to sell, and PGandE wishes to purchase, electric power from the Facility,
NOW THEREFORE, in consideration of the mutual covenants and agreements hereinafter set forth, the Parties agree as follows:

ARTICLE 1   SALE OF POWER

(a) Seller agrees to sell and deliver and PGandE agrees to purchase and accept delivery of the energy or energy and capacity as indicated below:
  1. ENERGY—_____ output; and,
                  [Net] or [Surplus]
  2. CAPACITY—_____
               [Capacity Rating in kW] or
               [Not Applicable]
  in accordance with Option selected in Article 3 and described in Appendix C—Schedule of Capacity Purchase Prices and Conditions.
(b) Seller may at any time reduce by any amount its capacity sale obligation by giving written notice thereof to PGandE, subject to possible repayments and payment adjustments as provided in Appendix D—Adjustment of Capacity Payments in the Event of Termination or Reduction.
(c) Seller's _____ kW Facility
             [Nameplate Rating of Generator(s)]
  located at _____ shall provide the energy or

  energy and capacity set forth above.

---

* Initial capitalization of words and phrases other than proper names indicates that these terms are defined in paragraph A-1 of Appendix A—Power Purchase General Terms and Conditions.

(d) The scheduled operation date of the Seller's facility is _____.

(date)

ARTICLE 2   TERM OF AGREEMENT

This Agreement shall be binding upon execution and shall remain in effect for a term of _____ years from the Operation Date; provided, that Seller may at any time terminate this Agreement by giving written notice thereof to PGandE, subject to possible repayments and payment adjustments as provided in Appendix D—Adjustment of Capacity Payments in the Event of Termination or Reduction.

ARTICLE 3   PURCHASE PRICE AND METHOD OF PAYMENT

(a) *Energy*
PGandE shall pay Seller for energy delivered and accepted in accordance with Appendix B—Schedule of Energy Purchase Prices.

(b) *Capacity*
Seller elects to be paid for capacity made available to PGandE according to: (check one)

_____   Option #1
_____   Option #2
_____   Option #3

as set forth in Appendix C—Schedule of Capacity Purchase Prices and Conditions. The applicable capacity price for purposes of computing payments under such option is $_____ per kilowatt-year, except as may be adjusted upward, as provided in Appendix C—Schedule of Capacity Purchase Prices and Conditions, or as otherwise provided in Appendix D—Adjustment of Capacity Payments in the Event of Termination or Reduction. [The capacity price is derived from Table 1, Appendix C.] PGandE's obligation to pay Seller for capacity furnished to PGandE shall commence as of the Operation Date.

ARTICLE 4   NOTICES

All written notices pursuant to Section A-19 of Appendix A—Power Purchase General Terms and Conditions shall be directed as follows:

to PGandE: _____
_____
_____

to Seller: _____
_____
_____

ARTICLE 5   DESIGNATED LOCATIONS

Where used in this Agreement, the following term shall mean:

Designated PGandE Switching Center:

_____Substation

ARTICLE 6   TERMS AND CONDITIONS

This agreement includes the following checked appendices which are attached and incorporated by reference herein:

(Check as appropriate)

_____X_____	Appendix A—Power Purchase General Terms and Conditions
_____X_____	Appendix B—Schedule of Energy Purchase Prices
_____X_____	Appendix C—Schedule of Capacity Purchase Prices and Conditions
_____X_____	Appendix D—Adjustment of Capacity Payments in the Event of Termination or Reduction
_____X_____	Appendix E—Simultaneous Purchase and Sale
_____X_____	Appendix F—Insurance
_____X_____	Appndix G—Special Facilities

IN WITNESS WHEREOF, the Parties hereto have caused this Agreement to be executed by their duly authorized representatives as of the last date hereinabove set forth:

SELLER

PACIFIC GAS AND ELECTRIC COMPANY

BY: _____

BY: _____

_____(Type Name)

_____(Type Name)

TITLE: _____

TITLE: _____

APPENDIX A
Power Purchase General Terms and Conditions

A-1.   DEFINITIONS
Whenever used in this Agreement, Appendices and attachments hereto, the following terms shall have the following meanings:

"Adjusted Capacity Purchase Price"—The $/kW-year purchase price from Table 1, Appendix C—Schedule of Capacity Purchase Prices and Conditions, of this Agreement for the period of Seller's actual performance.

"Alternate Fuel Generation"—For the purpose of this Agreement, generation from biomass, refuse-derived fuels and woodwaste, as generally provided in California Public Utilities Commission OII No. 26, Decision No. 91109 (December 19, 1979).

"Biomass Conversion"—The process of conversion of plant materials such as woodwaste, rice hulls, walnut shells, etc., into electricity or energy. (CPUC OII-26 Definition)

"Capacity Rating"—The maximum continuous ability of the Facility to generate electric energy, expressed in kilowatts, less station use and less step-up transformation losses to the high voltage bus at the generator site.

"Capacity Purchase Price Schedule"—The schedule of capacity prices published which sets forth the prices to be paid for capacity in $/kW-year. This schedule will be adjusted periodically.

"Capacity Sale Reduction"—A reduction in the amount of capacity provided or to be provided under this Agreement.

"Co-generation"—The sequential production of electricity and heat, steam or useful work from the same fuel source. (CPUC OII-26 Definition).

"Contract Capacity"—That capacity identified in Article 1(a)(2) of this Agreement except as otherwise changed as provided herein.

"Contract Capacity Price"—The price in $/kW-year set forth in Article 3(b) of this Agreement.

"Contract Termination"—The early termination of this Agreement.

"Current Capacity Price"—The $/kW-year capacity purchase price from the schedule of capacity purchase prices being published by PGandE at the time of termination or reduction of Contract Capacity.

"Designated PGandE Switching Center"—That Substation identified in Article 5 of this Agreement.

"Dispatchable"—That condition of the Facility whereby through engineering design, installed equipment, and operating conditions and procedures, the Facility may be called upon by PGandE, in a manner mutually agreed upon by the Parties, for operation at any time.

"Facility"—That generation facility described in Article 1 of this Agreement.

"Forced Outage"—Any outage caused by mechanical or electrical equipment failure that either fully or partially curtails the electrical output of the Facility.

"Interconnection Facilities"—All facilities required to be installed solely to interconnect and deliver power from Seller's generation to PGandE's system including, but not limited to connection, transformation, switching metering and safety equipment. Interconnection Facilities shall also include any necessary additions and/or reinforcements by PGandE to PGandE's system.

"Minimum Requirements"—Seller's requirements set forth in Paragraph C-2 of Appendix C—Schedule of Capacity Purchase Prices and Conditions.

"Net Energy Output"—The gross output a Seller's generating Facility produces in kilowatt hours, less station use and less step-up transformation losses to the high voltage bus at the generator site.

"Operation Date"—The day commencing at 0001 hours, following the day during which all features and equipment of Facility have reached a degree of completion and reliability, such that they are capable of operating simultaneously to deliver power continuously into PGandE's system; provided, that the Operation Date may occur only after such degree of completion and reliability has been demonstrated to PGandE's satisfaction by operation for a period not to exceed three months.

"Prudent Electrical Practices"—Those practices, methods and equipment, as changed from time to time, that are commonly used in prudent electrical engineering and operations to operate electric equipment lawfully and with safety, dependability, efficiency and economy.

"Refuse-Derived Fuels"—Fuels derived from municipal waste used as a fuel for electric energy production or low BTU gases from sewage treatment plants for use in turbines. (CPUC OII-26 Definition)

"Special Facilities"—Interconnection Facilities furnished by PGandE at Seller's request or because such facilities are necessary additions and/or reinforcements to PGandE's system.

"Surplus Energy Output"—The gross output a Seller's generating facility produces in kilowatt hours, less station use, less any other use by the Seller and less step-up transformation losses to the high voltage bus at the generation site.

"System Protection Facilities"—The equipment required to protect (1) PGandE's system and its customers from faults occurring at the Facility, and (2) the Facility from faults occurring on the PGandE system or on the systems of others to which it is directly or indirectly connected.

"Term of Agreement"—The period of time during which this Agreement will be in effect, as provided in Article 2 of this Agreement.

A-2. *ELECTRIC SERVICES SUPPLIED BY PGandE*

This Agreement does not provide for any electric services by PGandE to Seller. If Seller requires supplemental or standby services from PGandE, Seller shall enter into separate contract arrangements with PGandE in accordance with PGandE's applicable electric tariffs on file with and authorized by the California Public Utilities Commission.

A-3. *CONSTRUCTION*

A-3.1 *Land Rights:*

Seller hereby grants to PGandE for the term of this Agreement all necessary rights of way and easements to install, operate, maintain, replace and remove PGandE's metering and other Special Facilities, including adequate and continuing access rights on property of Seller and Seller agrees to execute such other grants, deeds or documents as PGandE may require to enable it to record such rights of way and easements. If any part of PGandE's facilities are to be installed on property owned by other than Seller, Seller shall, if PGandE is unable to do so without cost to PGandE, procure from the owners thereof, all necessary permanent rights of way and easements for the construction, operation, maintenance and replacement of PGandE's facilities upon such property in a form satisfactory to PGandE. In the event Seller is unable to secure them (i) by condemnation proceedings or (ii) by other means at such cost as may be agreeable to Seller, Seller shall reimburse PGandE for all costs incurred by PGandE in securing such rights.

A-3.2 *Facility And Equipment Design And Construction:*

Seller shall design, construct, install, own, operate and maintain the Facility and all equipment needed to generate and deliver energy or energy and capacity specified herein, except for any Special Facilities constructed, installed and maintained by PGandE pursuant to Appendix G—Special Facilities. Such Facility and equipment shall meet all requirements of applicable codes and all standards of Prudent Electrical Practice. Seller also agrees to meet reasonable PGandE requirements for Seller's Facility and equipment. Seller shall submit all its Facility and equipment specifications to PGandE for review prior to connecting its Facility and equipment to PGandE's system. PGandE's review of Seller's specifications shall not be construed as confirming nor endorsing the design nor as any warranty of safety, durability or reliability of the Facility or any of the equipment. PGandE shall not, by reason of such review or failure to review, be responsible for strength, details of design, adequacy or capacity of Seller's Facility or equipment, nor shall PGandE's acceptance be deemed to be an endorsement of any Facility or equipment. Seller agrees to change its Facility and equipment as may be reasonably required by PGandE to meet changing requirements of PGandE's system. Seller shall give notice to PGandE at three-month intervals of the estimated date of initial power deliveries.

A-3.3   *Interconnection Facility Construction And Meter Installation:*
  Seller shall construct, install, own and maintain Interconnection Facilities as required for PGandE to receive energy or energy and capacity from Seller's Facility. Seller's Interconnection Facilities shall be of a size to accommodate the delivery of the energy or energy and capacity designated in Article 1(a) of this Agreement. In the event it is necessary for PGandE to install Special Facilities or other Interconnection Facilities or to reinforce its system for purposes of this Agreement, Seller shall reimburse PGandE its costs in accordance with the terms and conditions of Appendix G—Special Facilities. At Seller's request, PGandE shall provide, install and maintain meters at a mutually agreed upon designated location to record and indicate the integrated demand for each hour and to measure kilowatt-hours. Meters for measurement of reactive volt-ampere hours shall be provided. PGandE may also install secondary meters, as appropriate, at a location within Seller's Facility, agreed to by both Parties, to enable Seller to make daily telephone reports to be delivered pursuant to paragraph A-4. All meter equipment, installation, ownership and administration costs therefor shall be borne by Seller, including costs incurred by PGandE for inspecting and testing such equipment, all as estimated by PGandE and quoted to Seller.

A-4.   *OPERATION*

A-4.1   *Facility And Equipment Operation And Maintenance:*
  Seller shall operate and maintain its Facility and equipment according to Prudent Electrical Practices and shall generate such reactive power as may be reasonably necessary to maintain voltage levels and reactive area support as instructed by PGandE's system dispatcher or his designated representative. If Seller is unable or unwilling to provide such reactive power, PGandE may do so at Seller's expense.

A-4.2   *Deliveries:*
  Seller shall deliver the energy or energy and capacity designated above, at the point where Seller's electrical conductors contact those of PGandE's at the transmission side of the high voltage disconnect, or at such other point as the Parties may agree.

A-4.3   *Communications:*
  PGandE and Seller shall maintain operating communications through PGandE's designated Substation. The operating communications shall include, but not be limited

to system paralleling or separation, scheduled and unscheduled shutdowns, equipment clearances and daily load reports.

A-4.4 *Meters:*

All meters used to determine the billing hereunder shall be sealed and the seals shall be broken only upon occasions when the meters are to be inspected, tested or adjusted.

PGandE shall, at Seller's expense, inspect and test all meters upon their installation and at least once every two years thereafter. If requested to do so by Seller, PGandE shall inspect or test a meter more frequently than every two years, but the expense of such inspection or test shall be paid by Seller unless upon being inspected or tested the meter is found to register inaccurately by more than two percent of full scale. Each Party shall give reasonable notice of the time when any inspection or test shall take place to the other Party, and that Party may have representatives present at the test or inspection. If a meter is found to be inaccurate or defective, it shall be adjusted, repaired or replaced, at Seller's expense, in order to provide accurate metering.

If a meter fails to register, or if the measurement made by a meter during a test varies by more than two percent from the measurement made by the standard meter used in the test, adjustment shall be made correcting all measurements made by the inaccurate meter for:

(1) the actual period during which inaccurate measurements were made, if the period can be determined, or if not

(2) the period immediately preceding the test of the meter equal to one-half the time from the date of the last previous test of the meter; provided, that the period covered by the correction shall not exceed six months.

Seller shall read the secondary meters daily and shall report the hourly readings and daily energy readings to PGandE's nearest switching center, currently the Substation designated in Article 5 of this Agreement, by telephone at an agreed upon time.

Seller with power deliveries greater than 10 MW shall telemeter the output information to PGandE's switching center designated in Article 5 of this Agreement.

Each Party, after reasonable notice to the other Party, shall have the right of access to all metering and related records.

A-5. *BILLING*

PGandE shall send a statement to Seller on or after the 20th day of the monthly billing period showing the kilowatt capacity, if any, and kilowatt-hours delivered to PGandE during the previous monthly billing period. Seller shall use this statement to compute charges for energy or energy and capacity delivered to PGandE. Seller shall then send a monthly billing statement to PGandE which states the energy or energy and capacity charges.

A-6. *PAYMENT*

PGandE shall make payment to Seller on or before the 15th day after the billing statement is received by PGandE. Where the 15th day falls on a Saturday, Sunday or holiday, the payment shall be due on the next following business day.

A-7. ADJUSTMENTS

In the event adjustments to billing statements are required as a result of corrected measurements made by inaccurate meters, the Parties shall use the corrected measurements described in paragraph A-4.4 to recompute the amounts due from or to PGandE for the energy or energy and capacity delivered under this Agreement during the period of inaccuracy. If the total amount, as recomputed, due from a Party for the period of inaccuracy varies from the total amount due as previously computed, and payment of the previously computed amount has been made, the difference in the amounts shall be paid to the Party entitled to it within 30 days after the paying Party is notified of the recomputation.

A-8. CHANGES IN CAPACITY RATING

Either Party may request, when it reasonably appears that the Capacity Rating of Seller's Facility may have changed for any reason including but not limited to a change in the steam supply, that a new Capacity Rating be determined.

A-9. CONTINUITY OF SERVICE

PGandE shall not be obligated to accept, and PGandE may require Seller to curtail, interrupt or reduce deliveries of energy or energy and capacity in order to construct, install, maintain, repair, replace, remove, investigate or inspect any of its equipment or any part of its system or if it determines that curtailment, interruption or reduction is necessary because of emergencies, forced outages, operating conditions on its system or as otherwise required by Prudent Electrical Practices. PGandE shall not be obligated to accept, and may require Seller to curtail, interrupt or reduce deliveries of energy or energy and capacity (1) whenever PGandE can obtain energy from another source, other than a PGandE fossil fueled plant, at a cost less than the price paid to Seller, (2) during any period when PGandE can generate or purchase an equivalent replacement amount of electric energy generated from renewable resources (including, but not limited to, solar, wind, biomass, geothermal, and hydro) or from plants designated for operation to minimize air pollution, or (3) during periods of minimum system operations; provided, that PGandE shall take or be prepared to take energy or energy and capacity from Seller for not less than 8,160 hours of each calendar year.

In the event of a force majeure, Seller shall not be obligated to deliver, and may curtail, interrupt or reduce deliveries of energy to PGandE and PGandE shall not be obligated to accept and may require Seller to curtail, interrupt or reduce deliveries of energy.

Except in case of emergency, in order not to interfere unreasonably with the other Party's operations, the curtailing, interrupting or reducing Party shall give the other Party reasonable prior notice of any curtailment, interruption or reduction, the reason for its occurrence and its probable duration. Seller always shall notify PGandE promptly of any complete or partial Facility outage.

A-10. *FORCE MAJEURE*

The term "force majeure" as used herein, means unforeseeable causes beyond the reasonable control of and without the fault or negligence of the Party claiming force majeure.

If either Party because of force majeure is rendered wholly or partly unable to perform its obligations under this Agreement, except for the obligation to make payments of money, that Party shall be excused from whatever performance is affected by the force majeure to the extent so affected provided that:

(a) the non-performing Party, within two weeks after the occurrence of the force majeure, gives the other Party written notice describing the particulars of the occurrence;

(b) the suspension of performance is of no greater scope and of no longer duration than is required by the force majeure;

(c) no obligations of either Party which arose before the occurrence causing the suspension of performance are excused as a result of the occurrence; and

(d) the non-performing Party uses its best efforts to remedy its inability to perform. This subparagraph shall not require the settlement of any strike, walkout, lockout or other labor dispute on terms which, in the sole judgment of the Party involved in the dispute, are contrary to its interest. It is understood and agreed that the settlement of strikes, walkouts, lockouts or other labor disputes shall be entirely within the discretion of the Party having the difficulty.

A-11. *INDEMNITY*

Each Party shall indemnify the other Party, its officers, agents, and employees against all loss, damage, expense and liability to third persons for injury to or death of person or injury to property, proximately caused by the indemnifying Party's construction, ownership, operation, or maintenance of, or by failure of, any of such Party's works or facilities used in connection with this Agreement. The indemnifying Party shall, on the other Party's request, defend any suit asserting a claim covered by this indemnity. The indemnifying Party shall pay all costs that may be incurred by the other Party in enforcing this indemnity.

A-12. *LIABILITY; DEDICATION*

Nothing in this Agreement shall be construed to create any duty to, any standard of care with reference to or any liability to any person not a Party to this Agreement.

Neither party shall be liable to the other for damages caused to the facilities of the other by reason of the operation, faulty operation, or nonoperation of the other's facilities.

No undertaking by one Party to the other under any provision of this Agreement shall constitute the dedication of that Party's system or any portion thereof to the other Party or to the public, nor affect the status of PGandE as an independent public utility corporation, or Seller as an independent individual or entity.

A-13. *SEVERAL OBLIGATIONS*

Except where specifically stated in this Agreement to be otherwise, the duties, obligations and liabilities of the Parties are intended to be several and not joint or collective. Nothing contained in this Agreement shall ever be construed to create an association, trust, partnership, or joint venture or impose a trust or partnership duty, obligation or liability on or with regard to either Party. Each Party shall be individually and severally liable for its own obligations under this Agreement.

A-14. *WAIVER*

Any waiver at any time by either Party of its rights with respect to a default under this Agreement, or with respect to any other matters arising in connection with this Agreement, shall not be deemed a waiver with respect to any subsequent default or other matter.

A-15. *ASSIGNMENT*

Neither Party shall voluntarily assign its rights nor delegate its duties under this Agreement, or any part of such rights or duties, without the written consent of the other Party, except in connection with the sale or merger of a substantial portion of its properties including Interconnection Facilities which it owns, and any such assignment or delegation made without such written consent shall be null and void. Consent for assignment will not be withheld unreasonably.

A-16. *CAPTIONS*

All indexes, titles, subject headings, section titles and similar items are provided for the purpose of reference and convenience and are not intended to be inclusive, definitive or to affect the meaning of the contents or scope of this Agreement.

A-17. *CHOICE OF LAWS*

This Agreement shall be construed and interpreted in accordance with the laws of the State of California, excluding any choice of law rules which may direct the application of the laws of another jurisdiction.

A-18. *GOVERNMENTAL JURISDICTION AND AUTHORIZATION*

This Agreement is subject to the jurisdiction of those governmental agencies having control over either Party or this Agreement. This Agreement shall not become effective until all required governmental authorizations and permits are first obtained and copies therof are submitted to PGandE; provided, that this Agreement shall not become effective unless it, and all provisions thereof, is authorized and permitted by such governmental agencies without change or condition.

This Agreement shall at all times be subject to such changes by such governmental agencies, and the Parties shall be subject to such conditions and obligations, as such governmental agencies may, from time to time, direct in the exercise of their jurisdiction. Both Parties agree to exert their best efforts to comply with all applicable rules and regulations of all governmental agencies having control over either Party or this Agreement. The Parties shall take all reasonable action necessary to secure all required governmental approval of this Agreement in its entirety and without change.

If after this Agreement becomes effective, any governmental agency having control over either Party or this Agreement requires any change in this Agreement, or imposes any condition or obligation on either Party, which either in its sole and absolute discretion, deems unreasonably burdensome, such Party may terminate this Agreement.

A-19. *NOTICES*

Any notice, demand or request required or permitted to be given by either Party to the other and any instrument required or permitted to be tendered or delivered.

APPENDIX B
Schedule of Energy Purchase Prices

PGandE shall pay Seller for energy delivered by Seller to PGandE at prices which will be based on PGandE's average quarterly cost of incremental fuel, which quarterly price shall be published by PGandE as provided by California Public Utilities Committion OII-26, Decision No. 91109 (December 19, 1979). The energy price so established shall be applied for periods as follows:

Table A

Average Cost Quarter Used	Months to Which Energy Price Applies
January—March	May—July
April—June	August—October
July—September	November—January
October—December	February—April

Energy prices will be applied to meter readings taken during the months indicated in the right-hand column above.

Table B

The energy prices to be applied to meter readings in August, September and October, 1981*, are:

	Period A	Period B
	August and September, 1981	October, 1981
On-Peak Period, per kWh	8.072 cents	7.752 cents
Partial-Peak Period, per kWh	7.765 cents	7.326 cents
Off-Peak Period, per kWh	6.685 cents	6.542 cents

* See Note on the next page.

Table C

	Monday through Friday*	Saturdays*	Sundays and Holidays
Period A (May 1 to September 30)			
On-Peak	12:30 P.M. to 6:30 P.M.		
Partial-Peak	8:30 A.M. to 12:30 P.M. 6:30 P.M. to 10:30 P.M.	8:30 A.M. to 10:30 P.M.	
Off-Peak	10:30 P.M. to 8:30 A.M.	10:30 P.M. to 8:30 A.M.	12:00A.M. to 12:00 A.M.
Period B (October 1 to April 30)			
On-Peak	4:30 P.M. to 8:30 P.M.		
Partial-Peak	8:30 P.M. to 10:30 P.M. 8:30 A.M. to 4:30 P.M.	8:30 A.M. to 10:30 P.M.	
Off-Peak	10:30 P.M. to 8:30 A.M.	10:30 P.M. to 8:30 A.M.	12:00 A.M. to 12:00 A.M.

* Except the following holidays: New Year's Day, Washington's Birthday, Memorial Day, Independence Day, Labor Day, Veteran's Day, Thanksgiving and Christmas Day, as said days are specified in Public Law 90–363 (U.S.C.A. Section 6103).

This table is subject to change to accord with the on-peak, partial-peak, and off-peak period as defined in PGandE's own rate schedules for the sale of electricity to its large industrial customers.

NOTE FOR READER

The following is for information purposes only. Other prices may be in effect in the future.

The energy prices, based on the quarterly average energy cost for April through June, 1981, if applied to both Period A and Period B, would be:

	Period A	Period B
	May 1 to September 30	October 1 to April 30
On-Peak	8.072¢/kWh	7.752¢/kWh
Partial-Peak	7.765	7.326
Off-Peak	6.685	6.542

APPENDIX C
Schedule of Capacity Purchase Prices and Conditions

C-1 *General*

This Appendix C shall apply if Seller has elected in Article 1(a) (2) of this Agreement to make available (deliver) Contract Capacity to PGandE. It establishes conditions and prices under which PGandE shall pay for such Contract Capacity.

C-2 *Minimum Requirements*

In order to qualify for a capacity payment, the following provisions must be met:

1. The Contract Capacity for payment purposes may not exceed the lowest Capacity Rating in any of the three peak months on PGandE's area system, which are presently the months of June, July and August.

2. The Contract Capacity must be available[1] for all of the on-peak hours[2] in the peak months on PGandE's area system, which are presently the months of June, July and August, subject to an allowance of 20 percent of those on-peak hours for forced outages.

3. Scheduled Outages must be performed between November and April unless otherwise agreed as provided in paragraph C-4 of this Appendix.

C-3 *Payment Options*

The Seller has three options for calculation of capacity payments, and Seller has made its selection in Article 3(b) of this Agreement. The three options are as follows:

Option #1
(Payment in Dollars per Kilowatt-month)

The Facility will be fully Dispatchable by PGandE. Seller must demonstrate that the Facility is fueled by a reliable fuel supply and adequate fuel storage is available to deliver power as requested by PGandE's system dispatcher. Payments will be

---

[1] As used herein "available" means either Dispatchable by PGandE or actually delivered to PGandE.

[2] On-peak, partial-peak and off-peak hours are defined on Table C, Appendix B—Schedule of Energy Purchase Prices.

made in twelve equal monthly amounts based on the Contract Capacity Prices set forth in Article 3(b) of this Agreement multiplied by the Contract Capacity.

Option #2
(Payment in Dollars per Kilowatt-month)

Payment each month will be based on the "monthly delivered capacity," but will not exceed the Contract Capacity.

As used herein, "monthly delivered capacity" shall be equal to FACTOR multiplied by the Contract Capacity. FACTOR is determined according to the following formula:

$$\text{FACTOR} = P \times \left(1.0 - \frac{M}{D}\right)$$

where
$$P = \frac{A}{C \times (B - S) \times (1 - F)}$$

where

A = Kilowatt-hours delivered on-peak and partial-peak
C = Contract Capacity in kilowatts
B = On-peak and partial-peak hours during the month
S = On-peak and partial-peak hours Facility is out of service on scheduled maintenance
M = The number of days during month Facility is out of service on scheduled maintenance
D = The number of days in the month
F = The fraction of on-peak and partial-peak hours allowed for Forced Outage. For the purpose of this Agreement, F equals 0.2.
P = Performance Factor (not to exceed 1.0)

The payment for a month is determined by multiplying the Contract Capacity Price as set forth in Article 3(b) of this Agreement by the FACTOR, as above determined, and then multiplying that product by the following allocation factor:

Allocation Factors*

Period A	Period B
0.1172	0.0588

Option #3
(Payment in cents per Kilowatt-hour)

Payments will be based on the time differentiated energy deliveries by Seller. The payment in cents per kilowatt hour is determined by multiplying the Contract Capacity Price, as set forth in Article 3(b) of this Agreement by the following allocation factors:

	Allocation Factors*	
	Period A	*Period B*
On-Peak	0.03357	0.02022
Partial-Peak	0.02323	0.00893
Off-Peak	0.00553	0.00476

Payment in any month will not exceed the maximum payment available under Option 2 assuming delivery of the full Contract Capacity.

GENERAL REQUIREMENTS AND INFORMATION

C-4 Scheduled Outages

To qualify for capacity payments, scheduled outages for maintenance must be performed within Period B (excluding the first month) unless otherwise agreed to by PGandE. Seller shall provide the PGandE system dispatcher with the scheduled outage dates (start and finish) at least six months in advance. These dates shall not be changed without written approval by PGandE.

Capacity payments will continue during the scheduled maintenance period and will be equal to the product of the average daily capacity payments of the preceding month and the number of days of actual outage for scheduled maintenance. Such payments during scheduled outage shall not be made for more than 35 days in any twelve-month period.

C-5 Adjustments to Contract Capacity

Seller may derate the Contract Capacity at any time. The derated capacity will be subject to Appendix D—Adjustment of Capacity Payments in the Event of Termination or Reduction.

Seller may increase the Contract Capacity with the approval of PGandE and receive payment for the additional capacity in accordance with the then applicable capacity purchase prices.

PGandE may derate the Contract Capacity as a result of appropriate tests, studies or prior performance. The derated capacity will be subject to Appendix D—Adjustment of Capacity Payments in the Event of Termination or Reduction.

C-6 Adjustment to Contract Capacity Price

The Contract Capacity Price will be adjusted upward to the highest capacity price for the scheduled operation date and agreement term published by PGandE between the date the Agreement is executed and the scheduled operation date set forth in Article 1(d) of this Agreement. The applicable Capacity Price schedule shall be attached to this Agreement and shall supersede Table 1 of Appendix C.

---

* These allocation factors will be subject to change by PGandE based on PGandE's marginal capacity cost allocation, as determined in general rate case proceedings before the CPUC. Periods A and B are defined in Table C, Appendix B—Schedule of Energy Purchase Prices.

Table 1. Capacity Price (Levelized $/kW-yr)

Operation Date Year	Term of Agreement												
	1	2	3	4	5	6	7	8	9	10	15	20	30
1980	—	—	—	55	56	57	59	60	61	62	68	73	81
1981	—	—	57	58	60	61	62	63	65	66	72	77	85
1982	—	59	60	61	63	64	65	67	68	69	75	81	89
1983	60	62	63	65	66	67	69	70	71	73	79	85	94
1984	63	65	66	68	69	71	72	74	75	76	83	89	98
1985	66	68	70	71	73	74	76	77	79	80	87	93	103

APPENDIX D
*Adjustment of Capacity
Payments in the Event of
Termination or Reduction*

D-1 *General Provisions*

A. This Appendix shall be applicable in the event there is a Contract Termination or a Capacity Sale Reduction and Seller is receiving capacity payments.

B. The Parties agree that the amount of the payment which PGandE is to make to Seller for capacity which Seller makes available to PGandE is based on the agreed value to PGandE of Seller's performance of his capacity obligations during the full period of the Term of Agreement. The Parties further agree that in the event PGandE does not receive such full performance by reason of a Contract Termination or a Capacity Sale Reduction, (1) PGandE shall be deemed damaged by reason thereof, (2) it would be impracticable or extremely difficult to fix the actual damages to PGandE resulting therefrom, (3) the refund and payments as provided in paragraphs D-2 through D-5, as applicable, are in the nature of adjustments in capacity prices and liquidated damages, and not a penalty, and are fair and reasonable, and (4) such refunds and payments represent a reasonable endeavor by the Parties to estimate a fair compensation for the reasonable losses that would result from such termination or reduction.

C. In the event of a Capacity Sales Reduction, the quantity by which the Contract Capacity is reduced shall be used to calculate the payments due PGandE in accordance with Paragraphs D-2 through D-5, as applicable.

D. Seller shall be invoiced by PGandE for all refunds and payments due under this appendix and Appendix G—Special Facilities, if applicable, and shall pay such amounts to PGandE within 30 days after receipt of said invoice.

E. PGandE shall have the right to offset any amounts due it against any present or future payments due Seller.

G. Notices of termination shall be made in accordance with Section A-19 of Appendix A—Power Purchase General Terms and Conditions.

D-2 *Termination Resulting from Governmental Action*

If either Party terminates either this Agreement, or all or a part of the Contract Capacity thereof, under the provisions of Section A-18 (changes in the Agreement required by any governmental agency), Appendix A—Power Purchase General Terms and Conditions, Seller shall refund to PGandE an amount equal to one-half the difference between the capacity payments already paid by PGandE (which were based on the original Term of Agreement) and the total capacity payments based on the period of Seller's actual performance using the Adjusted Capacity Purchase Price.

D-3 *Termination With Prescribed Notice*

In the event Seller terminates this entire Agreement, or all or part of the Contract Capacity thereof, with the following prescribed written notice:

Contract Capacity	*Length of Notice*
Under 25,000 kW	12 months
25,001 to 50,000 kW	36 months
50,001 to 100,000 kW	48 months
over 100,000 kW	60 months

A. Seller shall refund to PGandE an amount equal to the difference between the capacity payments already paid by PGandE (based on the original Term of Agreement) and the total capacity payments based on the period of Seller's actual performance using the Adjusted Capacity Purchase Price.

B. The Adjusted Capacity Purchase Price shall be calculated at the time the termination notice is received, and PGandE shall make capacity payments for the remainder of Seller's performance period at that price.

D-4 *Termination Without Prescribed Notice*

If Seller terminates that Agreement, or all or a part of the Contract Capacity thereof, without the notice prescribed in D-3:

A. Seller shall refund to PGandE an amount equal to the difference between the capacity payments already paid by PGandE (which were based on the original Term of Agreement) and the total capacity payments based on the period of Seller's actual performance using the Adjusted Capacity Purchase Price; and

B. Seller shall pay PGandE a one-time payment equal to the difference between the Current Capacity Price on the date of termination, for a term equal to the balance of the Term of Agreement, and the Contract Capacity Price, multiplied by the Contract Capacity. This payment will be pro-rated for the length of notice given, if any. The pro-ration factor shall be one minus the ratio of the actual number of months notice given and the prescribed length of notice (as set forth in D-3). In the event that the Current Capacity Price is less than the Contract Capacity Price, no payment under this paragraph D-4(B) shall be due either Party.

D-5 *Termination Due to Seller's Failure to Perform*

Except in the event of force majeure as defined in paragraph A-10 of Appendix A—Power Purchase General Terms and Conditions, in the event of failure of Seller to meet the Minimum Requirements set forth in paragraph C-2 of Appendix C—Schedule of Capacity Purchase Prices:

A. PGandE shall immediately suspend the capacity payments to Seller for a probationary period not to exceed 14 months.

B. If Seller meets or satisfies PGandE that it can meet its Minimum Requirements during the probationary period, PGandE shall make a retroactive capacity payment for the probationary period, and reinstate regular capacity payments.

C. If Seller fails to meet its Minimum Requirements during the probationary period, PGandE may derate the Contract Capacity appropriately or terminate the capacity purchases. In either case, the quantity by which the capacity is reduced shall be considered terminated without prescribed notice (as provided in paragraph D-4).

APPENDIX E
Simultaneous Purchase and Sale

PGandE shall purchase, except as provided in Paragraph A-9 of Appendix A—Power Purchase General Terms and Conditions, Seller's entire Net Energy Output and Contract Capacity, if any, at the prices stated in Appendix B—Schedule of Energy Purchase Prices and Appendix C—Schedule of Capacity Purchase Prices and Conditions as applicable; and, PGandE shall simultaneously supply all of Seller's electric service requirements under separate agreement at applicable filled rates.

During PGandE system emergencies, when Seller's Facility is operating, Seller agrees to provide its Contract Capacity to PGandE's system.

Provisions relating to curtailment or interruption as part of implementation of a statewide Electrical Emergency Plan approved by the California Public Utilities Commission will be observed by PGandE for all electrical service supplied by PGandE.

APPENDIX F
Insurance

Seller shall be required to maintain insurance as indicated below:

F-1 *WORKER'S COMPENSATION* _____
(Yes or No)

Seller shall furnish PGandE a certificate of workers' compensation insurance or self-insurance indicating compliance with the Labor Code of California and providing for 30 days' written notice to PGandE prior to cancellation of such insurance.

F-2 *LIABILITY* _____
(Yes or No)

Seller shall maintain in effect during the term of this Agreement insurance for both bodily injury and property damage liability, including automobile liability, in per occurrence limits of not less than $ _____.
Such insurance shall include:

_____assumption   of   contractual   liability,
_____an endorsement naming PGandE as an additional insured insofar   as   work   performed   under   this   Agreement   is   concerned,
_____a   severability   of   interest   clause,   and
_____provide that notice shall be given to *PGandE* at least 30 days prior to cancellation or material change in the form of such policies. Seller shall furnish PGandE, by delivering to the Manager, Insurance Department, Room 842, 77 Beale Street, San Francisco, California 94106, prior to commencing performance hereof but not less than 30 days before the scheduled date of initial power deliveries (Article 1 of this Power Sales Agreement), certificates of insurance together with the endorsements required therein. PGandE shall have the right to inspect the original policies of such insurance.

APPENDIX G
Special Facilities

[If Special Facilities, as provided in this Agreement, are required, Seller and PGandE shall enter into a separate agreement, which will be substantially in the form of PGandE's current Special Facilities agreement referred to in PGandE's Electric Rule No. 2, providing for Seller to reimburse PGandE its costs for such facilities. Such agreement will be deemed to become a part of this agreement as Appendix G.]

Appendix E

Selected Data for Capacity, Models, and Product Markets of the Copper Industry

This appendix contains 11 tables. Tables E-1, E-2, and E-3 present capacity data for 1975 for the three companies analyzed in Chapter 11. These data were summarized from each individual company's annual reports. Tables E-4 through E-9 present equations for the three econometric models discussed in Chapter 11. Tables E-10 and E-11 show product market data for the copper industry. Symbols used in these tables are explained in the related tables in Chapter 11.

Table E-1. The Capacity of Facilities of Company A.

Facility	Production Capacity (Thousand Short Tons)
Mine–mill	
I	283.3
II	59.8
III	36.2
IV	749.7
Smelter	
I	108.0
II	108.0
III	113.4
IV	270.0
Refinery	
I	186.0
II	276.0
III	103.0

Table E-2. The Capacity of Facilities of Company B.

Facility	Production Capacity (Thousand Short Tons)
Mine–mill	
I	44.8
II	29.6
Smelter	
I	259.2
II	155.5
III	162.0
Refinery	
I	420.0
II	168.0
III	156.0

Table E-3. The Capacity of Facilities of Company C.

Facility	Production Capacity (Thousand Short Tons)
Mine–mill	
I	117.0
II	298.1
III	94.2
Smelter	
I	202.5
Refinery	
I	252.0

Table E-4. Functional Type of Equations for ADL Model.

1. Y_0/Y_5 $= a_0 + a_1 X_6$

2. Y_1/X_5 $= b_0 + b_1 X_3/X_5 + b_2 D_1{}^a$

3. Y_2/X_5 $= c_0 + c_1 Y_8$

4. Y_3/X_5 $= d_0 + d_1 Y_2/X_5 + d_2 Y_5/X_1 + d_3 Y_7$

5. Y_4 $= e_0 + e_1 Y_0/X_5 + e_2 X_2 + e_3 (X_3/X_5) + e_4 (X_4/X_5) + e_5 Y_4(-1) + e_6 [Y_9(-1)]$

6. Y_5 $= Y_6 + Y_7$

7. Y_6 $= -Y_7 - Y_8 + Y_4 + Y_{12} + Y_9 + Y_{10} + Y_{11} + X_{10}$

8. $(Y_2 - Y_0)X_5$ $= h_0 + h_1 (Y_5/X_1) + h_2 [(X_3 - Y_0)/X_5]$

9. $(Y_3 - Y_0)/X_5$ $= k_0 + k_1 (Y_5/X_1) + k_2 [(X_3 - Y_0)/X_5]$

10. ΔY_9 $= l_1 X_9 + l_2 [Y_9(-1)] + D_1{}^a + D_2{}^a$

11. ΔY_{10} $= m_1 (Y_1/X_5) + m_2 Y_6 + m_3 (X_3/X_5) + m_4 [Y_{10}(-1)] + DD_1{}^a$

12. ΔY_{11} $= n_1 (Y_1/X_5) + n_2 Y_7 + n_3 [X_{11}(-1)] + n_4 D^a$

13. Y_{12} $= p_1 (Y_0/X_5) + p_2 (X_3/X_5) + p_3 (X_{13}/X_2) + p_4 D_1{}^a$

[a] D_1 and D_2 are dummy variables for strikes.

Table E-5. Functional Type of Equations for FCB Model.

1. $Y_0 = a_0 + a_1 Y_{29} + a_2 Y_{28}$

2. $Y_4 = b_0 + b_1 [Y_0(-1)/X_5(-1)] + b_2 [X_4(-1)/X_5(-1)] + b_3 X_2 + b_4 X_8 + b_5 X_8(-1) + b_6 Y_4(-1)$

3. $Y_{6.1} = c_0 + c_1 [Y_0/X_5] + c_2 Y_6(-1)$

4. $Y_9 - Y_9(-1) + Y_{10} - Y_{10}(-1) \quad Y_{6.1} + Y_{J5} + Y_{16} + Y_{27} - Y_4 + X_{10}$
   $Y_{11} - Y_{11}(-1) \quad =$

5. $Y_{15} = d_0 + d_1 Y_4$

6. $\ln(Y_{16}/X_{12} + \theta) = e_0 + e_1 \ln[Y_{16}(-1)/X_{12}(-1) + \theta] + e_2 \ln(Y_{28}/X_5)$

7. $Y_{17} = f_0 + f_1 (Y_0/X_5)^a + f_2 [Y_{17}(-1)]$

8. $Y_{18} = g_0 + g_1 (Y_0/X_{18}) + g_2 [Y_{18}(-1)]$

9. $Y_{19} = h_0 + h_1 X_{11}$

10. $Y_{20} = k_0 + k_1 (Y_{28}/X_5) + k_2 [Y_{20}(-1)]$

11. $Y_{21} = l_0 + l_1 [Y_{28}(-1)/X_{19}(-1)] + l_2 [X_4(-1)/X_5(-1)] + l_3 X_{15} + l_4 Y_{21}(-1)$

12. $Y_{22} = m_0 + m_1 [Y_{28}(-1)/X_{18}(-1)] + m_2 X_{14}$

13. $Y_{23} = n_0 + n_1 [Y_{28}(-1)/X_5(-1)] + n_2 [X_4(-1)/X_5(-1)] + n_3 X_{16} + n_4 Y_{23}(-1)$

14. $\ln(Y_{24}/X_{12} + \theta) = p_0 + p_1 \ln[Y_{24}(-1)/X_{12}(-1) + \theta] + p_2 \ln[(Y_{23}/X_{12} + \theta) + \ln(Y_{28}/X_5)]$

15. $Y_{25} - Y_{25}(-1) = Y_{17} + Y_{18} + Y_{19} + Y_{20} + Y_{24} - Y_{21} - Y_{22} - Y_{23} - Y_{27}$

16. $Y_{27} = q_0 + q_1 [Y_{29}/X_5 - Y_{28}/X_5) + q_2 [Y_4 - Y_{6.1}]^a$

17. $Y_{28}/X_5 = s_0 + s_1 [Y_{25}/Y_{23} - Y_{25}(-1)/Y_{23}(-1)] + s_2 [Y_{28}(-1)/X_5(-1)]$

18. $Y_{29}/X_5 = \alpha_0 + \alpha_1 [Y_9(-1) + Y_{10}(-1)]/Y_4(-1) + \alpha_2 [Y_{28}(-1) - Y_0(-1)]/X_5(-1) + \alpha_3 [Y_{29}(-1)/X_5(-1)]$

[a] The U.S. EMJ price is used instead of Chilean copper price because of its dramatic inflation.

Table E-6. Functional Type of Equations for CRA Model.

1. $\ln Y_0 = a_0 + a_1 \ln Y_0(-1) + a_2 \ln Y_2 + a_3 [\ln (X_7/X_7(-1))]$

2. $\ln Y_2 = b_0 + b_1 \ln X_3 + b_2 X_{11}$

3. $Y_4 = C_0 + C_1 Y_4(-1) + C_2 (Y_0/X_5) + C_3 (X_4/X_5) + C_4 X_2 + C_5 X_8 + C_6 X_8(-1)$

4. $\ln Y_{6.1} = d_0 + d_1 \ln Y_6(-1) + d_2 (Y_0/X_5) + d_3 X_1 + d_4 [X_6/X_6{}^{(74)}]$ b

5. $Y_9 = Y_9(-1) + 0.75 [Y_4(-1) - Y_{15}(-1) - Y_{16}(-1))]$

6. $(Y_9 + Y_{10})/Y_6 = e_0 + e_1 \left[1 - \dfrac{Y_2(-1) X_5}{Y_2 X_5(-1)} \right] + e_2 \left[\dfrac{[Y_9(-1) + Y_{10}(-1)]/Y_6(-1)}{} \right]$

7. $Y_{12} = Y_{21} + Y_{22} + Y_{23} + [Y_{26} - Y_{26}(-1)] - Y_{17} - Y_{18} - Y_{19} - Y_{20} - Y_{24}$

8. $Y_{13} = Y_{17} + Y_{18} + Y_{19} + Y_{20} + Y_{24} - Y_{21} - Y_{22} - Y_{23} - (Y_{25} - Y_{25}(-1))$

9. $Y_{14} = Y_{16} - Y_4 - [Y_9 - Y_9(-1)] - [Y_{10} - Y_{10}(-1)] - Y_{11} + Y_{15} + Y_{6.1} + Y_{13}$

10. $Y_{15} = f_0 + f_1 Y_4$

11. $\ln Y_{16} = g_0 + g_1 \ln (Y_2/X_{7.1}) + g_2 \ln Y_9 + g_3 \ln Y_{16}(-1)$

12. $\ln Y_{17} = h_0 + h_1 \ln [Y_0/X_{20}] + h_2 (\ln Y_{17}(-1) + \ln Y_{17}(-2) + \ln Y_{17}(-3))/3$

13. $\ln (Y_{18} + Y_{19} + Y_{20}) = k_0 + k_1 \ln [Y_{18}(-1) + Y_{19}(-1) + Y_{20}(-1)] + k_2 \ln [Y_0/(\alpha X_{17} + \beta X_{18} + \gamma X_{19})]$

14. $\ln (Y_{21} + Y_{22} + Y_{23}) = l_0 + l_1 \ln [Y_0/(\alpha X_{17} + \beta X_{18} + \gamma X_{19}] + l_2 \ln (X_4/(\alpha X_{17} + \beta X_{18} + \gamma X_{19}) + l_3 \ln [\alpha X_{14} + \beta X_{15} + \gamma X_{16}]$

15. $Y_{24} = m_0 + m_1 (Y_{21} + Y_{22} + Y_{23}) + m_2 [Y_0/(\alpha X_{17} + \beta X_{18} + \gamma X_{19})] + m_3 Y_{24}(-1)$

16. $\ln Y_{25} = \ln [Y_{21}(-1) + Y_{22}(-1) + Y_{23}(-1) + Y_{13}(-1)] + n_0 + n_1 (\ln Y_{25}(-1) - \ln (Y_{21}(-2) + Y_{22}(-2) + Y_{23}(-2) + Y_{13}(-2)] - n_2 [X_3/\alpha X_{17} + \beta X_{18} + \gamma X_{19}) + n_3 X_{11} + n_4 D_2{}^a$

17. $\ln Y_{26}(+1) = \ln (Y_{21} + Y_{22} + Y_{23} + Y_{13}) + p_0 + p_1 \ln X_3/(\alpha X_{16} + \beta X_{18} + \gamma X_{19}] + p_2 \ln X_{11} + p_3 D_2{}^a$

---

a D_1 and D_2 are dummy variables for strikes. b $X_6{}^{(74)} = X_6$ in 1974.

Table E-7. Re-estimated Equations for ADL Model.

		Sample period	Estimation method[c]	R^2 [e]	D.W. [f]
1.	$Y_0/Y_5 = -33.03 + 0.41 X_7/X_5$ [d] $\quad(-3.05)(8.31)$	1961–1969 1971–1973	OLSQ	0.874	3.4
2.	$Y_1/X_5 = 28.197 + 4.225 X_3/X_5 + 2.633 D1$ $\quad(1.48)\quad(1.90)\qquad(1.502)$	1964–1973	OLSQ	0.39	1.312
3.	$Y_2/X_5 = -21.73 + 0.074 Y_8 - 7.43 D3$ [a] $\quad(-0.36)\ (1.268)\quad(-1.258)$	1964–1973	INST	0.38	1.88
4.	$Y_3/X_5 = 73.795 + 1.498 Y_2/X_5 - 26.62 Y_5/X_1$ $\quad(2.096)\ (4.22)\qquad(-1.09)$ $\quad -0.2112\, Y_7$ $\quad(-1.49)$	1964–1973	INST	0.864	2.175
5.	$Y_4 = 2088.13 - 25.62 Y_0/X_5 + 15.15 X_2$ $\quad(2.94)\quad(-2.825)\qquad(3.472)$ $\quad -8.73 X_4/X_5 - 0.76 Y_9(-1) +$ $\quad(-0.58)\qquad(-0.51)$ $\quad 0.57\, Y_4(-1) - 119.9\, D1$ $\quad(2.87)\qquad(-5.59)$	1964–1973	INST	0.97	[b]
6.	$Y_5 = Y_6 + Y_7$				
7.	$Y_6 = -Y_7 - Y_8 + Y_4 + Y_{12} + Y_9 + Y_{10}$ $\quad + Y_{11} + X_{10}$				
8.	$(Y_2 - Y_0)/X_5 = -16.20 - 9.56\, Y_5/X_1 + 0.684(X_3 - Y_0) / X_5$ $\quad(-0.48)\ (-0.22)\qquad(3.93)$	1964–1973	INST	0.82	2.557

9. $(Y_3 - Y_0)/X_5$ =

$10.24 - 14.97 Y_5/X_1 +$
(0.257) (-2.85)
$0.625 (X_3 - Y_0)/X_5$
(3.03)

1964–1973 INST 0.71 1.782

10. ΔY_9 =

$525.83 - 0.133 X_9 - 0.605 Y_9(-1) -$
$(3.825)(-2.995)$ (-2.993)
$17.22 D1 + 16.33 D2$
(4.71) (4.19)

1964–1973 OLSQ 0.91 b

11. ΔY_{10} =

$0.044 Y_6 + 0.762 Y_1/X_5 - 0.61 X_3/X_5$
(1.6) (1.38) (-0.76)
$-1.27 Y_{10}(-1) + 78.34 D3^a$
(-10.50) (4.097)

1964–1973 INST 0.97 b

12. ΔY_{11} =

$5.966 + 0.301 X_3/X_5 - 0.187 Y_1/X_5$
(0.039) (0.604) (-0.3165)
$+ 0.087 Y_7 - 0.894 Y_{11}(-1) + 12.37 D3^a$
(0.130) (-1.03) (0.980)

1964–1973 INST 0.49 b

13. Y_{12} =

$1461.73 - 1.16 Y_0/X_5 + 5.06 X_3/X_5$
(5.90) (-0.256) (3.48)
$-1430.97 X_2/X_{13} - 35.69 X_{11} -$
(-9.50) (-4.39)
$33.35 D3^a$
(-0.66)

1964–1973 INST 0.992 2.90

Note: Numbers in the parentheses are student test values for respective coefficients.
[a] $D3$ is a dummy variable for 1972 and 1973's dramatic price change. [b] D.W. statistics are not shown here because they are biased due to the presence of lagged endogenous variables. [c] OLSQ = Ordinary least square scheme. [d] Due to the difficulty of computing operating costs of primary producers as defined by Arthur D. Little, the weekly wages of production workers in copper mining is used as a proxy. [e] R^2 = coefficients of determination. [f] D.W. = Durbin-Watson statistics to test auto-correlation in the error term.

Table E-8. Re-estimated Equations for FCB Model.

	Sample period	Estimation method	R^2	D.W.
1. Y_0/X_5 = $-0.224 + 0.995\,Y_{29} - 0.0007\,Y_{28}$ $\quad(-1.18)\quad(250.45)\quad(-0.47)$	1961–1969	INST	1.0	1.52
2. Y_4 = $217.56 + 92.765\,Y_0(-1)/X_5(-1) - 97.17\,X_4(-1)/$ $\quad(0.5)\quad(5.2)\quad(-3.92)$ $X_5(-1) + 45.305\,X_2 + 82.17\,X_8 -$ $\quad(4.3)\quad(3.3)$ $186.23\,X_8(-1) + 0.117\,Y_4(-1)$ $\quad(5.4)\quad(0.64)$	1961–1969	OLSQ	0.974	e
3. $Y_{6.1}$ = $180.73 + 18.27\,Y_0/X_5 - 80.63\,D1^a$ $\quad(1.10)\quad(4.6)\quad(-4.55)$ $+ 0.309\,Y_{6.1}(-1)$ $\quad(2.75)$	1961–1969	INST	0.95	e
4. $Y_9 - Y_9(-1) +$ $Y_{10} - Y_{10}(-1) + {}= Y_{6.1} + Y_{15} + Y_{16} + Y_{27} - Y_4 + X_{10}$ $Y_{11} - Y_{11}(-1)$				
5. Y_{15} = $190.1 + 0.323\,Y_4$ $\quad(0.97)\quad(2.74)$	1961–1969	INST	0.56	0.28
6. $\ln(Y_{16}/X_{12})$ = $-4.7 + 0.218\ln(Y_{28}/X_5) +$ $\quad(2.14)\quad(1.74)$ $0.315\ln(Y_{16}(-1)/X_{12}(-1))$	1961–1969	INST	0.83	e
7. Y_{17} = $172.74 + 6.20\,Y_0/X_5 + 0.27\,Y_{17}(-1)$ $\quad(1.88)\quad(2.0)\quad(0.81)$	1961–1969	INST	0.87	e

8. Y_{18}
$$= -52.766 - 37.925\, Y_0/X_{17} + 1.196\, Y_{18}(-1)$$
$$(-0.529)\quad (-0.162)$$
$$(6.032)$$
1961–1969   INST   0.91 e

9. Y_{19}
$$= 581.64 + 12.71\, X_{11}$$
$$(18.07)\quad (4.50)$$
1961–1969   OLSQ   0.67   1.85

10. Y_{20}
$$= -366.05 + 1.81\, Y_{28}/X_5 + 1.26\, Y_{20}(-1)$$
$$(-2.99)\quad (1.18)\qquad (15.41)$$
1961–1969   INST   0.97 e

11. Y_{21}
$$= 1405.3 + 21.93\, X_4(-1)/X_5(-1) +$$
$$(2.23)\quad (0.92)$$
$$36.01\, X_{15} - 4.44\, Y_{28}(-1)/X_{19}(-1)$$
$$(3.23)\qquad (2.45)$$
$$- 0.26\, Y_{21}(-1)$$
$$(-0.83)$$
1961–1969   OLSQ   0.96 e

12. Y_{22}
$$= 443.40 + 14.55\, X_{14} - 1.96\, Y_{28}(-1)/X_{18}(-1)$$
$$(4.63)\quad (11.66)\qquad (-1.59)$$
1961–1969, 1971–1973   OLSQ   0.95   2.38

13. Y_{23}
$$= 560.97 - 1.04\, Y_{28}(-1)/X_5(-1) -$$
$$(0.80)\quad (-0.45)$$
$$15.63\, X_4(-1)/X_5(-1) + 1.65\, X_{16} +$$
$$(-0.59)\qquad\qquad (0.153)$$
$$0.74\, Y_{23}(-1)$$
$$(1.57)$$
1961–1969   OLSQ   0.92   1.78

continued

Table E-8 (continued).

	Sample period	Estimation method	R^2	D.W.
14. $\ln(Y_{24}/X_{12}+\theta)^b = 0.26 + 0.42\ln(Y_{23}/(X_{12}+\theta^b))$ $\qquad (2.14)\quad (2.10)$ $\qquad + 0.241\ln\ Y_{24}(-1)/(X_{12}(-1)+\theta^b)^d$	1961–1969, 1971–1973	INST	0.92	1.78
15. $Y_{25}-Y_{25}(-1) = Y_{17}+Y_{18}+Y_{19}+Y_{20}+Y_{24}-Y_{21}-$ $\qquad Y_{22}-Y_{23}-Y_{27}$				
16. $Y_{27} = -272.1 + 2.9\,(Y_{29}/X_5 - Y_{28}/X_5) +$ $\qquad (-2.3)\quad (1.57)$ $\qquad 0.403\,(Y_4 - Y_{6,1})$ $\qquad (3.64)$	1957–1959, 1962–1972	INST	0.67	0.92
17. $Y_{28}/X_5 = 28.50 + -0.14[(\Delta Y_{25}^{C}/Y_{23} - \Delta Y_{25}^{C}(-1)/$ $\qquad (0.66)\quad (0.03)$ $\qquad Y_{23}(-1)]) \times 100 + 0.59\,(Y_{28}(-1)/$ $\qquad\qquad\qquad (1.7)$ $\qquad X_5(-1))$	1971–1973 1959–1965 1967–1973	INST	0.42	1.52
18. $Y_{29}/X_5 = 29.44 - 0.075\,[Y_{28}(-1) - Y_0(-1)]/$ $\qquad (4.43)\quad (-1.26)$ $\qquad X_6(-1) - 155.47\,[Y_9(-1) + Y_{10}(-1)]/$ $\qquad\qquad\qquad (-3.76)$ $\qquad Y_4(-1) + 0.83\,[Y_{29}(-1)/X_5(-1)]$ $\qquad\qquad\quad (8.13)$	1959–1965 1967–1973	OLSQ	0.90	1.68

[a] $D1$ is a dummy variable for strike. [b] θ is assumed as 140,000. [c] Due to the suitability of available data, ΔY_{25} is used instead of original Y_{25}. Symbols for the statistical criteria are the same as those in Table 2–6. [d] Y_{28} is not necessary in explaining this equation. [e] Since D.W. statistics are biased, they are not shown here.

Table E-9. Re-estimated Equations for CRA Model.

	Sample period	Estimated method	R^2	D.W.
1. $\ln Y_0 = -0.051 + 0.27 \ln Y_2 + 0.014 \ln (X_7/X_7(-1)) + 0.762\, Y_0(-1)$ $\quad(-0.19)\ (3.84)\qquad (0.98)\qquad\qquad (9.07)$	1961–1969 1971–1973	INST	0.96	c
2. $\ln Y_2 = 0.54 + 0.73 \ln X_3 + 0.009\, X_{11} - 0.02\, DD_1$ $\quad(1.28)\ (5.58)\qquad (0.77)\qquad (-0.15)$	1961–1969 1971–1973	OLSQ	0.91	3.79
3. $Y_4 = 1266.44 - 6.48\, Y_0/X_5 - 2246.94\, X_4/X_5 + 34.91\, X_2 - 3.925\, X_8 + 11.20\, X_8(-1) + 0.16\, Y_4(-1)$ $\quad(7.90)\ (-2.14)\qquad (-3.23)\qquad (9.47)\qquad (-0.60)\qquad (-1.82)\qquad (1.82)$	1957–1958 1961–1963 1969–1973	INST	0.998	c
4. $\ln Y_{6.1} = 6.45 + 0.015\, Y_0/X_5 - 0.0000\ /X_1 + 0.00006\, X_6/X_6{}^{(74)} + 0.022 \ln Y_{6.1}(-1) - 0.051\, DD_1$ $\quad(19.11)(2.29)\qquad (-0.03)\qquad (0.087)\qquad\qquad (0.25)\qquad\qquad (-2.85)$	1960–1966 1968–1969 1971–1973	OLSQ	0.92	c
5. $Y_9 = Y_9(-1) + 0.75\,(Y_4(-1) - Y_{15}(-1) - Y_{16}(-1))$				
6. $(Y_9 + Y_{10})/Y_{6.1} = 0.15 - 0.102\left(1 - \dfrac{Y_2(-1)\, X_5}{Y_2\, X_5(-1)}\right) + 0.06$ $\quad(6.95)\ (1.22)\qquad\qquad\qquad (0.38)$ $\quad (Y_9(-1) + Y_{10}(-1))/Y_6(-1)$	1961–1969 1971–1973	INST	0.39	c

continued

Table E-9. (continued).

		Sample period	Estimated method	R^2	D.W.
7.	$Y_{12} = Y_{21} + Y_{22} + Y_{23} + Y_{26} - Y_{26}(-1)$ $\quad - Y_{17} - Y_{18} - Y_{19} - Y_{20} - Y_{24}$				
8.	$Y_{13} = Y_{17} + Y_{18} + Y_{19} + Y_{20} + Y_{24} - Y_{21}$ $\quad - Y_{22} - Y_{23} - [Y_{25} - Y_{25}(-1)]$				
9.	$Y_{14} = Y_{16} - Y_4 - [Y_9 - Y_9(-1)] - [Y_{10} -$ $\quad Y_{10}(-1)] - Y_{11} + Y_{15} + Y_{6.1} + Y_{13}$				
10.	$Y_{15} = -262.8 + 0.34\, Y_4$ $\quad\quad (-4.4)\quad (16.51)$	1961–1969 1971–1973	INST	0.97	2.27
11.	$\ln Y_{16} = -1.16 - 0.0040\, DD_1 - 0.087\, \ln$ $\quad (-0.49)\ (-0.148)\quad\quad (-1.04)$ $\quad (Y_2/X_{7.1}) - 0.37 \ln Y_9 +$ $\quad\quad\quad\quad\quad (-0.94)$ $\quad 1.47 \ln Y_{16}(-1)$ $\quad (2.4)$	1960–1966 1968–1969 1971–1973	INST	0.67	c
12.	$\ln Y_{17} = 2.54 + 0.37 \ln Y_0/X_{20} + 0.386 \ln$ $\quad\quad (1.45)\ (0.94)\quad\quad\quad (0.77)$ $\quad [\ln Y_{17}(-1) + \ln Y_{17}(-2) +$ $\quad \ln Y_{17}(-3)]/3$	1961–1969	INST	0.86	1.97
13.	$\ln Y_{18} + Y_{19} +$ $\quad Y_{20}) = 0.347 + 9.38\, Y_0/(\alpha X_{17} + \beta X_{18} + \gamma X_{19})^a$ $\quad\quad\quad (0.98)\ (9.6)$ $\quad + 0.14 \ln [Y_{18}(-1) + Y_{19}(-1) + Y_{20}(-1)]$ $\quad\quad (1.9)$	1957–1967 1971–1973	INST	0.97	c

14. $\ln(Y_{21} + Y_{22} + Y_{23}) = 7.75 + 0.09 \ln[Y_0/(\alpha X_{17} + \beta X_{18} + \gamma X_{19})]^a + 0.43 \ln[X_4/(\alpha X_{14} + \beta X_{15} X_{16})]$

(48.65) (0.28)

(0.93)

$+ 0.011 \ln(\alpha X_{14} + \beta X_{15} X_{16})$

(4.75)

1961–1969   1971–1973   INST   0.95   1.67

15. $Y_{24} = -484.32 + 0.25(Y_{21} + Y_{22} + Y_{23}) + 111.6 Y_0/(\alpha X_{17} + \beta X_{18} + \gamma X_{19}) + 0.173 Y_{24}(-1)$

(-1.39) (4.0)

(1.3)

1961–1969   1971–1973   INST   0.94   1.57

16.b $\Delta Y_{25}/[Y_{21}(-1) + Y_{23}(-1) + Y_{13}(-1)] = -0.064 + 0.058 X_3/(\alpha X_{17} + \beta X_{18} + \gamma X_{19}) + 0.0002 X_{11} - 0.007 DD_1 + 0.69 DY_{25}(-1)/Y_{21}(-2) + Y_{22}(-2) + Y_{23}(-2) + Y_{13}(-2)$

(-1.48) (0.97)

(0.059) (-0.92)

(1.66)

1961–1969   1971–1973   INST   0.33

17.b $\Delta Y_{26}(+1)/(Y_{21} + Y_{22} + Y_{23} + Y_{13}) = -0.155 + 0.007 X_3/(\alpha X_{17} + \beta X_{18} + \gamma X_{19}) - 0.008 X_{11} - 0.006 DD_1$

(-4.64) (0.20)

(-0.34) (-1.11)

1961–1969   1971–1973   OLSQ   0.15   2.33

---

[a] γ, β, α are calculated by the following weights: OECD European countries = 0.629, Japan: 0.284, and rest-of-world: 0.087. Symbols for the statistical criteria are the same as those in Table 14-6. [b] Since some of the dependent variables were negative their logarithmic values were not possible. The straight lines were used. [c] D.W. statistics are not shown here.

Table E-10. Data for Exogenous Variables in the PCM.

Year	X_1	X_2	X_3	X_4	X_5	X_6	X_7	X_8	X_9	X_{10}
1957	2064.00	51.28	27.11	13.92	60.55	0.0	96.89	31.60	2214.00	79.00
1958	2081.49	44.90	24.57	12.83	61.89	0.0	94.19	29.89	2021.49	124.69
1959	2108.49	51.53	29.52	14.25	62.62	0.0	105.89	31.49	2383.00	4.49
1960	2309.00	52.44	30.52	13.36	62.69	0.0	116.79	31.95	2119.49	6.00
1961	2330.99	51.44	28.46	11.80	62.35	0.0	119.00	32.22	2213.00	-5.00
1962	2340.99	57.16	29.04	10.93	62.29	0.0	128.69	34.30	2423.00	-7.39
1963	2333.99	60.89	29.05	11.44	62.29	0.0	124.59	35.52	2540.49	-11.39
1964	2333.99	65.45	43.55	12.16	63.02	36.00	130.42	38.16	2863.00	-27.29
1965	2332.99	73.32	58.06	14.53	63.82	37.00	136.71	41.93	3052.49	-122.49
1966	2419.99	82.02	68.43	12.63	65.22	38.00	140.09	49.50	3398.49	-445.00
1967	2426.49	82.84	50.73	10.38	66.62	39.00	140.49	54.57	2903.00	-171.19
1968	2527.00	87.40	55.45	9.97	68.95	40.00	162.36	58.67	2911.00	-13.79
1969	2643.00	91.13	65.62	12.90	71.75	41.00	169.00	62.83	3274.00	-8.00
1970	2676.00	84.09	63.25	10.83	74.61	42.00	175.66	66.42	2820.00	0.0
1971	2676.00	82.35	48.78	8.25	77.94	43.00	178.45	65.61	2909.49	-1.79
1972	2793.00	89.80	48.06	7.44	80.67	44.00	192.18	69.80	3234.00	0.0
1973	2723.00	101.07	80.31	11.09	85.20	45.00	206.41	78.83	3596.49	-33.79
1974	2850.00	100.00	93.00	16.17	100.00	43.00	226.45	97.19	3097.00	-182.49
1975	0.0	0.0	0.0	0.0	110.45	0.0	248.13	95.16	0.0	0.0

Year	X_{11}	X_{12}	X_{14}	X_{15}	X_{16}	X_{17}	X_{18}	X_{19}
1957	1.00	924.09	14.32	43.84	0.0	57.18	57.48	45.05
1958	2.00	1257.69	14.00	44.49	37.49	56.55	55.38	44.33
1959	3.00	1392.79	16.90	47.09	0.0	58.02	54.82	43.84
1960	4.00	1277.09	15.48	51.92	42.44	57.56	54.87	44.76
1961	5.00	1412.89	20.07	54.54	44.27	55.65	54.59	45.95
1962	6.00	1563.49	22.17	56.77	47.84	53.71	54.41	47.44
1963	7.00	1727.89	23.96	59.21	51.05	54.30	53.77	48.29
1964	8.00	1765.29	30.03	67.27	55.64	55.26	53.23	45.14
1965	9.00	2060.99	31.51	70.36	58.45	56.12	53.18	46.73
1966	10.00	2422.29	50.81	72.24	62.74	57.69	53.39	48.50
1967	11.00	1884.69	45.44	72.29	64.66	58.77	52.62	47.64
1968	12.00	1944.29	56.10	77.27	70.00	60.09	52.91	46.30
1969	13.00	2077.89	65.52	84.82	75.29	62.32	54.07	50.77
1970	14.00	1884.09	77.82	89.52	77.96	65.88	56.02	56.52
1971	15.00	1977.39	78.59	89.92	82.28	69.52	57.05	60.26
1972	16.00	2241.59	82.02	93.01	88.53	74.06	66.08	66.54
1973	17.00	2386.09	99.06	100.17	97.52	81.54	85.04	83.56
1974	18.00	2685.59	100.00	99.99	99.99	100.00	100.00	99.99
1975	19.00	0.0	86.07	93.35	0.0	108.23	100.51	106.76

Table E-11.   Data for Endogenous Variables in the PCM.

Year	Y_0	Y_1	Y_2	Y_3	Y_4	Y_5	Y_6	$Y_{6.1}$	Y_7	Y_8
1957	29.28	31.83	19.88	32.75	2106.39	1676.09	1422.09	1320.65	254.00	808.69
1958	25.50	23.77	17.40	30.27	1974.59	1579.49	1373.09	1189.90	206.39	735.29
1959	30.86	28.31	22.32	33.21	2318.59	1331.49	1071.59	1002.18	259.89	883.09
1960	31.73	30.73	20.94	31.84	2079.29	1794.59	1559.59	1312.50	235.00	831.59
1961	29.62	27.40	21.56	32.45	2170.00	1813.89	1609.19	1415.78	204.69	778.39
1962	30.29	29.93	21.36	32.74	2361.09	1884.49	1677.59	1492.58	206.89	804.09
1963	30.29	28.24	21.96	33.35	2565.69	1884.39	1639.69	1474.11	244.69	849.00
1964	31.64	30.42	25.72	35.62	2775.00	1990.39	1740.69	1514.93	249.69	989.69
1965	34.66	35.53	34.14	47.01	2995.39	2214.70	1864.00	1642.39	276.79	1025.59
1966	35.80	53.84	44.21	57.08	3368.59	2183.29	1896.49	1736.56	286.79	1013.09
1967	37.85	51.59	32.82	45.69	2850.39	1526.59	1251.39	1159.29	275.19	932.09
1968	41.43	50.92	32.43	46.29	2813.49	1839.00	1546.29	1463.66	292.69	982.59
1969	47.05	48.26	42.44	56.30	3164.19	2214.89	1937.00	1876.78	277.89	1082.89
1970	57.12	52.18	39.05	52.91	2930.49	2242.69	1975.09	2089.54	267.59	978.49
1971	50.91	47.14	27.29	47.09	2956.00	1962.39	1733.29	1849.56	229.09	971.89
1972	50.11	48.79	38.63	58.43	3268.19	2258.49	2011.39	1998.53	247.09	1062.19
1973	58.51	51.40	49.91	71.78	3481.69	2312.59	2027.19	2087.35	285.39	1119.49
1974	0.0	81.63	54.07	79.87	3106.19	2136.49	1838.79	1936.32	297.69	1025.09
1975	0.0	0.0	0.0	0.0	0.0	0.0	0.0	1714.33	0.0	0.0

Year	Y_9	Y_{10}	Y_{11}	Y_{12}	Y_{13}	Y_{14}	Y_{15}	Y_{16}	Y_{17}	Y_{18}
1957	175.00	181.00	43.00	274.00	410.26	79.00	452.19	509.69	528.66	359.12
1958	179.00	80.69	49.00	287.39	349.29	124.69	433.69	489.29	512.45	345.12
1959	140.00	64.79	71.00	-32.89	347.95	4.49	509.29	556.89	600.53	395.28
1960	148.00	139.29	62.00	438.00	367.82	6.00	494.29	488.39	586.53	439.26
1961	152.00	79.79	43.00	480.49	382.12	-5.00	471.19	476.49	603.39	439.84
1962	157.00	117.39	59.00	269.79	373.31	-7.39	540.09	485.59	645.83	457.34
1963	149.00	76.89	47.00	228.39	416.17	-11.39	589.69	511.39	684.63	452.60
1964	159.00	45.59	53.00	267.69	436.21	-27.89	675.29	570.79	685.29	486.88
1965	168.00	60.79	49.00	256.09	353.97	-122.49	775.69	614.20	645.17	507.82
1966	231.00	65.69	52.00	135.09	389.70	-445.00	828.39	633.89	701.83	506.06
1967	182.00	55.39	48.00	-123.59	242.59	-171.19	683.89	578.69	727.73	613.31
1968	168.00	56.59	42.00	-72.69	317.53	-13.79	730.59	609.00	724.09	633.26
1969	178.00	45.89	57.00	130.69	271.43	-8.00	813.49	679.59	758.49	573.19
1970	242.00	160.59	68.00	168.89	188.90	0.0	744.00	620.09	762.34	672.73
1971	229.00	103.00	51.00	74.39	151.70	-1.79	760.79	548.00	780.75	721.45
1972	167.00	157.39	46.00	29.49	184.30	0.0	863.49	551.79	790.12	793.32
1973	167.00	49.09	42.00	72.59	166.00	-33.79	910.59	569.59	810.62	908.18
1974	0.0	0.0	0.0	-108.49	241.20	-182.49	868.89	562.49	994.38	905.42
1975	0.0	0.0	0.0	0.0	0.0	0.0	0.0	0.0	913.03	785.82

continued

Table E-11. (continued).

Year	Y_{19}	Y_{20}	Y_{21}	Y_{22}	Y_{23}	Y_{24}	ΔY_{25}	ΔY_{26}	Y_{27}	Y_{28}	Y_{29}
1957	480.27	718.82	2302.81	350.19	647.38	991.65	−358.13	−460.10	136.26	27.11	29.57
1958	441.02	770.63	2396.39	327.38	645.28	1087.08	−274.62	−545.63	61.89	24.57	25.76
1959	598.87	879.22	2402.34	395.72	708.06	1171.60	−241.17	−514.10	380.55	29.52	31.18
1960	635.36	1013.00	2784.07	555.55	695.72	1317.01	25.99	−460.30	−70.17	30.52	32.05
1961	633.49	1002.10	2893.53	664.57	719.88	1428.17	−73.41	−496.91	−98.37	28.46	29.82
1962	619.82	985.05	2789.03	585.31	761.26	1472.80	−58.27	−526.39	103.51	29.04	30.59
1963	648.26	1011.56	2910.94	652.67	841.37	1504.30	−291.40	−608.41	187.77	29.05	30.59
1964	696.98	1022.66	3171.31	846.01	1085.60	1795.55	−584.06	−907.30	168.51	43.55	31.95
1965	766.86	1000.35	3244.83	751.21	1147.04	2012.65	−308.08	−890.75	97.87	58.06	35.01
1966	687.17	1140.31	3062.73	783.18	1007.71	2076.66	3.80	−736.12	254.60	68.43	36.16
1967	730.82	1240.41	2976.42	973.32	1004.17	1923.66	−84.17	−859.32	366.19	50.73	38.22
1968	754.96	1288.44	3249.46	1122.35	1142.45	2173.35	−230.39	−966.60	290.23	55.45	41.84
1969	793.10	1317.45	3528.78	1265.32	1239.81	2482.93	−279.49	−1084.79	140.73	65.62	47.51
1970	754.08	1414.76	3608.91	1303.57	1232.50	2384.87	−177.20	−1014.65	20.00	63.25	57.70
1971	718.03	1576.03	3477.75	1276.13	1288.54	2259.11	−64.32	−799.83	77.30	48.78	51.43
1972	790.45	1833.16	3648.05	1439.93	1331.55	2267.21	−100.05	−948.38	154.80	48.06	50.61
1973	778.88	2069.29	4013.24	1789.91	1525.82	2596.99	−258.39	−1307.36	93.40	80.31	58.86
1974	769.40	2157.16	3941.59	1405.31	1644.00	2565.00	50.75	−901.99	349.70	93.00	76.64
1975	746.14	2095.31	3456.58	1249.78	3494.82	3082.57	−578.30	−74.41	0.0	0.0	63.53

REFERENCES

1. Arthur D. Little Inc. "Econometric Simulation and Impact Analysis Model of the U.S. Copper Industry, Technical Appendix to Economic Impact of Environmental Regulations on the U.S. Copper Industry," draft report to the U.S. Environmental Protection Agency, October 1976.
2. Fisher, F., Costner, P., and Baily, M. "An Econometric Model of the World Copper Industry," *Bell Journal of Economics and Management Science*, Vol. 3, No. 2, 568–609, Autumn 1972.
3. Charles River Associates, Inc. "Economic Analysis of the Copper Industry," report to the Property Management and Disposal Service, General Service Administration, March 1970.
4. ———. "An Econometric Model of the Copper Industry," report to the Property Management and Disposal Service, General Service Administration, September 1970.
5. ———. "The Effects of Pollution Control or the Nonferrous Metals Industries. Copper. Part III. The Economic Impact of Pollution Abatement on the Industry," NTIS, PB-207 163, December 1971.
6. Synergy, Inc. "A Forecasting System for Critical Imported Minerals," final report to Bureau of Mines, U.S. Department of the Interior, May 30, 1975.
7. American Bureau of Metal Statistics. *Yearbook of the American Bureau of Metal Statistics.* New York: The Bureau, 1975. Annual.
8. American Metal Market. *Metal Statistics.* New York: American Metal Market. Annual.
9. Copper Development Association, Inc. *Annual Data Copper Supply and Consumption.* New York: CDA.
10. *E/MJ.* New York: McGraw-Hill. Monthly.
11. U.S. Department of Commerce, Office of Business Economics. *Biennial Supplement to the Survey of Current Business.* Washington, D.C.: GPO.
12. U.S. Department of Labor, Bureau of Labor Statistics. *Employment and Earnings Statistics for the United States.* Washington, D.C., GPO. Annual.
13. ———. *Employment and Earnings and Monthly Report on the Labor Force.* Washington, D.C., GPO. Monthly.
14. ———. *Wholesale Prices and Prices Indexes.* Washington, D.C., GPO. Annual.
15. International Monetary Fund. *International Financial Statistics.* Washington, D.C.: The Fund. Monthly.
16. ———. *International Financial Statistics. Annual Supplement.* Washington, D.C.: The Fund. Monthly.
17. O.E.C.D. *Main Economic Indicators, Historical Statistics.* Biennial.
18. ———. *Main Economic Indicators.* Monthly.
19. Phelps Dodge Corporation. Annual Report.
20. United Nations. *Statistical Yearbook of the United Nations.* New York: The United Nations. Annual.
21. Metallgesellschaft, A. G. *Statistical Tables on Aluminum, Leach, Copper, Bine, Tin, Cadmium, Nickel, Mercury, and Silver.* Frankfurt am Main. Annual.

Appendix F

Pairwise Comparisons for Hierarchical Analysis

INTRODUCTION

This appendix presents a discussion of computing priority vectors (eigenvectors) among activities and computing consistency indices in comparisons. It also includes tables for the pairwise comparisons discussed in Chapter 11.

CALCULATION METHODS

As discussed in Chapter 11, pairs of objects are chosen for comparison from a set of them in order to rank their importance. These comparison values form a matrix (e.g., Table F-1). A priority vector can be obtained from this matrix by using the following equation:

$$AW^* = \lambda_{max} W^* \qquad \text{(F-1)}$$

where A is the matrix, W^* is the priority vector, and λ_{max} is the maximum eigenvalue. Note that W^* is normalized (i.e., $\sum_{i=1}^{n} W^* = 1$) for comparison convenience.

One of important properties in Equation F-1 is the following:

$$\det (\lambda I - A) = 0 \qquad \text{(F-2)}$$

where det is the symbol for determinant, and λ is an eigenvalue.

One difficult question for the above comparison and calculation is that the comparison values (or judgments) may not be consistent, and thus the priority vector for the matrix can become meaningless. For example, if cars are judged to be twice as preferable as houses and houses twice as preferable as children, and if in any case cars are not at least more preferable than children, then the resulting priority vector will become useless. Note that there need be no relationship between consistency and correctness of judgments. An individual may have excellent consistency in judgment but makes a series of wrong comparisions about reality.

A consistency index can be computed to measure the "closeness" to consistency of judgments in pairwise comparison. This index tells only how consistently an individual makes comparisons among objects. It can be expressed as follows:

$$\text{Index} = (\lambda_{max} - n)/(n - 1) \qquad \text{(F-3)}$$

where Index is the consistency index, λ_{max} is the maximum eigenvalue, and n is the number of objects compared in the matrix.

In general, if the value of a consistency index for a pairwise comparison matrix is less than 0.1, then we may conclude that judgments in the matrix are satisfactory with respect to consistency of comparisons.

An Example

Equations F-1, F-2, and F-3 are used to compute the maximum eigenvalue, priority vector, and consistency index for a comparison matrix. Table F-1 is used as an example.

Computing the Maximum Eigenvalue. Table F-1 is substituted for A in Equation F-2:

$$|\lambda I - A| = \begin{vmatrix} \lambda - 1 & -1 & -3 \\ -1 & \lambda - 1 & -3 \\ -1/3 & -1/3 & \lambda - 1 \end{vmatrix} = \lambda^2 (\lambda - 3) = 0$$

Therefore, λ_{max} is 3.

Computing the Priority Vector. According to Equation F-2, the priority vector is computed as follows:

$$\begin{pmatrix} 1 & 1 & 3 \\ 1 & 1 & 3 \\ 1/3 & 1/3 & 1 \end{pmatrix} \begin{pmatrix} W_1{}^* \\ W_2{}^* \\ W_3{}^* \end{pmatrix} = 3 \begin{pmatrix} W_1{}^* \\ W_2{}^* \\ W_3{}^* \end{pmatrix}$$

$$W^* = \begin{pmatrix} .4286 \\ .4286 \\ .1428 \end{pmatrix}$$

Computing the Consistency Index. Equation F-3 is used to compute the consistency index as follows:

$$\text{Index} = (3 - 3)/2 = 0$$

Therefore, the judgments in Table F-1 are consistent.

There are commercial packages available for use in computing the above values.

COMPARISON MATRICES FOR CHAPTER 11

The following symbols are used in Tables F-1 through F-13:

Symbol	Definition
ENV	Environmentalists
Policy I	Energy tax credit (13.9%)
Policy II	Tax rebates (1.8%)
Policy III	Aids (.4¢ per pound copper production)
λ^{max}	Maximum eigenvalue
$W*$	Dominant eigenvector (priority vector)
Index	Consistency index $(\lambda^{max} - n)/(n - 1)$
Eco.	Economic criterion
Pol.	Political criterion
Env.	Environmental criterion
GOV	Federal government

Table F-1. Importance of Policies with Respect to Economic Criterion under Environmentalists' (ENV) Consideration.

	Policy		
	I	II	III
I	1	1	3
II	1	1	3
III	1/3	1/3	1

$\lambda^{max} = 3.$ $W* = (.4286\ .4286\ .1428)$ Index $= 0$

Table F-2. Importance of Policies with Respect to Political Consideration for ENV.

	Policy		
	I	II	III
I	1	1/5	1/7
II	5	1	1/3
III	7	3	1

$\lambda^{max} = 3.065$ $W* = (.1884\ .08096\ .7306)$ Index $= .0325$

Table F-3. Importance of Policies with Respect to Environmental Considerations for ENV.

		Policy	
	I	II	III
I	1	1/3	1/3
II	3	1	1
III	3	1	1

$\lambda^{max} = 3.$ $W^* = (.1428\ .4286\ .4286)$ Index $= 0$

Table F-4. Importance of Policies from the Point of View of ENV.

		Criterion	
	Eco.	Pol.	Env.
Eco.	1	1/3	1/9
Pol.	3	1	1/7
Env.	9	7	1

$\lambda^{max} = 3.08$ $W^* = (.0549\ .2997\ .6554)$ Index $= 0.04$

Table F-5. Importance of Policies with Respect to Economic Consideration for Government (GOV).

		Policy	
	I	II	III
I	1	1/3	1/7
II	3	1	1/5
III	7	5	1

$\lambda^{max} = 3.065$ $W^* = (.08096\ .1884\ .7306)$ Index $= .0325$

Table F-6.   Importance of Policies with Respect to Political Consideration for GOV.

	Policy		
	I	II	III
I	1	7	3
II	1/7	1	1/2
III	1/3	2	1

$\lambda^{max} = 3.003$ $W^* = (.6817\ .1025\ .2158)$ Index $= .0015$

Table F-7.   Importance of Policies with Respect to Environmental Criterion for GOV.

	Policy		
	I	II	III
I	1	1/3	1/3
II	3	1	1
III	3	1	1

$\lambda^{max} = 3.$ $W^* = (.1428\ .4286\ .4286)$ Index $= 0$

Table F-8.   Importance of Criteria From the Point of View of GOV.

	Criterion		
	Eco.	Pol.	Env.
Eco.	1	1/5	3
Pol.	5	1	7
Env.	1/3	1/7	1

$\lambda^{max} = 3.065$ $W^* = (.1884\ .7306\ .08096)$ Index $= .0325$

Table F-9. Importance of Policies with Respect to the Economic Criterion for Industry.

	Policy		
	I	II	III
I	1	1/3	1/5
II	3	1	1/3
III	5	3	1

$\lambda^{max} = 3.039$ $W^* = (.1047 \ .2583 \ .637)$ Index $= .0195$

Table F-10. Importance of Policies with Respect to Political Criteria for Industry.

	Policy		
	I	II	III
I	1	3	1/3
II	1/3	1	1/5
III	3	5	1

$\lambda^{max} = 3.09$ $W^* = (.2583 \ .1047 \ .637)$ Index $= .045$

Table F-11. Importance of Policies with Respect to Environmental Criteria for Industry.

	Policy		
	I	II	III
I	1	1/3	1/3
II	3	1	1
III	3	1	1

$\lambda^{max} = 3$ $W^* = (.1428 \ .4286 \ .4286)$ Index $= 0$

Table F-12.   Importance of Criteria from the Point of View of Industry.

| | Criterion | | |
	Eco.	Pol.	Env.
Eco.	1	3	7
Pol.	1/3	1	3
Env.	1/7	1/3	1

$\lambda^{max} = 3.007$ $W^* = (.6694\ .2426\ .08795)$ Index $= .0035$

Table F-13.   Comparisons of Influence Among Interest Groups.

| | Interest Group | | |
	ENV	GOV	Industry
ENV	1	1/3	1/9
GOV	3	1	1/7
Industry	9	7	1

$\lambda^{max} = 3.039$ $W^* = (.2583\ .637\ .1047)$ Index $= .0195$

Acronyms and Abbreviations

ABMS	American Bureau of Metal Statistics
ac	acre
ac.	alternating current
ACS	Advanced Cogeneration System
ADL	Arthur D. Little
AES	Auto Efficiency Standards
AISI	American Iron and Steel Institute
AMM	American Metal Market
ASME	American Society of Mechanical Engineers
BA	budget authority
BACT	best available control technology
BAT	best available technology
bbl	barrel
B/C	benefit/cost
BCT	best conventional pollutant control technology
BEPS	Building Energy Performance Standards
BESOM	Brookhaven Energy Systems Optimization Model
BNL	Brookhaven National Laboratory
BOE	barrels (per day) of oil equivalent
BOF	basic oxygen furnace
BPT	best practicable control technology
Btu	British thermal unit
°C	degrees Centigrade
CAA	Clean Air Act
ca or cal	calorie
CBS	Current Business Survey
CCP	chance-constrained programming
CDA	Copper Development Association
CEC	California Energy Commission
CHP	combined heat and power generation
cm	centimeter

CO_2	carbon dioxide
COP	coefficient of performance
Cr	chromium
CRA	Charles River Associates
CTAS	Cogeneration Technology Alternatives Study
Cu	copper
db(A)	decibel
DEUS	Dual Energy Use System
DH	district heating
DMEA	Defense Minerals Exploration Administration
DOC	Department of Commerce
DOE	Department of Energy
DOT	Department of Transportation
DRI	Data Resources Inc.
ECDB	End-use Energy Consumption Data Base
ECPA	Energy Conservation and Production Act
EEA	Energy and Environmental Analysis, Inc.
EEI	Edison Electric Institute
EIA	Energy Information Administration
EMJ	*Engineering and Mining Journal*
EPA	Environmental Protection Agency
EPCA	Energy Policy and Conservation Act
EPRI	Electric Power Research Institute
ERA	Energy Regulatory Administration
ERDA	Energy Research and Development Administration
ERL	Environmental Research Laboratory
ESECA	Energy Supply and Environmental Coordination Act
ETA	Energy Tax Act
°F	degrees Fahrenheit
FBC	fluidized bed combustion
FCB	Fisher et al.
FEA	Federal Energy Administration
FERC	Federal Energy Regulatory Commission
FES	Facility Energy Utilization Data System
FEUDS	Facility Energy Utilization Data System
FGD	flue gas desulfurization
FPC	Federal Power Commission
ft	foot
FUA	Powerplant and Industrial Fuel Use Act
FWPCA	Federal Water Pollution Control Act
g	gram
gal	gallon
GAO	Government Accounting Office
GDP	gross domestic product
GEA	General Energy Associates
GNP	gross national product
gpm	gallons per minute
GT-HTGR	gas turbine high-temperature gas reactor

ha	2.47 acres
hp	horsepower
hr	hour
HUD	Department of Housing and Urban Development
IBM	International Business Machines
ICC	Interstate Commerce Commission
ICES	Integrated Community Energy System
ICOP	Industrial Cogeneration Optimization Program
IDHA	International District Heating Association
IEA	Institute of Energy Analysis
IFS	International Financial Statistics
IICP	Integrated Industry Cogeneration Program
in	inch
INCO	International Nickel Co.
IPEP	Industrial Plant Energy Profile
IPHDB	Industrial Profile Heat Data Base
ISTUM	Industrial Sector Technology Use Model
ITC	Intertechnology Corporation
kW	kilowatt
kWh	kilowatt-hour
LAER	lowest achievable emission rate
lb	pound
LEAP	Long-range Energy Analysis Package
LME	London Metals Exchange
LNG	liquefied natural gas
LP	linear programming
LWR	light water reactor
m	meter
mcf	millions of cubic feet
MEFS	Midterm Energy Forecasting System
MEI	Main Economic Indicators
MFBI	major industrial fuel-burning installation
MHD	magnetohydrodynamics
MIP	Mixed Integer Program
MMBtu	millions of British thermal units
MOPPS	Market Oriented Program Planning Study
MPS	mathematical program systems
MSE	mean squared error
MW	megawatt
MW(e)	megawatt (electric)
MW(t)	megawatt (thermal)
NBS	National Bureau of Standards
NEA	National Energy Act
NECPA	National Energy Conservation Policy Act
NEP	National Energy Plan
NEPA	National Environmental Policy Act
NGL	natural gas liquid
NO_2	nitrogen dioxide

NO_x	nitrogen oxide
NPDES	National Pollutant Discharge Elimination System
NSPC	Northern State Power Company
O&M	operation and maintenance
OECD	Organization for Economic Cooperation and Development
OIP	Office of Industrial Programs (DOE)
OPEC	Organization of Petroleum Exporting Countries
ORIM	Oak Ridge Industrial Model
ORNL	Oak Ridge National Laboratory
OSW	Office of Saline Water
PCM	Penn Composite Model
PIES	Project Independence Evaluation System
P.L.	public law
pp	payback period
ppm	parts per million
PSD	prevention of significant deterioration
PSE&G	Public Service Electric and Gas Company
psi	pounds per square inch
psia	pounds per square inch absolute
psig	pounds per square inch gauge
PUC	Public Utility Commission
PURPA	Public Utility Regulatory Policies Act
R&D	research and development
RFP	request for proposal
ROI	return on investment
SERI	Solar Energy Research Institute
SIC	standard industrial classification
SIP	State Implementation Plan
SMSA	Standard Metropolitan Statistical Area
SO_2	sulfur dioxide
SO_x	sulfur oxide
SRC	solvent-refined coal
T	absolute temperature
t	ton
T&D	transmission and distribution
TCF	tons per cubic foot
TETAS	Total Energy Technology Assessment Study
TVA	Tennessee Valley Authority
UNSY	United Nations Statistical Yearbook
UOM	University of Minnesota
USDA	United States Department of Agriculture
V	volt
W	watt
WAES	Workshop on Alternative Energy Strategies
yr	year

Index

absorption chillers, 453
absorptive refrigerator, 99
Advanced Cogeneration System (ACS), 384
air conditioning, 95
air emissions, 270
air pollution, 223–224
air preheaters, 311–312
alloy scrap, 109
aluminum
    direct reduction of, 456
    industry, 94–98
    smelter modification, 453–454
American Institute of Physics, 94
American Iron and Steel Institute (AISI), 458
American Metal Market, 237
anode baking process, 97
appliance standards, 211
aquatic waste, 414–415
Arthur D. Little (ADL) model, 255, 262–263
Auto Efficiency Standards, 39
automobile industry, 3–4, 444, 446

Bayer process, 97
beneficiation, 101–103, 158
best available technology (BAT), 431
best practicable control technology (BPT), 431
biofuels, 49
bio-gas production, 415
biological recycling, 410–416

blast furnace gasifier, 454
blended cement, 454
block heating power plant, 395–398
bottoming cycle, 332, 338
bottoming system (Rankine), 343, 345–346
Brayton cycle, 336, 337, 338, 381
Britain, 212
Brookhaven Energy Systems Optimization
    Model (BESOM), 31, 32
Building Energy Performance Standards, 39,
    41
buildings and community systems, 436
business energy conservation program, 210–
    211

California Energy Resources Conservation and
    Development Commission, 220–221
Canada, 26, 30, 33, 37, 212, 238
capacity production factor, 177
capital recovery factor, 449–450
carbon monoxide, 224
Carnot efficiency, 100
cement block drying, 455
chance-constrained programming (CCP)
    method, 196
Charles River Associates (CRA) model, 255,
    262–263
Clean Air Act, 200, 222, 225–226, 377
closed systems, 153–154, 391